37th International Symposium on Remote Sensing of Environment (ISRSE 37)

Earth Observation for Development and Adaptation to a Changing World

International Archives of the Photogrammetry, Remote Sensing and Spatial Information Sciences Volume XLII-3/W2

Tshwane, South Africa
8 - 12 May 2017

ISBN: 978-1-5108-7460-2

Printed from e-media with permission by:

Curran Associates, Inc.
57 Morehouse Lane
Red Hook, NY 12571

Some format issues inherent in the e-media version may also appear in this print version.

Printed by Curran Associates, Inc. (2019)

International Center for Remote Sensing of the Environment, Inc. (ICRSE)
1955 E. Sixth Street, Suite 115
Tucson, AZ 85719
USA

Phone: (520) 621-8567
Fax: (520) 621-7834

isrse@symposia.org

Additional copies of this publication are available from:

Curran Associates, Inc.
57 Morehouse Lane
Red Hook, NY 12571 USA
Phone: 845-758-0400
Fax: 845-758-2633
Email: curran@proceedings.com
Web: www.proceedings.com

TABLE OF CONTENTS

A RADIOMETRIC CALIBRATION FRAMEWORK FOR HIGH-RESOLUTION UAV-DERIVED IMAGERY FOR MAPPING INVASIVE ALIEN PLANTS. A HARRISIA POMANENSIS CASE STUDY

Mafanya M[1], Tsela P[1], Botai J[2], Manyama P[3], Chirima G[4], Monate T[5], Mnyengeza M[6]

[1]University Of Pretoria, [2]South African Weather Services: Applications, [3]South African National Biodiversity Institute: Invasive Species Programme, [4]Agricultural Research Council-Institute for Soil, Climate and Water: Geoinformatics Division, [5]CAD Mapping Aerial Surveyors, [6]Statistics South Africa

Aerial imagery derived from consumer grade digital cameras onboard unmanned aerial vehicles (UAVs) are increasingly being used for biodiversity monitoring and remote sensing of the environment. Studies on radiometric calibration of UAV derived mosaicked orthophotos for large scale mapping remain scarce. CMOS Mirrorless Interchangeable Lens Cameras (MILCs) mounted on UAVs are increasingly being used for land cover mapping applications, however radiometric calibration procedures for these sensors are not widely developed. Radiometric calibration improves image data quality and is important for consistent quantitative analysis and comparison across different calibrated image data sets. This study presents a radiometric calibration framework for (i) designing calibration targets, (ii) converting CMOS MILC Advanced Photo System type-C (APS-C) raw image digital numbers (DNs) to reflectance values, and (iii) performing accuracy assessment using in-situ spectral signatures. Three linear models of type -ln(reflectance) = m*DN+C were developed in this study for a band by band radiometric calibration of the high-resolution UAV imagery. In particular, these linear models were developed for a 24-bit Sony NEX-7 derived RGB mosaic. Results of this method showed a goodness of fit R^2 of 0.922 ($p<0.01$) between measured and estimated reflectance values for validation targets. These findings have significant implications for exploring options in developing radiometric calibration procedures for UAV derived imagery, particularly for large scale land cover monitoring applications. Furthermore, this work contributes to the recent research efforts in the quest for good-for-purpose calibration target materials and designs.

KEYWORDS: Digital Numbers; Reflectance Values; Digital Cameras; Empirical Line Method; Invasive Alien Plants

"SENTINEL-2 FOR AGRICULTURE" TO SUPPORT NATIONAL AGRICULTURE MONITORING: DEMONSTRATION IN MALI

Traore P[1], Traore S[2], Keita B[3], <u>Bellemans N[4]</u>, Bontemps S[4], Defourny P[4], Cara C[5], Nicola L[5], Udroiu C[5], Koetz B[6]
[1]ICRISAT, [2]Institut d'Economie Rurale, [3]Cellule de Planification et de Statistique, Ministere de l'Agriculture, [4]Earth and Life Institute, Université cath. de Louvain, [5]CS-Romania, [6]ESA-ESRIN

Launched in 2014, the ESA Sentinel-2 for Agriculture project aims to develop an operational, open-source system to transform Sentinel-2 (S2) data into agricultural monitoring products including (i) cloud-free composites, (ii) monthly cropland masks, (iii) cultivated crop type map for main crops and (iv) 10-day LAI and NDVI indicators.

The first phases of the project focused on algorithms selection, system design and implementation. The last phase started in March 2016 demonstrates the developed system by delivering S2-derived products over the 2016 growing season in near real-time. This demonstration is carried out over 3 national and 8 local sites over Europe, Africa and Asia. Mali is one of the national sites.

A large landlocked country with severe climatic and natural constraints, Mali faces recurring nutrition and food insecurities. Rapid population growth and urbanization exert significant pressure on a food production sector still largely dependent on smallholder agriculture, where limited inputs, low mechanization, lack of labor and market integration result in outstanding yield gaps and impede intensification. Improving national agricultural monitoring capabilities with timely data, information about crop status, crop area, and yield forecasts is considered a critical step towards strengthened food production information and market transparency.

The suite of agricultural products is generated by the project consortium and in parallel, by a Mali-based processing line operated by the National Agricultural Research Service, the Planning and Statistics Office of the Ministry of Agriculture and ICRISAT. The national team coordinates data collection campaigns for in-situ crop type (n=30,000) and condition (n=2,205). Derived information products focused on cotton, maize, millet, peanut, rice, and sorghum will be quantitatively and qualitatively evaluated to assess their usefulness for governmental and other entities. The final goal is the successful transfer of the system and integration of information produced into these different organizations.

KEYWORDS: Sentinel-2; agriculture monitoring; food security; operational; Mali

A COMPARATIVE ANALYSIS OF NFI-BASED AND LIDAR-BASED FOREST ABOVEGROUND BIOMASS MAPS IN MEXICO

Urbazaev M[1]
[1]*Friedrich-Schiller-University Jena*

Information on the spatial distribution of aboveground biomass (AGB) over large areas is needed for understanding and managing the processes involved in the carbon cycle, and supporting international policies for climate change mitigation and adaption. Furthermore, these products provide local stakeholders with important baseline data for the development of sustainable management strategies. Using remote sensing techniques it is possible to provide spatially explicit information of AGB from local to global scales. In this work we present a two-stage up-scaling approach to estimate forest aboveground biomass in Mexico at national scale based on multi-sensor remote sensing data. For this, we estimate firstly AGB along the airborne LiDAR transects using Mexican National Forest Inventory data collected by CONAFOR and very high resolution NASA G-LiHT LiDAR data. We calculated from discrete-return LiDAR data 88 LiDAR metrics that are then related to field-estimated AGB. In the next step, we calibrate active (ALOS PALSAR) and passive satellite imagery (Landsat) with LiDAR-based AGB estimates in a non-parametric Cubist model to create a national wall-to-wall AGB map. Finally, the generated AGB product is validated using independent Mexican National Forest Inventory (NFI) data that were not used for model training. Furthermore, we modelled AGB at national scale using satellite imagery and NFI data only and compared to the results from the two-stage up-scaling approach. The estimated AGB products showed similar goodness-of-fit statistics at different scales compared to the independent validation data set. However, we observed different AGB spatial patterns in two products, especially in regions where NFI data are not available, but where high AGB values occur.

A DEEP LEARNING-BASED METHOD FOR HIGH-ACCURACY DETECTION OF INLAND WATERWAY VESSELS USING HIGH-RESOLUTION OPTICAL IMAGERY

Xiao C[1,2], Chen N[1,2], Wen X[1], Chen Z[1,2]
[1]*State Key Laboratory for Information Engineering in Surveying, Mapping and Remote Sensing, Wuhan University,*
[2]*Collaborative Innovation Center of Geospatial Technology*

With the rapid development of inland water transportation, the quantity, tonnage, and speed of ships have increased progressively, which leads to more and more frequently happened emergencies. High-accuracy monitoring, especially detection of inland water vessels is of vital importance to shipping safety and traffic control, which, however, is still faced with challenges. To solve this problem, in this paper, a deep learning approach- Deep Neural Network (DNN), one of the state-of-the-art machine learning models, is taken advantage of to extract high-level features for classifying, as well as locating ships on high-resolution optical remote sensing images. Spatial information and spectral information are used collaboratively to improve the accuracy of detection. Experiments conducted in the area of Yangtze River Waterway, China, with 1-m-resolution panchromatic imagiery data and 4-m-resolution multispectral imagiery data from China recently launched Gaofen-2 (GF-2) satellite indicates that the proposed method provides more competitive performance than traditional related ones.

KEYWORDS: Deep learning, Ship detection, Classification, Optical remote sensing images, Gaofen-2 (GF-2)

A FREE HIGH RESOLUTION LANDCOVER ON THE SMALL INDIAN OCEAN ISLANDS, AN EXAMPLE OF ITS USE WITH THE STUDY OF VECTOR-BORNE DISEASES

Révillion C[1], Attoumane A[2], Herbreteau V[2]
[1]UR/UMR 228 Espace Dev (IRD, UAG, UM, UR), [2]IRD/UMR 228 Espace Dev (IRD, UAG, UM, UR)

The southwest of the Indian Ocean comprises numerous islands of less than 3000 sq km (Comoros, Seychelles, Mascarene Islands). These small island territories have very fragmented and diversified environments. Satellite imagery and products at low and medium spatial resolution (usually from 4 kilometers to 250 meters) are not or poorly suited to the study of these areas. This is particularly the case for ecological studies such as the study of vector-borne diseases where the evaluation of interactions at a fine scale is critical to understand their spatial dynamics.

To overcome this need, we realized a homogeneous land cover mapping of these small islands, by analysing SPOT 5 satellite images acquired between July 2013 and July 2014 by SEAS-OI Station. We used an object-based image analysis method to identify the 12 major classes of land cover / land use of these tropical islands. This methodology together with a good knowledge of the field has enabled us to achieve an overall accuracy of 85%, making it an operational product.

In order to illustrate the importance of this dataset, we present its use in health research to describe thelocation of health observations: whether the place of residence of patients or the points of capture of disease vectors through several examples in Reunion, Mayotte, the Union of the Comoros and the Republic of Seychelles. In these studies, high resolution allows to provide very fine information on the surrounding environment, while little data is usually available at this scale. In the statistical analyzes, these data were mainly used to differentiate between infected and non-infected individuals.

This high resolution landcover is available on a web portal to allow a wider distribution to researchers and thematicians using land cover information in their field.

KEYWORDS: landcover, remote sensing, object-based image analysis, environment, health

A HANDS-ON APPROACH TO HUMAN CAPITAL DEVELOPMENT USING NEXT GENERATION SMALL SATELLITES

Busakwe T[1]
[1]*Space Advisory Company*

The global economy is currently experiencing a shifting paradigm from a resource-based economy towards a more knowledge-based economy. The development of skills and human capital is seen as the key mechanism towards strengthening the position of economies in the global market. Space is one of the sectors that has seen on-going rapid changes and providing a country with a good understanding of Space science and technology together with a labour force with adaptive skills to deal with these changes is essential for any thriving economy.

With the above realisation, several training institutions offer theoretical space training programmes as mechanisms for science and technology advancement and to support the space aspirations of participating nations. However, there is a need to supplement the theoretical training with hands-on industry training.

In South Africa, 23 years of experience has been gathered in this domain. The lessons learned from a number of Human Capital Development (HCD) programs conducted in prior years provide a backdrop against which current HCD programs are structured and implemented.

This paper will explore the benefits of a hands-on industry training program in the Space sector. It will evaluate the successes and failures of past HCD programs implemented in South Africa and present a new HCD model not only for South Africa, but for the African continent to support societies that can compete and contribute to the rest of the world as we head forward into the modern knowledge-based economy. It will demonstrate how the Phoenix Hyperspectral satellite is used as a training platform to provide the required hands on experience. It will further be demonstrated how this model fosters collaboration and partnership, especially with and between developing countries, to ensure sustainable capacity building efforts.

A MULTI SCALE REMOTE SENSING APPROACH FOR MONITORING AND ASSESSING ECOSYSTEM REHABILITATION OF MINE SITES

Bartolo R[1], Erskine P[2], Whiteside T[1], Esparon A[1]
[1]Department Of The Environment And Energy-ERISS, [2]Sustainable Minerals Institute, The University of Queensland

Developing completion criteria for ecosystems created on mined land typically relies upon indicator values derived from reference systems using a traditional plot or transect approach. However, these methods tend to sample a very small portion of a site (typically <1%) and can be problematic on rehabilitated sites where the instability of newly constructed landforms restricts access. In order to provide leading practice monitoring we have developed a multi-scaled remote sensing approach for the assessment of ecosystem rehabilitation against closure criteria.

We present two case studies demonstrating the application of Remotely Piloted Aircraft System (RPAS) derived imagery to monitor revegetation in a mine site setting. The first case study is at Jabiluka mine, located within the Jabiluka mineral lease within Kakadu National Park. The 12 ha site is now in care and maintenance with rehabilitation works commencing in 2013. Revegetation to date has occurred in two stages with planting of tubestock 12 months apart (2013 and 2014). RPAS flights have been conducted regularly to collect multispectral imagery for the purpose of monitoring the revegetation success at the plant scale across the entire site. The second case study follows a rehabilitated mine site in central Queensland which has been monitored for 20 years prior to being exposed to a controlled fire. RPAS flights have provided us with a very high spatial and temporal data set impossible to collect using other remote sensing methods due to cost, clouds and smoke. We used a Geographic object-based image analysis (GEOBIA) framework to identify plants on the two sites across the temporal record as well as the proportional cover. We then applied a number of landscape fate metrics to quantify plant fate including growth, mortality and recruitment monitor the progress of the revegetation effort at the sites.

KEYWORDS: UAV, UAS, rehabilitation, revegetation, landscape metrics

A MULTI-SENSOR AND MULTI-TEMPORAL APPROACH TO VEGETATION STRUCTURE MAPPING – RESULTS FROM THE ARS AFRICAE EXPERIMENT AT THE SKUKUZA FLUX TOWER

Schmullius C[1], Onyango V[1], Berger C[1], Lueck W[2], Huettich C[3], Baade J[1]
[1]University Jena, Department for Earth Observation, [2]PCI Geomatics, [3]JenaOptronik

Ars AfricaE aims to deepen the knowledge on the functional role of structural elements in ecosystems such as species content or biomass. One of the objectives is to combine a multi-agent-based simulation model with information about spatio-temporal landsurface dynamics derived from remote sensing, historical aerial photography and Terrestrial Laser Scanner in-situ data. For the Skukuza flux tower, a space-time cube is being generated with a lateral dimension of 5 x 5 sqkm and a temporal dimension reaching back to the first available remotely sensed scenes (i.e. aerial photographs from the 1920ies, Corona spy satellite images from the 1970ies).

This presentation describes results from a lidar campaign in September 2015 with the full-waveform RIEGL Terrestrial Laser Scanner VZ-1000 and synergistic evaluation using 1) digital surface models retrieved from Worldview images, 2) vegetation height and woody cover retrieved from DMC-scenes, 3) above-ground biomass from PALSAR radar data, and 4) vegetation volume retrieved from Sentinel-1 radar data. Up- and downscaling techniques for the different spatial and spectral resolutions of data and products over time are being tested, adapted and/or developed to innovatively exploit the physical interactions behind reflectance and backscatter characteristics from the observed land surfaces.

The overall goal is to support distinction between 'disturbances' (variation with subsequent recovery) and 'degradation' (sustained loss of function) with operational monitoring systems. The final procedures contribute to the Ars AfricaE management concept which is being developed in close cooperation with SANPark's Scientific Services.

KEYWORDS: RIEGL TLS, PALSAR, vegetation structure, Skukuza

A MULTI-TEMPORAL APPROACH TO BURN SCAR DETECTION USING SYNTHETIC APERTURE RADAR DATA

Engelbrecht J[1], Theron A[1], Vhengani L[1]
[1] *CSIR*

The impact of wild fires are severe, affecting the built- and natural environment and threatening health and safety. However, for some ecosystems, including the Fynbos biome in the Western Cape region of South Africa, fires are beneficial since they stimulate species diversity and influence nutrient cycles. In fact, without periodic fires, these shrublands cannot persist. Effective fire management strategies are required to keep the land in a healthy condition. The optimal fire frequency is dependent on the specific species but should not be too often nor too infrequently. Burnt area identification over time has been performed to determine veld age in remote regions. If the time interval since the preceding fire becomes too long, a controlled burn to ensure optimal ecosystem functioning may be considered. In addition to informing optimal fire management practices, burnt-area mapping is also important for disaster recovery and disaster relief.

Satellite data has been used for the monitoring of active fires as well as post event assessments. However, successful burn-scar mapping using optical and multispectral data is frequently limited by smoke or cloud cover at the time of data acquisition. Furthermore, spectral overlaps between burnt and unburnt classes can negatively affect classification accuracy. To address these limitations, the use of multi-temporal Sentinel-1 SAR data for burn scar identification was considered. One pre-burn and one post-burn scene were subject to H-A-alpha decomposition to determine the proportion of vegetation scattering. A Normalised Difference alpha Index (NDαI) was calculated to determine the change in vegetation scattering before and after the burn. The resulting image was subject to image segmentation for burn scar identification. The results suggest that burn scars could be identified with a high degree of confidence confirming that SAR data can provide a complimentary source of information for operational burnt area mapping.

KEYWORDS: Burn scar mapping, SAR

A NEW ERA FOR LAND CHANGE MONITORING: THE USGS LAND CHANGE MONITORING, ASSESSMENT, AND PROJECTION (LCMAP) PROJECT

Obregón A, Loveland T[1], Woodcock C[2]
[1]U.S. Geological Survey, [2]Boston University

The U.S. Geological Survey is implementing a new land cover and land change mapping and monitoring capability titled LCMAP – Land Change Monitoring, Assessment, and Projection. LCMAP enables the continuous tracking and characterization of changes in land cover, use, and condition for the purpose of translating such information into assessments of current and historical processes of change. Basic elements of LCMAP include (1) an "analysis-ready" Landsat archive organized to support time series analysis, (2) land cover mapping and land change monitoring, (3) land change assessment, (4) scenario-based land cover projections, and (5) the systems and services that enable these capabilities. The LCMAP mapping and monitoring system is based on the Continuous Change Detection and Classification (CCDC) algorithm developed by Boston University researchers (Zhu, Z. and C. Woodcock, 2014, Continuous change detection and classification of land cover using all available Landsat data: Remote Sensing of Environment 144:152–171). CCDC is used to detect and characterize historical land change at any point in the 1984-present Landsat record and includes the capability to detect land change in near real time. CCDC also includes tools for classifying land cover at any point in the Landsat record. Initially, science-quality annual United States land cover and land change products including maps, statistics, and estimates of uncertainty are being generated. This presentation focuses on the transformation of the Landsat archive into an "analysis ready" structure, the land cover and land change processing capabilities, and initial 30 year set of land cover and land change results.

KEYWORDS: Land cover, land change, Landsat, monitoring

A NEW HORIZON FOR QUANTIFYING THE SPATIAL AND TEMPORAL VARIATIONS OF C3 AND C4 GRASS SPECIES ABOVEGROUND BIOMASS

Shoko C[1], Mutanga O[1]

[1]Discipline of Geography, School of Agricultural, Earth and Environmental Sciences, University of KwaZulu-Natal, Private Bag X01, Scottsville

C3 and C4 grasses represent a fundamental species functional type of grassland ecosystems which play fundamental role in the regulation of carbon, water balance, biodiversity, food security and veld condition. The ability of these grasses to provide ecosystem goods and services varies spatially and temporarily, due to the influence of climatic variables and topographic features. However, the current and projected environmental changes are anticipated to compromise the distribution and productivity of these ecosystems with substantial effects on their ability to provide the required services. Monitoring the temporal variations of C3 and C4 grasses aboveground biomass (AGB) therefore becomes key in understanding their response to environmental changes and their contribution to the provision of services. Emerging remote sensing sensors with better earth imaging characteristics present key data sources to monitor the seasonal variations of C3 and C4 grasses AGB, which has been limited by the scarcity of suitable sensors. The innovative sensors (i.e. Sentinel 2 MSI and Landsat 8 OLI), with high revisit frequency, more and unique spectral bands, as well as large geographic coverage, provide better opportunity for accurate and spatial characterization of C3 and C4 grasses AGB over space and time, in a cost-effective manner. This study therefore attempts to characterize the seasonal variations of C3 and C4 grass species AGB using emerging multispectral sensors. It is anticipated that the advanced sensors will provide a better temporal and spatial variations of AGB of the target species, which might be difficult to achieve, when using the available broadband multispectral sensors.

KEYWORDS: Climate change, food security, seasonal variations, remote sensing, ecosystem goods and services

A NEW SAR CLASSIFICATION SCHEME FOR SEDIMENTS ON INTERTIDAL FLATS BASED ON MULTI-FREQUENCY POLARIMETRIC SAR IMAGERY

Wang W[1,2], <u>**Gade M**</u>[1]
[1]Universität Hamburg, [2]Institute of Remote Sensing and Digital Earth, CAS

We present a new classification scheme for muddy and sandy sediments on exposed intertidal flats, which is based on synthetic aperture radar (SAR) data, and use ALOS-2 (L-band), Radarsat-2 (C-band) and TerraSAR-X (X-band) fully polarimetric SAR imagery to demonstrate its effectiveness. Four test sites on the German North Sea coast were chosen, which represent typical surface compositions of different sediments, vegetation, and habitats, and of which a large amount of SAR is used for our analyses. Both Freeman-Durden and Cloude-Pottier polarimetric decomposition is utilized, and an additional descriptor called Double-Bounce Eigenvalue Relative Difference (DERD) is introduced into the feature sets instead of the original polarimetric intensity channels. The classification is conducted following Random Forest theory and the results are verified using ground truth data from field campaigns and an existing classification based on optical imagery. The use of Kennaugh elements for classification purposes is demonstrated using both fully and dual polarimetric multi-frequency and multi-temporal SAR data. Our results show that the proposed classification scheme can be applied for the discrimination of muddy and sandy sediments using L-Band SAR imagery, while shorter microwaves (X- and C-band) can be used to also detect more detailed features such as bivalve beds on intertidal flats.

A NOVEL UNSUPERVISED OBJECT-BASED SEA ICE CLASSIFICATION OF POLARIMETRIC SYNTHETIC APERTURE RADAR DATA

Mahdianpari M[1], **Salehi B[1]**, Mohammadimanesh F[1]
[1]*C-CORE, and Memorial University of Newfoundland*

PolSAR images are complex multi-dimensional data which can be extensively used in sea ice monitoring and classification. The discrimination capability of PolSAR data makes them a unique source of information with a significant contribution in tackling problems concerning environmental applications especially in the Arctic regions. Moreover, the increasing temperature of the Earth due to climate change, combined with the degradation of ice thickness in the Arctic region increases the navigation of shipping routes more than ever. Therefore, sea ice monitoring is one of the most important applications of remote sensing tools since it is time and cost efficient in comparison with extensive field campaigns.

This paper presents an unsupervised object-based classification method for PolSAR data, which allows the use of multiple sources of statistical evidences. Expectation-Maximization (EM) classifier is proposed to estimate sea ice classes from the scattering matrix that is also modified using Fisher Linear Discriminant Analysis (FLDA). In fact, the FLDA reduces the dimension of a given statistical model by defining a projection that reduces the within-class diversity while increasing the between-class scatter. In fact, FLDA transformation is used to make the Most Discriminated Features (MDF) space for image classification. The proposed method has three major steps, first is the initialization for defining equally probable proportions. Second step is the (Expectation) E-STEP, which updates the posterior probabilities of each object, and finally (Maximization) M-STEP updates the parameters estimation. Our proposed classification method is applied to a dual polarimetric L-band dataset, ALOS-1 PALSAR acquired from Baffin Bay/Labrador Sea, Northern Canada. We evaluate the classification performance of the proposed approach and compare it with other well-known classic methods, and the results will be presented in the full paper.

KEYWORDS: Polarimetric Synthetic Aperture Radar (PolSAR), Sea ice, Object-based classification, Expectation-Maximization (EM), Fisher Linear Discriminant Analysis (FLDA)

A PILOT STUDY TO DELIMITATE TSETSE TARGET POPULATIONS IN ZIMBABWE

Chikowore G[1], Dicko A[2], Chinwada P[3], Zimba M[3], Shereni W[1], Roger F[4], Bouyer J[2], Guerrini L[4,5]

[1]Tsetse Control Division, Department of Livestock and Veterinary Services, [2]2 Centre de Coopération Internationale en Recherche Agronomique pour le Développement, Unité Mixte de Recherche Contrôle des Maladies Animales Exotiques et Emergentes, [3]Department of Biological Sciences, University of Zimbabwe, [4]Centre de Coopération Internationale en Recherche Agronomique pour le Développement, Unité de Recherche Animal et Gestion Intégré des Risques, [5]Centre de Coopération Internationale en Recherche Agronomique pour le Développement, Unité de Recherche Animal et Gestion Intégré des Risques, Research Platform Production and Conservation in Partnership, University of Zimbabwe

Tsetse (Glossina sensu stricto) are cyclical vectors of human and animal trypanosomoses, that are presently targeted by the Pan African Tsetse and Trypanosomiasis Eradication Campaign (PATTEC) coordinated by the African Union. Tsetse distribution in Zimbabwe is not accurately delimited throughout the country, which is a pre-requisite to achieve effective control. Presented is a methodology that could help doing so throughout the country, in a pilot study in the Masoka area, Mid-Zambezi valley in Zimbabwe, and targeting two savannah species, Glossina morsitans morsitans and Glossina pallidipes. First, we used the presence data to study habitat suitability of both species based on climatic and environmental data derived from MODIS and SPOT 5 satellite images. Factors influencing distribution were studied using an Ecological Niche Factor Analysis (ENFA) whilst habitat suitability was predicted using a Maximum Entropy (MaxEnt) model at a spatial resolution of 250 m. Area Under the Curve (AUC), an indicator of model performance, was 0.89 for G. m. morsitans and 0.96 for G. pallidipes. We then used the predicted suitable areas to calculate the probability that flies were really absent from the grid cells where they were not captured during the study based on a probability model using a risk threshold of 0.05. Apart from grid cells where G. m. morsitans and G. pallidipes were captured, there was a high probability of presence in an additional 104 km2 and 0 km2 respectively. The modelling process proved to be useful in optimizing the outputs of presence/absence surveys and allowed defining accurately the limits of tsetse infested areas. This model can be extended to all the tsetse infested parts of Zimbabwe and may also be useful for other PATTEC national initiatives in other African countries.

KEYWORDS: Glossina, tsetse distribution, habitat suitability, Ecological Niche Factor Analysis, Maximum Entropy, MODIS, SPOT5

A PROPOSED APPROACH TO IMPLEMENT THE GEO USER NEEDS AND GAP ANALYSIS PROCESS

Mlisa A[1], Rum G[1]
[1]*Group On Earth Observations (GEO)*

The *GEO Strategic Plan 2016-2025: Implementing GEOSS* makes clear that GEO will do much more than just deliver data:

GEO will play a key role in systematically...delivering tools, skills and services to allow the intelligent exploitation of the data by the user communities...This end-to-end process of identifying needs, ensuring the availability of data with which to develop information to address societal challenges, and transforming that information into knowledge through the generation of products and services for end-users, defines the scope of GEO.

To date GEO has mostly focused on data rather than tools and services that transform it into higher level products--information and knowledge products--that enable users to address societal challenges. Thus there remains a gap between users and data that hinders its exploitation and application. This gap must be bridged. Fortunately, GEOSS has now matured enough so that bridging is possible and community portals--in particular, those that provide tools and services--are the key bridging component. Three types of portals can be defined: simple websites, data portals, and what might best be called "exploration platforms". Such platforms host the tools and services that allow a user to explore, understand, and apply the data, information, and knowledge that it provides access to. These tools and services include, for example, visualization tools for temporal and spatial trends, predictive models that can address "what if" questions, generation of derived products, and links to explanatory variables that provide insight into why another variable is changing. The GEO Energy Portal http://webservice-energy.org will be demonstrated as an example. There is a tight relationship between a community portal and user needs, portals typically being the physical manifestation of those needs. Consequently, portal development and user needs can be developed together, iteratively, a topic that needs further discussion.

A PUBLIC FREE AND OPEN TOOL FOR MAPPING HUMAN SETTLEMENTS

Kemper T[1], Politis P[2], Maffenini L[3], Pesaresi M[1], Mudau N[4], Sehn Körting T[5]
[1]European Commission, Joint Research Centre (JRC), [2]ARHS Development (under contract of JRC), [3]GFT (under contract of JRC), [4]South African National Space Agency, [5]Instituto Nacional de Pesquisas Espaciais

The access to up-to-date information on human settlements is crucial for any kind of spatial analysis related to human and physical exposure to threats such as natural disasters and conflicts or environmental contamination and degradation, to the impact of human activities on ecosystems and to the human access to resources. This paper presents a public, free and open tool for the automatic extraction of settlement information from high and very high spatial resolution satellite imagery. Based on the experiences gained during the production of the Global Human Settlement Layer (GHSL) the tool uses the symbolic machine learning (SML) classifier to map settlements with the help of coarse resolution settlement or land cover information. The SML classifier is a multi-purpose classifier that uses in the case of the settlement mapping radiometric, textural and morphological features as input for the classifier. Although the current workflows are optimized for settlement mapping the may as well be used for general land cover mapping provided there is an appropriate training data set available. The tool allows the parameterization of the workflow, which allows the adaptation to support any high or very high spatial resolution sensor. Currently the tool includes ready-to-use workflows for SPOT-5, SPOT-6/7, RapidEye and CBERS-4, but it was also tested with GeoEye-1, WorldView-2/3, Pléiades, Deimos-2 and Quickbird. The software is designed for the processing of massive data sets. This will be demonstrated with examples based on SPOT-5 for the Middle East and South Africa as well as RapidEye for Brazil.

A RECOVERY OBSERVATORY FOR HURRICANE MATTHEW IMPACT IN HAITI

De Boissezon H[1], Leitmann J[2], Proy C[1], **Hosford S[1]**, Eddy A[3], Durand A[4], Mellucci C[5], Chalifoux S[6], Petiteville I[7], Ito M[8], Green D[9], Zoffoli S[10], Jones B[11], Danzeglocke J[12]

[1]Cnes, [2]GFDRR World bank, [3]Athena Global, [4]Icube-SERTIT, [5]UNDP, [6]CSA, [7]ESA / ESRIN, [8]JAXA, [9]NASA/Godard Space Flight Center, [10]ASI, [11]USGS, [12]DLR

Large populations living in vulnerable areas have led to record damages and substantial loss of life in disasters ranging from the 2004 Indian Ocean tsunami to the 2011 Tohoku tsunami. More recently, Hurricane Matthew swept through the Caribbean causing widespread damage and loss of life. In Haiti alone, over 1,000 people died, large towns in the southwest were devastated; a month after the hurricane, some 1.4 million people were still in desperate need of humanitarian assistance. Recovery from catastrophes such as these lasts years. While Earth Observation satellite imagery is used after many disasters to support damage assessment, use of EO for recovery tracking is rare.

In 2015, the Committee on Earth Observation Satellites (CEOS) launched an initiative to bring together satellite operators, international donors, and stakeholders from the international disaster risk management community, to create a pilot where satellite data and value added products could be used to support recovery planning and monitoring over a four-year period. A Recovery Observatory (RO) Oversight Team, co-chaired by the French space agency CNES and the World Bank Global Facility for Disaster Reduction and Recovery, was created, and oversaw in 2016 a series of demonstrator showcase activities. CEOS is currently engaged in the triggering process of the RO for Hurricane Matthew impact in Haiti. A full-scale RO in Haiti would be a unique EO contribution to disaster recovery.

The principal aims of this project are to:

1. Demonstrate in a high-profile context the value of using satellite Earth Observations to support Recovery: near-term (baseline for recovery) and long-term (recovery planning and monitoring);
2. Work with the recovery community to define a sustainable vision for increased use of satellite Earth observations;
3. Establish institutional relationships between CEOS and recovery stakeholders;
4. Foster innovation around high-technology applications.

KEYWORDS: Earth Observation, disaster, recovery, planning, monitoring

A REVIEW OF SINKHOLE PRECURSOR DETECTION THROUGH SAR INTERFEROMETRY

Theron A[1, 2], Engelbrecht J[1], Kemp J[2]
[1]CSIR Meraka Institute, [2]Department of Geography and Environmental Studies, Stellenbosch University

Sinkholes are an unpredictable geohazard that endangers life and structures globally. So far they have remained an underestimated risk in relation to other geohazards. However, due to the recent explosive population growth and urbanisation trends experienced worldwide, susceptible terrains are rapidly being developed upon. Furthermore, sinkhole events are known to be caused by anthropogenic actions. These factors lead to increased risk due to an increasing frequency of sinkhole events as well as greater exposure of populations. There is currently no way of providing a reliable early warning to sinkhole formation. However, early signs of sinkhole development, such as tension cracks in the ground and infrastructure are often seen due to ground subsidence. Such subsidence, thought to occur before sinkhole formation, is referred to as precursory deformation and is key to an operational early warning system. However, current deformation monitoring techniques are not capable of monitoring large areas for small-scale, unpredictable, deformation associated with sinkholes. Satellite remote sensing, specifically Differential Interferometric Synthetic Aperture Radar (DInSAR), is able to frequently monitor large areas for millimetre scale deformation. It has proven to be valuable for many applications and is increasingly being explored for sinkhole precursor detection. DInSAR has successfully provided empirical evidence of sinkhole precursors across the world. Results have been reported on from diverse environments using a wide variety of SAR systems and processing methodologies. However, there is a clear need for an improved understanding of DINSAR's capabilities and limitations as well as the physical characteristics of sinkhole precursors. This review aims to show that, while challenging, predicting imminent sinkhole formation based on precursor subsidence detected from space is now within reach. There is, however, still an urgent need to develop this capability further towards the ultimate goal of an operational sinkhole early warning system.

KEYWORDS: Differential Interferometry, Sinkhole, Geohazard, SAR, Review

A SPATIAL-TEMPORAL METHOD FOR ASSESSING THE ENERGY BALANCE DYNAMICS OF PARTIALLY SEALED SURFACES

Pipkins K[1], Kleinschmit B[1], Wessolek G[1]
[1]*Technische Universität Berlin*

The effects of different types of sealed surfaces on the surface energy balance have been well-studied. However, these studies typically aggregate these surfaces into continuous units. The proposed method seeks to disaggregate such surfaces into paving and seam areas using spatial methods, and to consider the temperature dynamics under wet and dry conditions between these two components, for different levels of surface sealing. The work is undertaken using a thermal camera to record a time series of images over two lysimeters with differing levels of surface sealing. The images are subsequently decomposed into component materials using object-based image analysis and compared on the basis of both the surface materials as well as the spatial configuration of materials. Finally, a surface energy balance method is used to estimate evaporation rates from the surfaces, both separately for the different surface components as well as using the total surface mean. These results are validated using the output of the weighing lysimeter. Our findings will determine whether the explicitly spatial method is an improvement over the mean aggregate method.

KEYWORDS: Sealed surfaces, Evapotranspiration, Surface energy balance, Thermal imaging, Lysimeter

ACCESS TO TERRASAR-X AND TANDEM-X MISSION DATA FOR SCIENCE

Rossner G[1], Bock M[1], Danzeglocke J[1], Roth A[2], Hajnsek I[3], Hoffmann J[1]
[1]DLR Space Administration, [2]DLR German Remote Sensing Data Center , [3]DLR Microwaves and Radar Institute

The German earth observation missionsTerraSAR-X and TanDEM-X have been built on basis of a public-private-partnership. The private investment was made on the basis of a business models that included first of all the perspective to sell data to commercial users worldwide, while the public investment of Germany was done to support the scientific exploitation of the mission.

The twin satellites TerraSAR-X and TanDEM-X were launched in 2007 and 2009 respectively, providing each high resolution X-Band SAR data in five different modes, ranging from staring spotlight data of 0.25m resolution to wide scan SAR products with a scene width of 54000km² and 40m resolution. In bi-static constellation the satellites provide a high quality Global DEM data product of 2m vertical and 12 m horizontal resolution, as well as additional experimental modes, that are acquired during special mission phases with varying across-track and along-track baselines. Both satellites are beyond their expected lifetime, but with the systems in excellent state further data exploitation until 2019 and even beyond is expected.

The basic data access mechanism for each mission is a general announcement of opportunity (AO) permanently open. Users need to register and provide a proposal with a description of the research envisioned including schedule and data requirements as well as information on the team composition.

To support the various international GEO Research and Monitoring initiatives, suitable approaches for EO data provision are agreed with other space agencies within dedicated CEOS Working groups as e.g. the GFOI Spatial data Coordination Group, the Disasters Working Group or the Polar Space Task Group.

KEYWORDS: science access, commercial EO data, TerraSAR-X, TanDEM-X, GEO, CEOS, Open data

ACCURACY ASSESSMENT OF TANDEM-X I-DEMS OVER THE UK

Feng L[1], Muller J[1]
[1]*Mullard Space Science Laboratory, University College London*

Many global DEMs have been produced from spaceborne EO sensors, such as from SAR (SRTM, TerraSAR-X, TanDEM-X), stereo-photogrammetry (ASTER, SPOT, PRISM and IRS-3P) and lidar (ICESat). From the TanDEM-X mission (bistatic X-Band interferometric SAR), globally consistent and edited Digital Elevation Model (DEM) will be available from 2017. However, its accuracy has until now not been independently validated. Through a data grant (IDEM_CALVAL0207) from DLR an intermediate first-pass only TanDEM-X DEM (called i-DEM) was made available covering different regions at 1/3rd arc-second (≈12.5m) and 1 arc-second (≈30m) over the UK along with 3 arc-second i-DEM of selected locations around the world.

This 30m and 90m i-DEM tiles have been assessed using both airborne photogrammetrically derived DTMs and kinematic GPS. Photogrammetric DTMs at 5m spacing were produced commercially by Blue Sky and released for academic research use through the LANDMAP project for the assessment of the i-DEM tiles. The LANDMAP project also provides us with Kinematic GPS data covering the major road networks over the British Isles, with an accuracy of about 2.5m in X, Y and Z at 20m spacing that can be used to assess the quality of the i-DEM data. In addition, the Ice, Cloud and land Elevation Satellite (ICESat) provides globally-distributed elevation data of high accuracy (≤1m) that is well-suited for evaluating large area DEMs.

This report will demonstrate why the DEMs need to be co-registered after surface matching before they can be properly compared and water bodies removed using Ordnance Survey® vector datasets. After co-registration, GIS spatial statistical analysis and 2D and 3D visualization analysis methods and difference map and statistics are shown at 12m, 30m and 90m IDEM data. Furthermore, the true horizontal resolution of the i-DEM will be studied and the results will be described.

KEYWORDS: TanDEM-X; Accuracy Assessment; digital elevation model (DEM); IDEM; ICESat

ADAPTIVE CANOPY HEIGHT MODEL PROCESSING OF LIDAR DATA FOR ESTIMATING FRACTIONAL CANOPY COVER IN SAVANNAS

van den Bergh F[1], Wessels K[1], Mathieu R[1], Main R[1], Naidoo L[1]
[1]*CSIR*

Fractional canopy cover is a metric frequently used to monitor woody vegetation structure across large areas with satellite data. We use airborne LiDAR data to estimate canopy cover in order to generate training and validation data for national scale SAR-based monitoring systems.

Following LiDAR point classification (ground, vegetation, etc.) and subtraction of the ground surface model, a LiDAR point cloud can be rasterized at a fixed ground sampling distance (say, a 1 m GSD) to produce a Canopy Height Model (CHM). Provided that the LiDAR point density is sufficiently high, fractional cover can be expressed as the count of CHM cells with a height exceeding 1 m (e.g., for general woody vegetation), divided by the count of non-empty cells, of those CHM cells falling within the extent of a coarser resolution cell (say, 25 m GSD) of the canopy cover estimate.

The choice of CHM GSD is significant: a 2 m CHM can lead to canopy cover overestimation exceeding 20%, relative to the cover estimate derived from a 1 m CHM. On the other hand, if the point density drops to 0.5 points/m^2, a 1 m CHM will contain many empty cells, leading to severe underestimation (again in excess of 20%), especially in open savanna landscapes with isolated trees.

We propose a method of blending cover estimates from two CHMs (e.g., 1 m and 2 m GSDs), guided by point density, on a per 25 m cell scale. Through simulated thinning of high point density data we show that the blending method produces unbiased estimates that reduces the error to less than 4% over a density range of 0.5 to 12 points/m^2. The proposed blending method allows consistent canopy cover estimates to be obtained from LiDAR data sets with varying point densities within this range.

ADDRESSING INTERCONNECTIONS BETWEEN SUSTAINABLE DEVELOPMENT GOALS AND SUPPORTING POLICY DEVELOPMENT WITH AGENT-BASED MODELS: THE EXAMPLE OF GENDER EQUALITY AND SUBSISTENCE FARMING

Jules-Plag S[2], **Plag H**[1]
[1]Old Dominion University, [2]Tiwah UG

The Sustainable Development Goals (SDGs) pose wicked problems to society that require straddling the interests of different policy making departments at national levels. Making progress towards the SDGs requires Earth observation and science support to address core questions, including the question of what policies and actions can facilitate progress towards the targets. We assess the capability of Agent-Based Models (ABMs) to provide support for policy development using the example of SDG 5 "Achieve gender equality and empower all women and girls." In societies with a large fraction of subsistence farming, policy options to increase gender equality are those that impact land ownership and access to financing.

ABMs provide an environment for the incorporation of domain expertise in a rigorous manner to explore wicked problem solutions when not all of the relevant issues are totally clear. The iterative process of AMB development including formalizing rules, experimenting with various scenarios, test assumptions, assessing internal model consistencies, and addressing sensitivities to data availability and quality helps to identify the most relevant aspects and the core issues impacting the goal and also guides data collection. Once developed, the ABMs can be generalized and applied over different domains.

The exploratory system ABM simulates the seasonal cycle of buying seeds (including getting financing, if needed), seeding, growing, harvesting, and utilizing (marginal) gains. Agents are the male and female farmers; land-owner societies; markets to buy seeds and fertilizers and sell produce; traditional financial sources (banks) and micro-financing actors. The model was run over the seasonal cycle for a period of 100 years for many different scenarios. The results indicate that decoupling land ownership from financing and introducing quota systems are two potential policy options to increase gender equality. Additionally, the results help to identify gaps in the current formulation of SDG targets and indicators.

ADDRESSING THE RELATIONSHIPS BETWEEN LAND COVER CHANGE AND FIRE IN MIOMBO FORESTS IN GURUÉ, CENTRAL MOZAMBIQUE

Soares M[1], Mahamane M[1], **Ribeiro N**[1]
[1]*Eduardo Mondlane University*

The District of Gurué has had a reduction in its woodlands, with an estimated deforestation rate of 1.57% between 2000-2014. Land clearing for agriculture is the main driver of deforestation in Mozambique and fires play an important role in that process. Gurué has had an increase in agricultural production, especially for soy plantations, which have contributed to deforestation. This study presents the land use and land cover (LULC) change for Gurué between 2000 and 2015 and the role fire plays in those changes. The study was conducted using remote sensing data (Landsat images and MODIS data) and field data. A supervised classification with maximum likelihood of LULC was conducted using Landsat images. Changes in a vegetation index were also investigated. The fire regime was analysed using two MODIS sensor products: active fires and burned area. Temporal and spatial distribution of fires was analysed. The LULC classification accuracy was 80.3, 79.0 and 77.6% for 2000, 2005 and 2015 respectively. The deforestation rate in the district between 2000 and 2015 was 1.9%. The vegetation index indicated an overall decrease in plant biomass in the same period. The period between August and October comprised 96% of the active fires, being more frequent in western and southern Gurué. Fire occurrence was independent of LULC class, and fire frequency was not correlated with changes in LULC or plant biomass reduction. Fires in Gurué appear to be a tool in LULC change, rather than a driver of deforestation and forest degradation.

ADVANCING SATELLITE-BASED PHENOLOGY MONITORING: A CASE IN SOUTHERN AFRICA

Dubovyk O[1], Parplies A[1], Landmann T[2], Erasmus B[3], Schellberg J[4]
[1]Center for Remote Sensing of Land Surfaces, University Of Bonn, [2]International Centre of Insect Physiology and Ecology, [3]University of Witwatersrand, [4]Agrar- und Produktionsökologie, University of Bonn

Monitoring land surface phenology from satellites is essential for characterizing vegetation dynamics over large areas. Availability of very high spatial resolution image time series from RapidEye satellites constellation allow the phenology analysis also at field level, which is especially important for fragmented landscapes such as in South Africa. A bi-weekly time series of the multispectral 5m RapidEye images, covering the 2012-2013 growing season, were used to derive key phenometrics for a study area located in the Free State Province of South Africa.

Based on a noise-reduced Normalized Difference Vegetation Index (NDVI) time series, a set of quantitative metrics were extracted that summarized the phenology of vegetation at field, farm and regional scales. Vegetation parameters included both phenology metrics (start, end, length of growing season) and productivity metrics (amplitude and the small integral). For the study area, we detected one growing season that started around November and ended in July. Further statistical analysis showed that the spatial patterns of phenometrics were correletaed (R2=..?) with different farming management systems. The set of key phenometrics including length, amplitude and small integral proved to be beneficial for comparing and detecting variances across the investigated area in the grassland ecosystem. It was also possible to differentiate productivity changes at field level over the analysed period of time. The applied satellite-based approach yielded promising results for the analysis of vegetation dynamics and land surface phenology at high level of spatial details. The elaborated approach is deemed to be useful for precision agriculture applications and could be applied for similar cases in South Africa and beyond.

AEROSOL-RADIATION-CLOUD INTERACTIONS IN THE SOUTH-EAST ATLANTIC: FIRST RESULTS FROM NASA'S ORACLES-2016 DEPLOYMENT AND PLANS FOR FUTURE ACTIVITIES

Redemann J[1], Wood R[2], Zuidema P[3], Haywood J[4,5], Piketh S[6], Formenti P[7], Knox N[8], Holben B[9], Abel S[5], Science Team O

[1]NASA Ames Research Center, [2]University of Washington, [3]University of Miami, [4]University of Exeter, [5]Met Office, [6]North-West University, [7]LISA, CNRS, Université Paris Est Créteil et Université Paris Diderot, [8]Namibia University of Science and Technology, [9]NASA Goddard Space Flight Center

Southern Africa produces almost a third of the Earth's biomass burning (BB) aerosol particles. Particles lofted into the mid-troposphere are transported westward over the South-East (SE) Atlantic, home to one of the three permanent subtropical stratocumulus (Sc) cloud decks in the world. The SE Atlantic stratocumulus deck interacts with the dense layers of BB aerosols that initially overlay the cloud deck, but later subside and may mix into the clouds. These interactions include adjustments to aerosol-induced solar heating and microphysical effects, and their global representation in climate models remains one of the largest uncertainties in estimates of future climate. Hence, new observations over the SE Atlantic have significant implications for regional and global climate change predictions.

Our understanding of aerosol-cloud interactions in the SE Atlantic is severely limited. Most notably, we are missing knowledge on the absorptive and cloud nucleating properties of aerosols, including their vertical distribution relative to clouds, on the locations and degree of aerosol mixing into clouds, on the processes that govern cloud property adjustments, and on the importance of aerosol effects on clouds relative to co-varying synoptic scale meteorology.

We describe first results from various synergistic, international research activities aimed at studying aerosol-cloud interactions in the region:

• NASA's airborne ORACLES (ObseRvations of Aerosols Above Clouds and Their IntEractionS) deployment in August/September of 2016,

• the DoE's LASIC (Layered Atlantic Smoke Interactions with Clouds) deployment of the ARM Mobile Facility to Ascension Island (June 2016 – October 2017),

• the ground-based components of CNRS' AEROCLO-sA (Aerosols Clouds and Fog over the west coast of southern Africa), and

• ongoing regional-scale integrative, process-oriented science efforts as part of SEALS-sA (Sea Earth Atmosphere Linkages Study in southern Africa).

We expect to describe experimental setups as well as showcase initial aerosol and cloud property distributions. Furthermore, we discuss the implementation of future activities in these programs in coordination with the UK Met Office's CLARIFY (CLoud-Aerosol-Radiation Interactions and Forcing) experiment in 2017.

AFRICA DATA INTENSIVE RESEARCH CLOUD: EARTH OBSERVATION DATA PLATFORM IN SUPPORT OF THE AFRICAN SPACE PROGRAMME AND AFRIGEOSS INITIATIVE

Mlisa A[1], Munsami V[2], Newby T[3], Hugo W[4], Sithole H[5]
[1]Group On Earth Observations (geo), [2]Department of Science and Technology, [3]Agriculture Research Council (ARC), [4]South African Environmental Observation Network (SAEON), [5]Center for High Performance Computing

Earth observations (both space and in-situ) data and information are already being used in Africa for decision making in different socio-economic and natural resource areas, including, health, energy, water, agriculture and transport, thanks to a number of initiatives implemented by different actors at national, sub-regional and continental levels. Through the support of the Group on Earth Observations (GEO) Data Sharing Principles the Earth observation community has at its disposal a suite of freely available satellite data sets, which include low to high-resolution satellite data. However, the infrastructure challenges relating to data storage, processing and bandwidth limits the "free and open" access to these datasets in Africa. The AfriGEOSS Initiative and the African Union's recently approved African Space Policy and the African Space Strategy, call for the integration of space related efforts and infrastructure across the continent to enable the use of space science and technology for improving the quality of life of its citizens. Current efforts on developing the requisite infrastructure tend to be based on single data streams for a specific domain, tailored for a single institution, country or region, and at most is project based. Africa may currently lack adequate data infrastructure, but already there is an emerging capacity that can be tapped into and built upon. A number of African countries have or are developing centers of high performance computing, which should be explored for use in the Earth observations domain. The Africa Data Intensive Research Cloud (ADIRC) is developing an infrastructure that potentially will facilitate access to Earth Observations data for the continent. The pilot stage of the ADIRC Earth Observation platform envisages to establish an open access African Agricultural Data Cube. The platform will enable access to data and information on agriculture by both government and private sector agricultural stakeholders for a food secure Africa.

AFRICA LAND COVER MAPPING AT 10 M WITH SENTINEL-2: CURRENT ACHIEVEMENTS OF THE LAND COVER COMPONENT OF THE ESA CLIMATE CHANGE INITIATIVE

Defourny P[1], Achard F[2], Boettcher M[3], Bontemps S[1], Brockmann C[3], Eberenz J[4], Gamba P[5], Georgievski G[6], Herold M[4], Hagemann S[6], Hartley A[7], Kirches G[3], Lamarche C[1], Lisini G[5], MacBean N[8], Moreau I[1], Ottlé C[8], Peylin P[8], Riedel T[9], Salentinig A[5], Santoro M[10], Schmullius C[9], Vittek M[1], Ramoino F[11], Arino O[11]
[1]UCLouvain-Geomatics, [2]Joint Research Center, [3]Brockmann Consult, [4]Wageningen University, [5]University of Pavia, [6]Max Planck Institute, [7]Met Office, [8]Laboratoire des Sciences du Climat et de l'Environnement, [9]Jena University, [10]Gamma RS, [11]European Space Agency

Essential Climate Variables were listed by the Global Climate Observing System as critical information to further understand the climate system and support climate modelling. The European Space Agency launched its Climate Change Initiative to provide an adequate response to the requirements for long-term satellite-based products for climate.

Within this program, the land cover (LC) project aims at addressing the requirement of a high spatial resolution LC map expressed by the climate science community. To reach that goal, the project will generate a 10-m prototype map over Africa based on Sentinel-2 and supplemented by Landsat-8 with a legend of 10 classes. Taking advantage of the Sentinel era, the potentiality to derive a new water body product based on Sentinel-1 mission is analyzed.

The increase in spatial resolution requires significant methodological adjustments and innovations to the processing chains developed for medium spatial resolution imagery at global scale. For the pre-processing, for example, the topography and adjacency effects have to be taken into account in the atmospheric correction. In addition, due to the lower revisiting capacity of high spatial resolution sensors such as Sentinel-2 and Landsat-8, the spatial consistency of surface reflectance between few images becomes a critical aspect in the production of high spatial resolution composites.

For the LC classification, challenges are of a different nature. The decametric resolution captures the landscape elements diversity and distinct evolution through time due to slightly different seasonality and ecological gradients. Specific effort is made to ensure the consistency between this decametric map and the medium resolution global LC maps already developed within the project. While the rich literature on LC mapping at high resolution supports the processing chain development, the data flow provided by Sentinel-2 forces to revisit the classification strategy to map the LC consistently over space and time.

AFRICAN TEMPORAL GEO-SPATIAL DATA, APPRAISE THE PAST TO EVALUATE PLANS FOR THE FUTURE

Becker R[1,2,3]
[1]*Geodyn Technology,* [2]*Gistec,* [3]*GeoMark Systems*

African Temporal Geospatial Data

Appraise the Past to evaluate plans for the future

To plan for the future, it is essential to understand the present as well as the past. It is hindsight that is the most valuable asset when examining how to proceed. Apart from natural catastrophes, conditions do not happen, they develop. To change or influence them, not only must their present state be considered, but it is essential to investigate what triggered them in the first place. Just as a doctor makes an anamnesis before prescribing a treatment.

This requires access to records that might have to go back many years. In Africa, such records are hard, if not impossible, to retrieve. Moreover, whatever can be located is likely to have been compiled by a great variety of agencies, each presenting their own version of events and, consequently, often contradictory.

To overcome such problems, it is proposed to use aerial photography coverages that have been captured and – hopefully – archived over the past decades. Throughout this period, these Temporal Geospatial Data trace the course, interaction and consequences of events. Combined with Satellite Imagery, they provide unbiased evidence of yesterday's development that can now be analyzed and assessed based on the knowledge of today.

This presentation deals with the advanced data conversion and automated processing procedures required for convenient and user-friendly information extraction, as well as with Data Mining and Big Data Processing.
The aspect of Temporal Aerial Photography to document National Heritage is not covered by this presentation.

KEYWORDS: Aerial photography, Temporal geospatial data

AFRIGEOSS INITIATIVE: STRENGTHENING USE OF EARTH OBSERVATIONS AND BRINGING GEOSS TO AFRICA

Mlisa A[1]
[1]*Group On Earth Observations (geo)*

The development and uptake of Earth observation (EO) data, information and knowledge is critical to improving the socio-economic status of the African continent. The Group on Earth Observations (GEO)'s AfriGEOSS initiative recognizes the need to improve and coordinate observation systems across the Societal Benefit Areas (SBAs) in Africa, in particular, in water resource management, food security and agriculture, sustainable urban planning and development and sustainable forest management. Strong advocacy of open data-sharing policies and practices, improved infrastructure, as well as for increased use of EO data and information in policy and decision-making, are the foundation of moving forward in these vital SBA areas. Similarly, focusing significant effort on building human, institutional and technological capabilities will ensure the African continent benefits from better access, understanding and use of EO data, products and services. AfriGEOSS provides a robust framework for coordination and connecting relevant stakeholders, institutions and agencies across Africa and with Africa's international partners to initiate mutual activities within the scope of GEO and in support of key environmental and related African agendas, the recently adopted African Space Programme and the African prioritized sustainable development goals.

To advance the progress made in meeting the AfriGEOSS objectives, ownership – illustrated by commitment of both in-kind and financial resources - by the African community is crucial. Secondly, success is largely dependent on access to the information of Who is doing What Where. This information needs to come from within Africa, to strengthen partnerships on the continent; from international partners, to enable participation and linkages of African activities with all relevant GEO activities, thus enabling AfriGEOSS to fulfill its objective of being the gateway to Africa for EO activities, thereby reducing duplication of efforts, and leveraging all investments being made in or for Africa.

ALGORITHM FOR IMPROVED TREE COUNTING AND DETECTION OF HIGH VALUE CROPS IN MISAMIS ORIENTAL THROUGH ADAPTIVE MACHINE LEARNING APPROACH WITH THE INTEGRATION OF WATERSHED TRANSFORMATION AND LOCAL MAXIMA ANALYSIS

Pelayo J[1], Villar R[1]
[1]*Phil-LiDAR 2.B.11*

The Philippines is long considered as a valuable producer of high value crops globally. The country's employment and economy have been dependent on agriculture, thus increasing its demand for efficient agricultural mechanism. Remote sensing and geographic information technology proven to effectively provide applications for precision agriculture through image-processing technique considering the development of the aerial scanning technology in the country. Accurate information concerning the spatial correlation within field is very important for precision farming of high value crops, specially. The availability of height information and high spatial resolution images obtained from aerial scanning together with the development of new image analysis methods are offering relevant influence to precision agriculture techniques and applications. In this study, an algorithm was developed and implemented to detect and count high value crops simultaneously through adaptive scaling of support vector machine (SVM) algorithm subjected to object-oriented approach combining watershed transformation and local maxima criterion. The methodology is compared to cutting-edge template matching algorithm procedures to demonstrate its effectiveness on a demanding tree counting recognition and delineation problem. Since common data and image processing techniques are utilized, thus can be easily implemented in production processes to cover large agricultural areas. The algorithm are tested on high value crops like Palm, Mango and Coconut located in Misamis Oriental, Philippines - showing a good performance in particular for young adult and adult trees, significantly 90% above. The process can also be used as support tools for agricultural inventories or database updating, allowing for the reduction of field work and manual interpretation tasks.

KEYWORDS: Machine Learning, LiDAR, OBIA, Precision Agriculture

ALL FORESTS ARE NOT EQUAL: QUANTIFYING THE BIODIVERSITY VALUE OF MOUNTAIN FORESTS IN JAVA, INDONESIA

Higginbottom T[1], **Symeonakis E[1]**, Collar N[2], Marsden S[1]
[1]Manchester Metropolitan University, [2]Birdlife International

Tropical forests are the most biodiverse ecosystems on the planet and a range of policies in place for their protection. Although the last decade has seen major progress in mapping forest characteristics, such as cover and biomass, quantifying biodiversity value remains an elusive goal. With human and climate-related pressures expected to increase over the coming decades, mapping biodiversity value is more critical than ever for ensuring conservation efforts are deployed as effectively as possible. Mapping the value of forest for biodiversity is a complicated endeavour, as it requires quantification of various interlinking parameters. Unfortunately, no single Earth-observation sensor can provide a full inventory of these parameters. However, recent developments in open-data allow multi-sensor approaches. On the one hand, optical sensors, e.g. Landsat, offer a means for 2-dimensional landcover mapping, but provide no context on 3-dimensional properties. Conversely, radar data, in particular L-band wavelengths, can accurately capture the 3-d properties for biomass and height quantification.

We implement a multi-sensor approach for quantifying the biodiversity value of highland forests in west Java, Indonesia. We combine Landsat-derived land cover maps, ALOS-PALSAR-based biomass estimates and habitat fragmentation metrics as complementary measures of forest health and biodiversity value. Our results indicate that large areas of forest has relatively little biodiversity value, due to low biomass and high fragmentation, attributable to selective logging and encroachment from smallholder agriculture. Protected areas, especially Halimun Salak National Park, contain the most productive forests, due to protection of older trees. Further protection is provided by steep mountainous areas that are difficult to access and unsuitable for exploitation. These areas have intact forests with little fragmentation, but are not suitable for supporting high biomass due to the steep slopes. These results are of value to biodiversity conservation efforts in Java, and will be distributed through the dryad repository (http://datadryad.org/) .

ALOS-2 AND ITS FOLLOW-ON SATELLITE

Hayashi M[1], Motohka T[1], Suzuki S[1], Shimada M[2,1]
[1]*Japan Aerospace Exploration Agency (JAXA),* [2]*Tokyo Denki University*

Advanced Land Observing Satellite-2 (ALOS-2) is a Japanese earth observation satellite launched in 2014 for disaster mitigation, environmental monitoring such as forest and sea/land ice, land monitoring such as agriculture and natural resources, and technology development. ALOS-2 observes the earth surface with the Phased Array type L-band Synthetic Aperture Radar-2 (PALSAR-2) that has many improved performances compared to previous Japanese L-band SAR satellites. ALOS-2 data are systematically acquired according to the basic observation scenario (BOS) and are effectively used for many applications, for example, global forest/non-forest maps and JICA-JAXA Forest Early Warning System in the Tropics (JJ-FAST). To keep and enhance the applications using ALOS-2 data, JAXA plans to launch a successor satellite to the ALOS-2 in JFY 2021. The concepts of the ALOS-2 follow-on are expanding swath width and increasing observation frequency while keeping highly spatial resolution of the ALOS-2 (e.g., 200 km swath at 3 m resolution) for improving the response of disaster monitoring, early detection of anomalies on the earth surface, and enabling time-series analysis. We present the status of ALOS-2, its applications for environment, and the ALOS-2 follow-on satellite.

KEYWORDS: ALOS-2, PALSAR-2, L-band SAR, disaster monitoring, forest monitoring

AMERIGEOSS COMMUNITY PLATFORM: AN EARTH OBSERVATION DELIVERY FRAMEWORK THAT BRINGS TOGETHER SOCIAL, ECONOMIC AND ENVIRONMENTAL DATA FROM A GLOBAL COMMUNITY OF CONTRIBUTORS TO SUPPORT UNDERSTANDING AND DECISION-MAKING

Frazier E[1,2,3], DeLoatch I[1,2,3]
[1]Federal Geographic Data Committee, [2]Group on Earth Observation , [3]U.S. Geological Survey

AmeriGEOSS is a framework that seeks to promote collaboration and coordination among Group on Earth Observation (GEO) Members of the American continent, "to realize a future wherein decisions and actions, for the benefit of the region, are informed by coordinated, comprehensive and sustained Earth observations and information". The AmeriGEOSS Community Platform was established as a capacity building capability in 2016 to increase regional capacity to acquire, share, store, maintain and utilize Earth Observation data and information. The user-driven community platform is envisioned to provide local, national and global open data accessible, tools, applications, products and services that support understanding and decision-making.

The AmeriGEOSS community platform brings together social, economic and environmental data from a global community of contributors to support communal access, discovery and usability. The AmeriGEOSS Community Platform provides tools and applications that support data visualization, curation, harmonization, analytics, collaboration, co-creation of products and services and other resources that can support capacity building and accelerate understanding and decision-making.

In this session Mr. Frazier will provide accomplishments, lessons learned, best practices and future plans for the AmeriGEOSS Community Platform.

AMERIGEOSS: PROGRESS AND PRIORITIES FOR EARTH OBSERVATION-INFORMED DECISIONS IN THE AMERICAS

Searby N[1], Gutierrez A[2], Quimbay Valencia D[3], Medico A[4], Cumberbatch C[5], Ferreira H[6], Mills L[7]
[1]NASA Earth Sciences' Applied Sciences Program, [2]NOAA Office of Water Prediction, [3]IDEAM, [4]CONAE, [5]Belize Hydromet Service, [6]INPE, [7]Environment and Climate Change Canada

The AmeriGEOSS initiative promotes collaboration and coordination among the Group on Earth Observations (GEO) members of the American continent, "to realize a future wherein decisions and actions, for the benefit of the region, are informed by coordinated, comprehensive and sustained Earth observations and information." The Americas Caucus members (Argentina, Bahamas, Belize, Brazil, Canada, Chile, Colombia, Costa Rica, Ecuador, Honduras, Mexico, Panama, Paraguay, Peru, the United States, and Uruguay) have recognized the value of working together to meet regional development and sustainability objectives. Since October 2014, the initiative has focused on four priority GEO Societal Benefit Areas, prioritized by the Americas Caucus members: food security and sustainable agriculture, disaster resilience, water resources management, and biodiversity and ecosystem sustainability. Foundational activities, e.g. data infrastructure and capacity building, are also a focus. Needs have been identified by member countries in each of these areas to leverage and tailor GEO global initiatives and foundational activities such as the GEO Global Agricultural Monitoring System (GEOGLAM), the GEO Global Water Sustainability (GEOGLOWS), GEONETCast, and the new Pole-to-Pole Marine Biodiversity Network (mBON), to regional needs. Workshops and training events, including AmeriGEOSS Week 2016, have begun to address these regional needs. In 2016, the initiative focused on the development of the AmeriGEOSS Community Platform to bring EO resources together and to provide the environment for regional coordination and collaboration. In 2017, in coordination with the priority area working groups, test cases on the use of the platform will be developed for demonstration purposes and future platform improvements. Efforts will also focus on the development of better software to support the use of the GEONETCast system in the region, and will continue to implement capacity building activities and projects amongst its GEO members. Regional GEO principals invite non-Americas Caucus members to join the efforts of AmeriGEOSS.

AN ABOVE-GROUND BIOMASS MAP OF AFRICAN SAVANNAHS AND WOODLANDS AT 25 METERS RESOLUTION DERIVED FROM ALOS PALSAR

Bouvet A[1], Mermoz S[1], Villard L[1], Mathieu R[2], Le Toan T[1], Naidoo L[2]
[1]CESBIO, [2]CSIR

Savannahs and woodlands are among the most important biomes in Africa: they cover half of sub-Saharan Africa, provide vital ecosystem services to the rural communities, and play a major part in the carbon budget. Despite their importance and their fragility, they are much less studied than other ecosystems like rainforests. In particular, the distribution and amount of the above-ground woody biomass (AGB) is largely unknown . In this paper, we produce the first continental map of the AGB of African savannahs and woodlands at a resolution of 25 meters. The map is built from the 2010 L-band PALSAR mosaic produced by JAXA, along the following steps: a) stratification into wet/dry season areas in order to account for seasonal effects, b) development of a direct model relating the PALSAR backscatter to AGB, with the help of in situ and ancillary data, c) Bayesian inversion of the direct model. A value of AGB and its uncertainty has been assigned to each pixel. This approach allows estimating AGB until 85Mg.ha-1 approximately, while dense forests and non-vegetated areas are masked out using the ESA CCI Land Cover dataset. The resulting map is validated using a cross-validation approach and a comparison with AGB estimates obtained from LiDAR datasets, leading to an RMSE of 12 to 17 Mg.ha-1 . Finally, carbon stocks for savannahs in Africa and 34 countries are estimated and compared with estimates by FAO.
The approach developed in this study can be applied to to open forests of other continents, provided high quality in situ AGB estimates are available, and to similar L-band mosaics produced at later epochs, for example the ALOS-2 PALSAR-2 mosaics that are being produced on a yearly basis since 2015.

AN ASSESSMENT OF URBAN SPRAWL USING ARCHIVAL SPOT IMAGERY IN THE CITY OF TSHWANE, SOUTH AFRICA

Magidi J[2], Ahmed F[2]
[1]Tshwane University of Technology, [2]University of Witwatersrand

Gaining an understanding of the extent and dynamics of urban sprawl is quite crucial in sustainable land use management and spatial planning. Information on urban sprawl is an important input for predicting future urban land use changes and sustainable urban planning. Influx of people into urban areas due to migration and natural population increase has led to rapid increase in urban areas and encroachment of urban like environments into non-urban areas. One of the most observable effects of urbanisation is the controlled and uncontrolled urban sprawl and morphological change of urban areas. This study is aimed at assessing and monitoring the similarities and differences in urban sprawl dimensions, patterns and trends in the four main suburban areas (two former townships and two former suburbs) in the City of Tshwane (COT) which are: Northern Townships, Eastern Townships Pretoria Moot and Pretoria East Suburbs. Classified Spot 5 and Spot 6 images of 2005, 2010, and 2015 respectively were used in the analysis and quantification of spatio-temporal variations in urban sprawl in the four study areas. Landscape metrics revealed significant changes in urban sprawl between these four regions in the City of Tshwane and the use of remotes sensing in helping decision maker with relevant planning information.

KEYWORDS: urban Sprawl, SPOT, GIS, land use change, comparison, landscape metrics

AN EVALUATION OF SENTINEL-2 BASED LAI AND CANOPY LEAF CHLOROPHYLL PRODUCTS FOR AGRO-ECOLOGICAL APPLICATIONS

Adjorlolo C[1], Mashiyi N[1], Mangara P[1], Kganyago M[1], Odindi J[2]
[1]Sansa, [2]University of Kwa-Zulu Natal

Accurate estimation of vegetation biophysical parameters is necessary for a large variety of agro-ecological applications. The recent availability of free and open high spatial resolution data from Sentinel-2 sensors should offer better capabilities for retrieving specific vegetation variables frequently over wide-swath. The retrieval of canopy leaf area index (LAI) and chlorophyll content in the leaf for specific crop types or heterogeneous grasslands using sentinel-2A top of canopy (TOC) reflectance is investigated. The LAI and canopy leaf chlorophyll content retrieval algorithms were implemented by means of the ESA's SNAP Sentinel-2 biophysical processor, for some test sites in South Africa. The algorithms are based on well-established methods that have been proven efficient for measuring biophysical variables using other remote sensing data, such as, MERIS, SPOT, and LANDSAT sensors. The current study sets out to evaluate LAI and leaf chlorophyll content products retrieval using Sentinel-2 images along with actual field measurements – by means of Plant Canopy Analyzer and Chlorophyll Content Fluorescent Meter instruments. To validate the performance of the models' inversion based on Sentinel-2 TOC reflectances, the RMSE and R2 between the field measurements that coincide with the Sentinel-2 image acquisition dates and the estimated parameters will be tested. Results of this study should demonstrate the performance of the Sentinel-2 data and associated retrieval algorithm for the estimation of LAI and chlorophyll content in the leaf.

AN INTERFEROMETRIC SYNTHETIC APERTURE RADAR (INSAR) ANALYSIS FOR WATER LEVEL MONITORING: A CASE STUDY IN NEWFOUNDLAND AND LABRADOR

Mohammadimanesh F[1], **Salehi B[1]**, Motagh M[2], Brisco B[3], Mahdianpari M[1]
[1]C-CORE, and Memorial University of Newfoundland, [2]Helmholtz Center Potsdam, GFZ German Research Center for Geosciences, [3]Canada Centre for Mapping and Earth Observation

Wetland ecosystems are one of the most productive and desirable habitat for a variety of plant and animal species. Despite their benefit, they are not well maintained and understood. Radar and optical satellite imagery is widely used for monitoring vegetation patterns and wetlands, especially for classification of different types of vegetation. However, monitoring of wetland water level using satellite imagery is not investigated that much. Interferometric Synthetic Aperture Radar (InSAR) technique has been proved to be an efficient technique for displacement measurements of solid earth changes such as earthquake, landslide, and subsidence. However, the use of such techniques in monitoring water level changes is still in its early stages and need more investigation to be done. In fact, using InSAR technique for water level changes is challenging due to low(or zero) backscatter from water surface and that is why water surface appears dark in a SAR image. However, in case of double-bounce scattering between water surface and flooded vegetation, interferometric coherence will be maintained high and water level change is observable in phase data. InSAR technique provides high resolution spaced-map of water surface fluctuation which is time and cost efficient in comparison with field campaign. In this study, the capability of this technique for water level changes are investigated using TerraSAR-X data in Newfoundland and Labrador province, Canada. Our preliminary results are promising, and final results will be demonstrated in the full paper.

KEYWORDS: Water level, Wetland, InSAR, interferogram, TerraSAR-X

AN OPEN-SOURCE, MULTI-SCALAR APPROACH FOR MONITORING AND REPORTING ON LAND DEGRADATION

Zvoleff A[1], Gonzalez-roglich M[1], Andelman S[1]
[1]*Conservation International*

Sustainable Development Goal (SDG) target 15.3 calls for countries to "combat desertification, restore degraded land and soil, including land affected by desertification, drought and floods, and strive to achieve a land degradation-neutral world". To track progress towards this goal, countries must be able to measure changes in the area of land that is degraded. However, a lack of standardized methods and capacity for implementing monitoring hampers progress towards these goals.

The GEF-Land Degradation Monitoring Project is a Global Environment Facility (GEF)-funded project responding to these challenges by providing guidance on robust methods and an open-source toolbox for assessing, monitoring status, and estimating trends in land degradation using remote sensing. The project is using a multi-scalar approach, drawing on a range of imagery, from AVHRR (8 km), to MODIS (250 m), Landsat (30 m), and very high resolution commercial imagery (50 cm), to enable countries to track and understand degradation at a fine scale and (at lower resolutions) over a 30-year time series.

The project is piloting products and tools in four countries (Kenya, Tanzania, Uganda, and Senegal) through a partnership between Vital Signs at Conservation International, Lund University and the National Aeronautics and Space Administration (NASA). The project will provide an open source, open access data platform with tools and data products that will be available to all countries as a global public good. The guidance and toolbox will be available by December 2017 and can be employed directly to facilitate the calculation of the sub-indicators for SDG target 15.3, as well as to inform land management and investment decisions and improve reporting to the United Nations Convention to Combat Desertification (UNCCD) and to the GEF.

AN OPERATIONAL PROCESSING CHAIN FOR MEDIUM RESOLUTION EARTH OBSERVATION DATA TO ESTABLISH DATACUBES FOR ENVIRONMENTAL MONITORING IN SOUTHERN AFRICA

Röder A[1], Frantz D[1], Hill J[1], Schneibel A[1], Stellmes M[2]
[1]Trier University, Dept. of Environmental Remote Sensing and Geoinformatics, [2]Freie Universität Berlin, Remote Sensing and Geoinformatics

Monitoring environmental processes, such as deforestation or land degradation, requires time series of data at sufficient quantitative quality. Reflecting the need to cover long time periods, Landsat data are one of the most valuable resources for Earth observation, which may be complemented by Sentinel-2 for recent dates. Open data policy provides unique opportunities to cover large areas with a level of geometric detail appropriate for many applications; however, data scarcity during wet seasons, a very low signal level in sparsely vegetated areas and the WRS structure are obstacles for operational interpretation approaches, and limit the suitability of existing data production systems such as the LEDAPS. We developed a processing chain that addresses all relevant aspects from cloud-/cloud shadow-detection, individual retrieval of water vapor and aerosol optical depth for radiometric correction for every acquisition date, correction of topography-induced illumination variation, BRDF-correction and projection to a gridded database irrespective of WRS-2 conventions.

Based on this, a number of second- and third-level products can be supplied. On the one hand, this includes the full time series of reflectance data of any existing image for a given location (or seasonal/temporal subsets thereof), which enables application of time series applications such as trend analysis or temporal segmentation. Further to this, we developed a method to fuse MODIS phenology and Landsat reflectance data to generate Landsat-resolution phenology layers. These can serve as valuable information products by themselves, but are also employed in the generation of phenology-adaptive, pixel-based composites that support large-area applications requiring multi-temporal datasets, for instance to support regional or national reporting requirements under international programs.

The derived data products have been utilized in a number of case study applications on rangeland dynamics, deforestation or agricultural expansion in different countries of Southwestern Africa, such as Angola, Zambia, Botswana and Namibia.

AN OPERATIONAL RPAS PROGRAM FOR ENVIRONMENTAL MONITORING IN NORTHERN AUSTRALIA

Bartolo R[1], Whiteside T[1], Sparrow B[2]
[1]Department Of The Environment And Energy-eriss, [2]The University of Adelaide

The Supervising Scientist Branch (SSB), within the Australian Government's Department of the Environment and Energy, is developing an environmental monitoring program using remotely piloted aircraft systems (RPAS). The RPAS program is key in providing data to support scientific research related to monitoring the Alligator Rivers Region (including the World Heritage listed Kakadu National Park (KNP)) for potential impacts of uranium mining. The SSB also undertakes landscape assessment work in KNP in collaboration with park management. With the closure of Ranger uranium mine scheduled for 2026, methods for assessing and monitoring closure criteria, particularly for revegetation and landform, are the focus of the RPAS research.

SSB operate a range of RPAs (both fixed wing and multi rotors) and associated sensors. The most used platform is the Swampfox fixed wing due to its ability to cover larger areas than the multi rotor platforms. SSB's sensor suite includes standard and modified DSLR cameras, a 5 band multispectral sensor, a hyperspectral sensor (100 spectral bands), a thermal sensor, live link video and a LiDAR system.

This paper summarises the results of some of the key RPAS applications SSB have been working on (such as mine site revegetation assessment over time and surface water turbidity monitoring of inland water bodies). Specifically, we present a case study on deriving vegetation structure and composition information in savanna ecosystems. The case study highlights the use of multispectral RPAS imagery acquired in KNP in conjunction with detailed ground survey data undertaken for the Terrestrial Ecosystem Research Network's AusPlots ecological surveillance monitoring program. Combining RPAS analysis with field-based monitoring data enables information on composition and structure to be mapped over larger areas than is possible with field data alone. This assists park managers in determining appropriate management practices and where to implement them.

KEYWORDS: UAV, UAS, savanna, mine site monitoring

AN OVERVIEW OF OPERATIONAL EARTH OBSERVATIONS DATASETS FOR MONITORING SUSTAINABLE DEVELOPMENT GOALS IN SOUTH AFRICA

Newby T[1], Abd Elbasit M[1], Chirima G[1]
[1]*Agricultural Research Council*

Operational and easily accessible Earth Observations datasets are becoming available for monitoring terrestrial environments. Seventeen Sustainable Development Goals with 160 targets have been published by the United Nations to guide the public, private and volunteer sectors in global development. Monitoring progress towards these goals and targets requires suitable systems and data collection strategies, many of which have still to be developed. National governments require national data collection systems. The Group on Earth Observation (GEO) has over the past 12 years made significant strides in encouraging and facilitating open access to operational earth observation datasets. Many of these datasets are suitable for monitoring progress towards achievement of the SDG's. Specifically, access to course and medium resolution satellite image datasets from sensors on board satellites such as NOAA, MODIS, ProbaV, CBERS, Landsat and the Sentinels make the monitoring of many of the SDG goals feasible. Considering the monitoring of the SDG's that relate to terrestrial natural resources, climate change, food security & agriculture (goals 2, 6. 13 and 15) South Africa has the opportunity of utilising these cloud free image data because of its geographical advantage. The country is largely located in a climatic region where cloud cover makes optical remote sensing feasible for monitoring terrestrial landscapes. RADAR datasets are also increasingly becoming accessible. Further the temporal, spectral and spatial resolution of open access image datasets available and suitable for national monitoring for the country further motivate the use of these data for SDG monitoring. The paradigm shifts in using cloud computing for image processing whereby the algorithms are sent to the data as opposed to bringing the data to the algorithm makes the need for large band width and computing infrastructure obsolete. These factors make Earth Observation datasets and technology a feasible means of monitoring selected SDG's in South Africa.

AN OVERVIEW OF THE UNITED NATIONS URBAN SDG TARGETS AND INDICATORS: IMPLICATIONS FOR SPATIAL DATA NEEDS

Ndugwa R[1]
[1]*Un-habitat*

Sixty per cent of the global population will live in cities by 2030, with 90% of urban growth in coming decades likely to occur in low- and middle-income countries. By 2030, a majority of people in every region will live in urban areas, including in Africa and Asia, which are currently the least urbanized regions. The steady trend towards urbanization will influence virtually every facet of human endeavor in the coming years, including health, economic, social, and environmental. Towards this aim, the world has embraced a series of targets to address the global urban agenda and this is well reflected in the stand-alone SDG 11 and other urban related indicators in other SDGs.

Several key principles underpin the monitoring approaches to cities performance over the next 15 years. Monitoring cities as units of analysis and reporting performance at national levels requires ambitious and transformative shifts at all levels in the way data has been collected in the past, including an intensified focus on deploying new spatial analysis technologies.

This presentation will share updates on the design of monitoring tools based on the concept of national sample of cities and how this is being used to study the in-depth aspects of SDGs and other urban agenda targets. Links to the City Prosperity Initiative under UN-Habitat will be made, and demonstrations will be provided on how cities have been studied and compared using a uniform set of indicators in selected countries leading to strengthening of the capacities of national governments to create a multilevel coordination of national/local monitoring and reporting, and reinforce inter-linkages of Goal 11 indicators and other SDGs with an urban component, including the systematic dis-aggregation of information along key dimensions of urban development and spatial city forms.

KEYWORDS: Spatial data, SDGs, City prosperity

ANALYSIS OF FOREST FRAGMENTATION BY REMOTE SENSING APPLICATION IN NEW CALEDONIA

Despinoy M[1], Mangeas M[1]
[1]*Ird*

Tropical moist forests on ultramafic rocks encompass about 1380 species of vascular plants, 82% of which are endemic to New Caledonia (Jaffré et al., 2009). Human pressure has considerably fragmented and reduced the area originally occupied by this ecosystem. Today, these forests remain only in the form of small fragments (<10% of the Caledonian vegetation).

The analyzes of the DYNAMIC project are based on the effect induced by the fragmentation itself on the durability of the forest fragments. However, due to the heterogeneity of trees species, the distinction between forest classes is difficult and it is more pertinent to look for structural limits between vegetation classes in the canopy than for indices of floristic composition.

In this paper, we propose a innovative approach based on satellite imagery (QuickBird, Pleiades) which provides structural criteria aiming at characterizing the new-caledonian forest. The processing consists firstly in calculating the spatial autocorrelation allowing to estimate the spatial dependence between the values of the same variable (vegetation indices, brightness, variance...) in different geographical sites. Then, this study provides a pixel classification (SVM) according to their probability (ranging from 0 to 1) to belong or not to a forest structure for which boundaries have been demarcated by botanists. The best obtained model implements FV-ML55 and NDVI-ML55 (FV for Vegetation cover and ML for Moran Local with a 55 pixels window). It is validated with an AUC score of 0.83 corresponding to the ROC curve.

Then, the resulting "forest gradient", analyzed between 2004 and 2014, shows forest dynamics (mainly regressions). Finally, a study on connectivity was made bringing different scenarios on the relationships between patches.

These results allowed to assess the degree of connectivity between the forest patches, to identify the ecological corridors and to propose a method of prioritization in the conservation and restoration measures.

ANALYSIS OF TEMPORAL STATISTICS AND LONG TERM CLIMATE OBSERVATIONS FOR DERIVING THE PREDISPOSITION OF FORESTS TO STRESS EVENTS

Reichmuth A[1], Pinnel N[2], Rogge D[2], Heiden U[2]
[1]University Of Würzburg, [2]German Aerospace Center

Forest ecosystems are affected by stress induced changes in various ways. Environmental factors that affect trees negatively can be distinguished between biotic and abiotic factors. Abiotic factors are non-living such as drought, storm, frost, etc. Biotic factors are of living kind such as fungi or insects. Tree species react to stress in terms of activating their repair process and/or long-term adaptation of their morphology and metabolism. Depending on the strength of stress events this can lead to resistance and repair or severe damages and even plant death. However, with regard to water or nutrient supply, tree species respond very differently. Especially for coniferous tree species bark beetle infestations are a consequence of primary damage in form of drought and unfavourale conditions for trees. Therefor it is crucial to analyse the predisposition of forests to stress events. Long-term temporal statistics of Landsat data will be analysed for change of forest state and linked to temporal climate records and soil-moisture data. Especially longterm and repeated drought periods result in lower vitality of forests prolonging for several years after the drought event. The preliminary results of a case study will be presented and an outlook for further research will be given.

ANALYZING FIRE BEHAVIOUR FROM SPACE USING MEDIUM AND HIGH RESOLUTION IR SENSORS

Rücker G[1], Tiemann J[1], Leimbach D[1], Tanpipat V[2], Lorenz E[1]
[1]Zebris Gbr, [2]Chulalongkorn University, [3]DLR

Fire behaviour is well described by a fire's direction and rate of spread, and its energy release rate. Fire intensity as defined by Byram (1959) is the most commonly used term describing fire behaviour in the wildfire community. It is, however, difficult to observe from space.

Here, we assess fire spread and fire radiative energy release rate (fire radiative power) using infrared sensors with different spatial, spectral and temporal resolutions, focusing on sensors providing a spatial resolution higher than 500 m. The sensors used offer either high spatial resolution (Sentinel 2, Landsat 8) for fire detection with low temporal resolution, moderate spatial resolution and fire radiative power retrievals with moderate temporal resolution over selected sites (FIreBird satellites), or moderate spatial resolution, high temporal resolution, but no fire radiative power retrievals (S-NPP VIIRS I-band active fire product), or moderate to low spatial resolution and fire radiative power retrievals at high temporal resolution (MODIS).

We extracted fire fronts from Landsat and Sentinel 2 (using the Shortwave Infrared bands) and use the available fire products for S-NPP VIIRS and MODIS. We analyzed rate of spread and energy release rate for a sample of fire events in South Africa, Thailand, Brazil and Indonesia.

From these results we derive preliminary metrics on fire behaviour in our study areas.
For a number of observed fire events, we run fire behaviour models to compare observed and modelled rates of spread and energy release rates. We furthermore relate our results to published values on fire intensity.

KEYWORDS: Fire spread, fire radiative power, FireBird, Sentinel 2, S-NPP VIIRS

ANNUAL CROP-TYPE CLASSIFICATION FROM MULTITEMPORAL LANDSAT-8 AND SENTINEL-2 DATA BASED ON DEEP-LEARNING

Karakizi C[1], Vakalopoulou M[1], Karantzalos K[1]
[1]*National Technical University Of Athens*

Agricultural monitoring is of significant importance in order to distinguish, identify and measure the main crop production areas and estimate the production as early in the year as possible. To this end, multitemporal satellite remote sensing data are employed along with pixel-based or object-based classification techniques towards the accurate land cover mapping of agricultural areas. However, crop-type classification and identification is not a trivial task since the spectral signatures between the different crop types and grassland may be quite similar due to phenology, climate and cultivation practices. In this paper, we address the annual crop-type classification task based on multitemporal Landsat-8 and Sentinel-2 data depending on the year and availability. After the atmospheric corrections, the cloudy pixels are interpolated through a standard multitemporal analysis and then all data are simplified based on anisotropic morphological leveling. Based on a deep learning architecture and the manually collected extensive ground truth dataset from the main agricultural regions in Greece, a hierarchical feature learning procedure is taking place. Deep hierarchies of non-linear features are stacked and efficiently exploited by the classifier. Experimental results for three years namely 2014, 2015 and 2016 are promising. A comparative evaluation with standard SVM and random forest classifiers indicated higher overall accuracy rates for the developed methodology.

ISRSE37 TSHWANE 2017

ANNUAL GLOBAL LAND COVER MAPS FROM THE 1990S TO 2015 FOR CLIMATE MODELLING: THE LAND COVER COMPONENT OF THE ESA CLIMATE CHANGE INITIATIVE

Defourny P[1], Achard F[2], Boettcher M[3], Bontemps S[1], Brockmann C[3], Eberenz J[4], Gamba P[5], Georgievski G[6], Herold M[4], Hagemann S[6], Hartley A[7], Kirches G[3], Lamarche C[1], MacBean N[8], Moreau I[1], Ottlé C[8], Peylin P[8], Radoux J[1], Riedel T[9], Salentinig A[5], Santoro M[10], Schmullius C[9], Vittek M[1], Ramoino F[11], Arino O[11]

[1]Earth and Life Institute, Université cath. de Louvain, [2]Joint Research Center, [3]Brockmann Consult, [4]Wageningen University, [5]University of Pavia, [6]Max Planck Institute, [7]Met Office, [8]Laboratoire des Sciences du Climat et de l'Environnement, [9]Jena University, [10]Gamma Remote Sensing, [11]ESA-ESRIN

Essential Climate Variables were listed by the Global Climate Observing System as critical information to further understand the climate system and support climate modelling. The European Space Agency launched its Climate Change Initiative to provide an adequate response to the set of requirements for long-term satellite-based products for climate.

Within this program, the Land Cover project aimed at revisiting all algorithms required for the generation of global LC products that are stable and consistent over time. To this end, the LC concept was revisited to deliver a set of three consistent global LC products at 300 m spatial resolution for three 5-year epochs centered on the years 2010, 2005 and 2000. It also delivered climatological 7-day time series representing seasonal dynamics of the land surface and 7-day surface reflectance time series for the whole archive of MERIS data .

In January 2017, the project will deliver an updated dataset made of 300m global annual land cover maps from the 1990s to 2015. The entire MERIS archive from 2003 to 2012 is first classified into a unique 10-year baseline LC map, which is then back- and up-dated using (i) AVHRR time series from 1992 to 1999, (ii) SPOT-VGT time series from 1998 to 2012 and (iii) PROBA-V time series from 2013 to 2015. This method avoids independent classifications from year to year, thus ensuring temporal and spatial consistency between successive maps.

Another significant output of the project is a global map of open permanent water bodies at 150 m spatial resolution derived from Envisat ASAR images and ancillary data.

All products are delivered with an aggregation tool, enabling re-projection and re-sampling as well as the translation from land cover classes into Plant Functional Types for different climate models.

KEYWORDS: Global land cover mapping, climate, annual time series, 2015

ANTHROPOGENIC HEAT FLUX ESTIMATION FROM SPACE: RESULTS OF THE SECOND PHASE OF THE URBANFLUXES PROJECT

Chrysoulakis N[1], Marconcini M[2], Gastellu-Etchegorry J[3], Grimmond C[4], Feigenwinter C[5], Lindberg F[6], Del Frate F[7], Klostermann J[8], Mitraka Z[1], Esch T[2], Landier K[3], Gabey A[4], Parlow E[5], Olofson F[6]

[1]Ffoundation for Research and Technology Hellas (FORTH), [2]German Aerospace Center (DLR), [3]Centre d'Etude Spatiale de la Biosphère (CESBIO), [4]University of Reading, [5]University of Basel, [6]University of Goeteborgs, [7]GEO-K s.r.l., [8]ALTERRA

The H2020-Space project URBANFLUXES (URBan ANthrpogenic heat FLUX from Earth observation Satellites) investigates the potential of Copernicus Sentinels to retrieve anthropogenic heat flux, as a key component of the Urban Energy Budget (UEB). URBANFLUXES advances the current knowledge of the impacts of UEB fluxes on urban heat island and consequently on energy consumption in cities. In URBANFLUXES, the anthropogenic heat flux is estimated as a residual of UEB. Therefore, the rest UEB components, namely, the net all-wave radiation, the net change in heat storage and the turbulent sensible and latent heat fluxes are independently estimated from Earth Observation (EO), whereas the advection term is included in the error of the anthropogenic heat flux estimation from the UEB closure. The Discrete Anisotropic Radiative Transfer (DART) model is employed to improve the estimation of the net all-wave radiation balance, whereas the Element Surface Temperature Method (ESTM), adjusted to satellite observations is used to improve the estimation the estimation of the net change in heat storage. Furthermore the estimation of the turbulent sensible and latent heat fluxes is based on the Aerodynamic Resistance Method (ARM). Based on these outcomes, QF is estimated by regressing the sum of the turbulent heat fluxes versus the available energy. In-situ flux measurements are used to evaluate URBANFLUXES outcomes, whereas uncertainties are specified and analyzed. URBANFLUXES is expected to prepare the ground for further innovative exploitation of EO in scientific activities (climate variability studies at local and regional scales) and future and emerging applications (sustainable urban planning, mitigation technologies) to benefit climate change mitigation/adaptation. This study presents the results of the second phase of the project and detailed information on URBANFLUXES is available at: http://urbanfluxes.eu

APPLICATION OF A SEMI-EMPIRICAL MODEL IN THE ESTIMATION OF SAVANNA BIOMASS, USING HYPER-TEMPORAL C-BAND DATA

Main R[1,2], Mathieu R[1,2], Cho M[1], Wessels K[3,2], Asner G[5], Naidoo L[1,2], Kleynhans W[3,4]

[1]Ecosystems Earth Observation, Natural Resources and the Environment, CSIR, [2]Department of Geography, Geomatics and Meteorology, University of Pretoria, [3]Remote Sensing Research Unit, Meraka Institute, CSIR, [4]Department of Electrical, Electronic and Computer Engineering, University of Pretoria, [5]Department of Global Ecology, Carnegie Institution for Science

The estimation of woody biomass in southern African deciduous savannas is important given a) the critical role of trees and shrubs as a source of energy and timber products in rural populations, b) the need to monitor processes such as bush encroachment, and c) the combined effect of these interactions on regional savanna carbon stocks and fluxes. SAR-based biomass estimation using empirically-based methodologies experience shortcomings in terms of their generalisation and robustness across time series. Therefore this study investigated the applicability and performance of a SAR backscatter-based semi-empirical Water Cloud Model (WCM), and its subsequent inversion, in the estimation of above ground biomass, using hyper-temporal (n=14) C-band data (i.e. Sentinel1). Field biomass estimates (n=38, 1ha plots) were up-scaled to produce a ~63 000ha LiDAR-based biomass product, which together with multi-temporal Sentinel-1 backscatter data served to parameterise the WCM model. The model achieved encouraging accuracies, with dry season images producing R2 of ~0.65, RMSE of ~7.8 (Tonnes/Ha) and rRMSE values of ~44%. Wet season images produced poorer accuracies, and multi-temporal combinations of the two seasons reduced extreme values and helped smooth the effects of seasonal or phenological variability. The research indicates that it is possible to estimate savanna woody biomass using hyper-temporal C-band SAR backscatter coupled with a semi-empirical modelling approach; that the model is sensitive to seasonal changes in landscape moisture; and that there is potential for routine monitoring of savanna biomass using Sentinel1 data.

KEYWORDS: Biomass, Sentinel-1, C-band, Semi-empirical, Water cloud model

APPLICATION OF EARTH OBSERVATION IN MAPPING MANGROVE COVER WITHIN WESTERN INDIAN OCEAN

Ouko E[1]
[1]*Rcmrd*

Mangroves are important components of the world's coastal ecosystems, providing vital ecosystem services to coastal communities. However, these systems are continuously under threat from anthropogenic and natural factors, e.g. the expansion of human settlements, expansion of coastal aquaculture, the impacts of tidal waves and storm surges, and climate change. Earth observation offers one of the most effective tools that can support monitoring of changes in mangrove cover and diversity. Application of higher resolution data can significantly improve interpretation of mangrove cover maps, thus enhancing species specific monitoring and recovery efforts. This study compares mangrove cover maps generated from a medium resolution satellite (30m LANDSAT 7 ETM+) and a high resolution satellite (8m FORMOSAT- 2), and WORLVIEW-2 with a view to improving cover detection and interpretation of such maps. The satellite images were pre-processed and analysed in ENVI and ArcGIS software. A classification scheme was developed followed by image classification (supervised algorithm), and a post classification smoothing under-taken through a low-pass mean frequency filter. Results from classification statistics for the same area showed differing above ground biomass for LANDSAT and FORMOSAT- 2 imageries due to variations in the areas covered by clouds. For a smaller cloud free area (Sii Island), FORMOSAT- 2 identified a cover of 1.96 km2, WORLDVIEW-2 1.94 km2 while LANDSAT 8 indicated 1.86 km2. The slight difference in coverage could be attributed to refined FORMOSAT- 2 and WORLDVIEW-2 grids, which can enhance area and boundary delineations.

KEYWORDS: Mangroves, FORMOSAT- 2, LANDSAT 7 ETM+, LANDSAT 8 OLI, WORLDVIEW-2

APPLICATION OF REMOTE SENSING TECHNOLOGY IN EXPLORING THE ORIGIN OF CHINESE CIVILIZATION

NIE Y[1], Yu L[1], zhu j[1], Ren Y[2], Bai X[2], Enwei L[3]
[1]*Institute Of Remote Sensing And Digital Earth,* [2]*Center of Conservation and Restoration of Cultural Heritage,* [3]*Agency of Cultural Relics Management*

The project of "Research on the Application of Remote Sensing Technology in Exploring the Origin of Chinese Civilization" is a major national scientific research project in China's 12th Five-year Plan period. It is an important part of the project "Exploration of the Origin of Chinese Civilization and Related Technology Research for the Protection of Cultural Heritage". The interdisciplinary project was set up by the State Administration of Cultural Heritage. The remote sensing project focused on three ancient city settlements related to the origin of Chinese civilization, including Liangzhu, Taosi and Erlitou. Multiple remote sensing techniques were combined to extract the weak information, detect the anomalous areas and interpret the archaeological sites from RS images in order to determine the optimal sensor and band combination for identifying buried archaeological features in different regions. The research tentatively proposed the theoretical framework and methodology of weak archaeological information mining from different remote sensing images. It is vital to understand the physical basis of remote sensing for future archaeological research especially the physical characteristic and corresponding spectral properties of archaeological features. The spectral characteristics of historic sites in central China were collected to establish the archaeological spectral library. The library can not only assist with the comparability and sharing of the archaeological spectral data, but also be used to determine the effective detecting bands for different ancient sites. Furthermore, it provided parameters for the interpretation and identification of the surface and sub-surface historic sites.

KEYWORDS: Remote sensing for archaeology,Origin of Chinese Civilization, Weak information,Archaeological spectral library, Detection of historic sites

APPLICATION OF SENTINEL DATA TO IMPROVE THE WATER RESOURCE MANAGEMENT OF RWANDA

Csorba Á[1], Bukombe B[2], Szegi T[1], Naramabuye F[2], Uwiragiye Y[2], Waltner I[1], Michéli E[1]
[1]*Szent István University,* [2]*University of Rwanda*

The Rwandan agriculture strongly relies in the dry seasons on the water stored in artificial reservoirs of various sizes for irrigation purposes. Furthermore, the success of irrigation depends on a wide range of soil properties which directly affect the moisture regime of the growing medium. By integrating Sentinel and auxiliary data the objectives of our study are to monitor the water level fluctuation in the reservoirs, estimate the volume of water available for irrigation and to combine this information with soil property maps to support the decision making for sustainable irrigation water management in a study area in Southern Rwanda. For water level and volume estimation a series of Sentinel-1 (product type: GRD, acquisition mode: IW, polarizations: HH and VH) data were obtained covering the study area and spanning over a period of two years. To map the extent of water bodies the Radar-Based Water Body Mapping module of the Water Observation and Information System (WOIS) was used. High-resolution optical data (Sentinel-2) were used for validation in cloud-free periods. To estimate the volume changes in the reservoirs, the information derived from the water body mapping procedure and digital elevation models were combined. To support sustainable irrigation water management, digital soil property maps were developed by the application of wide range of environmental covariates related to soil forming factors. To develop covariates which represent the land use a time series analysis of the 2 years of Sentinel-1 data was performed. As auxiliary soil data, the ISRIC-WISE harmonized soil profile database was used. The developed digital soil mapping approach is planned to be integrated into a new WOIS workflow.

KEYWORDS: Irrigation, Rwanda, Time-series analysis, Digital soil mapping, Sentinel

ARCHAEOLOGICAL SURVEYS ON THE GERMAN NORTH SEA COAST USING HIGH-RESOLUTION SYNTHETIC APERTURE RADAR DATA

Gade M[1], Kohlus J[2], Mertens C[3]
[1]Universität Hamburg, [2]LKN Schleswig Holstein, Nationalparkamt, [3]Jürgen Kost & Cornelia Mertens Nationalpark-Wattführer GbR

We show that high-resolution space-borne Synthetic Aperture Radar (SAR) imagery with pixel sizes reaching below 1 m^2 can be used to complement archaeological surveys in areas that are difficult to access. After major storm surges in the 14th and 17th centuries, vast areas on the German North Sea coast were lost to the sea. What was left of former settlements and historical land use was buried under sediments for centuries, but when the surface layer is driven away under the permanent action of wind, currents, and waves, they appear again on the Wadden Sea surface. However, the frequent flooding and, thereby, the strong erosion of the intertidal flats make any archaeological monitoring a difficult task, so that remote sensing techniques appear to be an efficient and cost-effective instrument for any archaeological surveillance of that area. Spaceborne SAR images clearly show remnants of farmhouse foundations and of former systems of ditches, dating back to the 14th and to the 16th/17th centuries. In particular, the very high-resolution acquisition ('staring spotlight') mode of the German TerraSAR/ TanDEM-X satellites allows for the detection of various kinds of residuals of historical land use with high precision. In addition, we also investigate the capability of SARs working at lower microwave frequencies (on Radarsat-2 and on ALOS-2) to complement our archaeological survey of historical cultural traces, some of which have been unknown so far.

ASSESSING FOREST COVER DYNAMICS AND ECOSYSTEM SERVICES PERCEPTION IN THE ATLANTIC FOREST OF PARAGUAY; COMBINING REMOTE SENSING AND HOUSEHOLD LEVEL DATA

Da Ponte Canova E[1,3], Fleckenstein M[2], Parker A[2], Oppelt N[3], Kuenzer C[1], Gessner U[1]
[1]German Aerospace Center (DLR), [2]World Wildlife Fund, [3]Kiel University

The Upper Parana Atlantic Forest (BAAPA) in Paraguay is one of the most threatened tropical forests in the world. The rapid growth of deforestation has resulted in the loss of 91% of its forest cover endangering not only their continuity but the biodiversity within them. In order to halt the continuous advancement of deforestations activities (e.g. agricultural crops, cattle ranching, and illegal logging), many strategies and programs have been initiated, e.g. Payment for Ecosystem Services (PES) schemes. While the approach of ecosystem services (ES) has been wildly applied by policy makers, it is not perceived strongly by the direct users of the forest. Therefore, this study provides a comprehensive understanding how landowners in the BAAPA region perceive the ES derived from the forest and what is their influence on forest conservation. The results were obtained from the combination of Earth Observation-based data and an extensive household survey carried out at the BAAPA region from January to February 2016. Remotely sensed data acquired from Landsat images from 2000 until 2016 were utilized in order to derive the extent of the forest cover and annual deforestation rates over the last 16 years. Household surveys provided a comprehensive understanding of the perception of the ecosystems service influence on the preservation of the forest in regards to a mixture of landowners, such as indigenous communities, small, medium and large farmers-scale farmers. Common to all is the understanding of the high ecological value of the forest. A strong dependency on forest-related products was observed for small and medium landowners whereas large-scale farmers considered forest's main value to be mainly recreational and cultural. PES appears to be well accepted by forest owners however a stronger advertising must be given. Understanding the social value given to ecosystem services is a valuable contribution towards to conserving natural resources.

ASSESSING MANGROVE DIEBACK IN NORTHERN AUSTRALIA USING LIDAR AND ULTRA-HIGH RESOLUTION IMAGERY

Erskine P[1], Bartolo R[2], Lucas R[3], Asbridge E[3], Woodroffe C[4]
[1]Sustainable Minerals Institute, The University of Queensland, [2]Department of the Environment and Energy-ERISS, [3]School of Biological, Earth and Environmental Sciences, University of New South Wales, [4]School of Earth and Environmental Sciences, University of Wollongong

Extensive mangrove dieback has been observed across northern Australia since late 2015. The unprecedented dieback was first reported in the Gulf of Carpentaria. The most recent estimates indicate that 10,000 ha have been affected in a region where there had previously been both seaward and landward expansion of mangroves. It has been hypothesised that the dieback is climate driven due to a combination of higher than normal temperatures and low rainfall wet season. In September 2016 we conducted an airborne survey of mangroves in the World Heritage listed Kakadu National Park and discovered extensive dieback of mangroves, particularly in the landward mangrove zone dominated by Avicennia marina.

We acquired ultra high resolution multispectral imagery (<20 cm) using a Sony Nex5 and MicaSense RedEdge sensor (blue, green, red, red edge, near infrared and 12-bit data) on-board both a Swampfox fixed wing remotely piloted aircraft system (RPAS) and a R44 helicopter. The focus area for this paper is the West Alligator River mangroves. We compare the structure (height) and survival of the mangroves obtained from the ultra high resolution imagery with a previous LiDAR survey undertaken in 2011 to investigate effects of dieback on mangrove structure. By combining these techniques we are able to map spatially important features of mangrove forests and demonstrate which areas of this habitat have been degraded. We also summarise the challenges in acquiring image data in such a remote location and the role RPAS can play in conducting vegetation surveys in such locations.

KEYWORDS: UAV, UAS, mangrove structure, salinity.

ASSESSING THE ACCURACY OF NON-PARAMETRIC INVERSION OF RADIATIVE TRANSFER MODELS FOR ESTIMATING LEAF AREA INDEX OF AFRICAN RANGELAND USING LANDSAT 8 OLI DATA

Masemola C[1], Ramoelo A[1], Cho M[1]
[1]*CSir,* [2]*University of South Africa*

The regional monitoring of rangeland relies on accurate estimation of essential vegetation variables (EBVs) such as the leaf area index (LAI). LAI is a standard EBV that can be retrieved from Earth observation imagery using both physical based radiative transfer models (RTMs) and statistical models. This paper investigated the predictive power of Radiative transfer models inverted using machine learning regression algorithms (MLRAs). Several MLRAs algorithms were assessed in this study such as, decision trees (TREE), Bagging trees (BAGTREE), Gradient Boost & Adaboost, Extreme Learning Machines (ELM), Naive Bayes, Kernel Ridge Regression (KRR), Relevance Vector Machine (RVM), Gaussian Process Regression (GPR), Variational Heteroscedastic Gaussian Process Regression (VHGPR). The accuracy of the algorithms has been compared to conventional spectral indices and Look-up table RTM inversion approaches. Overall, we observed high accuracy when using machine learning regression algorithms (MLRAs) as compared to vegetation indices and LUT approaches. In terms of accuracy and robustness MLRAs outperformed conventional spectral indices and Look-up table approaches. The study demonstrated the possibility to map grass LAI with high accuracy using Landsat 8 OLI at a regional scale on condition that non-parametric techniques are used.

ASSESSING THE SPATIO-TEMPORAL CHARACTERISTICS OF WETLANDS ALONG THE KLIP RIVER USING REMOTE SENSING TECHNIQUES

Zwedzi L[1], Tesfamichael S[1]

[1]*Department Of Geography, Environmental Management And Energy Studies, University Of Johannesburg Kingsway Campus*

Wetlands play a vital role by providing ecosystem services and regulating the environment; however they are under severe stress due to natural and anthropogenic factors. Sustaining wetlands through efficient management requires timely and accurate monitoring that can be achieved through the use of remote sensing techniques. This paper compared the utility of Landsat and SPOT data in characterizing wetlands. Specifically, the study assessed the effect of spatial resolution of remotely-sensed on delineation of wetland boundaries. In addition, the study aimed at performing time-series analysis of wetlands using Landsat data. Several wetlands found in the south of the City of Johannesburg, South Africa were used in the study. These wetlands are highly influenced by anthropogenic activities, given their vicinity to residential and industrial sites of the city. Reference data consisting of wetland boundaries were obtained from an existing database and confirmed and modified by using high-spatial resolution Google Earth™ imagery. Unsupervised classification of Landsat and SPOT images was used to identify wetlands that were subsequently compared with the reference data. Inter-annual dynamics of wetland extents were analyzed by comparing post-classified Landsat images acquired in 1991, 2001, 2009 and 2015. Results comparing the effect of spatial resolution showed the dependence of accuracy on spatial resolution, indicating the need for using high spatial resolution for accounting small wetlands. Dynamics of the wetlands were rather mixed showing expansion and contraction of at different times. These phenomena are explained in relation to image characteristics and other variables such as climate and settlement related patterns.

KEYWORDS: Classification techniques, Satellite imagery, Wetlands delineation, Remote sensing, Dynamics of wetlands

ASSESSING THE SUCCESS OF ASBESTOS MINE REHABILITATION USING VEGETATION DYNAMIC AND EARTH OBSERVATION DATA

AdamNeuvhirwa N[1], Tengela W[2], **Adam E[2]**, Watson I[1], Cornelissen3 H[3], Malefetse3 T[3]
[1]Centre for Sustainability in Mining and Industry, University of The Witwatersrand, Johannesburg,, [2]School of Geography, Archaeology and Environmental Studies, University of The Witwatersrand, Johannesburg, [3]Mintek, Johannesburg, South Africa

Asbestos mines in South Africa have been of great concern since the 1980s due to the health risks posed by asbestos fibres to humans. efforts have been placed in order to rehabilitate derelict and ownerless asbestos mines. The South African derelict and ownerless mines rehabilitation programme is one of the government initiatives to address the asbestos mines problem. It is currently overseen by the Department of Mineral Resources (DMR) in partnership with Mintek a provider of minerals processing and metallurgical engineering products and services.

An important aspect of this program is maintenance and monitoring of the progress of the rehabilitated sites towards the rehabilitation aims or objectives. This is done through post-closure management plans which are at a localised scale. Since this is a developing field in South Africa, these rehabilitation management plans have been limited to field measures which are often tedious and require extensive labour.

The efficiency of these methods has also been questioned in relation to assessing progress at rehabilitated sites. Therefore, there is a need for another approach to quantify the success of rehabilitation programs. Remote sensing is a modernised tool for managing mine rehabilitation plans and offers a platform which translates ground based surveys into a medium that allows variable assessments at different scales. The aims of this research is to test the use of remote sensing techniques in monitoring the process and the success of asbestos mine rehabilitation using vegetation species and density as indicator. Landsat data and high resolution data from Worldview 2 and RapidEye were tested in mapping vegetation species around the mine sites to understand the vegetation composition and diversity, While NDVIs were calculated from of Landsat images to quantify the vegetation recover and density after the rehabilitation process.

KEYWORDS: Asbestos Mines; mines rehabilitation; Remote sensing; vegetation species

ASSESSING THE VERTICAL AND HORIZONTAL MOVEMENT OF ICE AND SNOW FACIES ALONG THE BALTORO GLACIER BY SAR TECHNOLOGY

Wendleder A[1], Friedl P[1], Mayer C[2], Schmitt A[3]

[1]German Aerospace Center, [2]Bavarian Academy of Sciences and Humanities, [3]University of Applied Sciences Munich

In the semi-arid countries of central Asia like Pakistan the livelihoods of the people relies heavily upon the use of river water mainly for irrigated agriculture and hydropower generation. Additionally, the increasing population and higher living standards induces considerably higher water consumption in the next few decades. Hence, water availability, especially in dry season, will be an increasing problem. The glaciers of central Asia contribute substantially to the stream-off, especially in the dry season. The Karakoram Range in northern Pakistan and adjacent India and China is one of the main glaciered areas of the region, together with the Pamirs and Tien Shan. The Baltoro glacier is located in the eastern part of the Karakorum and, with its length of more than 60 km, is one of the world's largest valley glaciers. Climate change and variability in the Karakorum Range raise the question about the spatio-temporal variability in glacier contribution to river runoff and potential changes in the runoff components. These temporal variations and changes can be monitored and mapped the help of satellites equipped with Synthetic Aperture Radars (SAR) with accurate high resolution information with. The advantage of SAR is their all-weather and day and night observation capability.

The objective is to analyze the vertical and horizontal change of the Baltoro glacier of the last decades with respect to seasonal and temporal variations. Therefore, three different approaches are applied: (1) intensity tracking for the monitoring of the glacier velocity, (2) comparison of different Digital Elevation Models (DEM) for the monitoring of vertical changes and (3) classification of snow facies for the monitoring of horizontal changes. All theses analyzes base on a time series of divers SAR data (ERS-1/2, Envisat ASAR, Radarsat-1/2, SRTM, ALOS PALSAR-1/2, TerraSAR-X, TanDEM-X and Sentinel-1).

ASSESSMENT OF CLIMATE EFFECT ON CROP WATER PRODUCTIVITY USING LANDSAT AND SENTINEL DATA

Akdim N[1], Jaquot P[2], El Ghandour F[1], Iannini L[2], Herrero Huerta M[2], Lagüela S[3], Labassi K[1], Menenti M[2], Alfieri S[2]

[1]Faculty of Sciences, Chouaib Doukkali University, [2]Department of Geosciences and Remote Sensing, Delft University of Technology, [3]Department of Cartographic and Terrain Engineering, University of Salamanca

The climate variability effect reflected on crop water productivity (CWP) in the Doukkala irrigated area (Morocco) is being investigated by looking at the CWP response in wetter, average and dryer years. This investigation is a contribution to the ESA-TIGER Bridge Program. The growing season 2016-2017 was selected for monitoring purposes in order to exploit the potentialities of the new generation Sentinel data and contribute to the Sen2Agr project.

We plan to analyze the 2016/2017 cropping season using Sentinel data. In preparation of that, we have evaluated the CWP for the cropping season 2014-2015, identified as an average year by analyzing the ERA-INTERIM data.

A composite time series of Landsat 7 TM and Landsat 8 OLI-TIRS image data has been corrected for stripes and clouds and gaps filled by applying the HANTS algorithm in order to reproduce the temporal resolution of Sentinel 2 (A and B) data. The availability of thermal bands allowed the estimation of real evapotranspiration maps with SEBAL over the study area. A data fusion processing chain has been developed and applied to add a thermal band to the Sentinel 2 image data. Unsupervised multi-temporal classification was performed using ISODATA to produce the crop maps necessary to estimate crop yield. Crop yield was estimated using Sentinel 1 data to obtain CWP in combination with the SEBAL actual ET.

The combination of Landsat 7 and Landsat 8 showed that image acquisition at high temporal resolution is necessary to capture the temporal variability of actual ET. Using ground data on the actual yield of sugarbeet a semi – empirical relationship to estimate crop yield with a time series of Sentinel 1 data has been calibrated and evaluated to obtain CWP. This method provides an alternate solution to the estimate crop yield by means of multi–spectral data (Sentinel 2 or Landsat).

ASSESSMENT OF COASTAL VEGETATION INTACTNESS IN FALSE BAY, SOUTH AFRICA, USING WV-2 IMAGERY

Lück-Vogel M[1,2], Mbolambi C[1,2]
[1]CSIR, [2]Stellenbosch University

Vegetated coasts around the world fulfil an important role. They act as buffers against the impacts of driving oceans forces such as waves and wind, and also protect the coastal hinterland from wind-blown sand and coastal erosion.

However, due to the increasing pressure those coastal environments are experiencing from urban encroachment, pollution and direct usage impacts such as trespassing, the intactness of coastal vegetation is frequently being diminished. As a result, its capacity to function as a coastal protection barrier decreases. The monitoring and management of coastal vegetation belts is therefore of utmost importance, particularly for the urbanised and industrialised areas where the hinterland hosts infrastructure of enormous economic value and importance.

However, the assessment of coastal vegetation condition is often difficult, due to the large extent and frequent inaccessibility of the zone. Furthermore, in many countries such as South Africa, there is a scarcity of coastal vegetation expertise and resources to perform the task locally.

Remote sensing therefore holds huge potential for coastal vegetation condition assessment. In the work presented here, the use of WorldView-2 imagery was investigated for assessing the intactness of a stretch of coastal vegetation in the Municipal area of Cape Town. In an object-based approach, a decision tree classification was developed to identify pristine, slightly, moderately and severely degraded Cape Flats Dune Strandveld vegetation as well as areas infested with alien Acacia species in the vicinity of Khayelitsha, one of the fastest growing informal settlements in the region.

Overall accuracies of >80% show that the results hold promise for being a useful tool for assessing the degree of intactness of coastal Strandveld vegetation. This would provide a valuable tool to environmental managers for monitoring and managing natural coastal zones and their functionality.

KEYWORDS: Coastal vegetation, degradation, WorldView-2, False Bay

ASSESSMENT OF THE RELATIONSHIP BETWEEN THE ABUNDANCE OF MOSQUITO-BORNE DISEASES' VECTORS AND MICROCLIMATIC INDICATORS DERIVED FROM EARTH OBSERVATION IMAGES

Kotchi S[1], Ludwig A[1], Brazeau S[1]
[1]*Public Health Agency Of Canada*

The emergence and re-emergence of several diseases including mosquito-borne diseases (MBD) are among others favored by increasing temperatures and climate variability in many parts of the world. Prevention and management of these diseases at local scale require a good knowledge of the environmental conditions associated to their vectors' habitats. The objective of this study was to evaluate the influence of microclimatic conditions and micro-habitat characteristics on the abundance of some MBD' vectors in southern Quebec, Canada.

Earth observation images from MODIS and AVHRR sensors acquired between 2010 and 2015 were used to estimate microclimatic indicators (MCI) related to vegetation quantity and condition, surface temperature, surface moisture (Temperature/vegetation dryness Index, TVDI), near surface air temperature (Modified temperature/vegetation index, MTVX), and hydro-thermal stress conditions of vegetation (Vegetation heat stress Index, VHSI). These indicators were used to perform a dynamic spatialization of precipitation data derived from weather stations. SPOT-6, Landsat-8 and Radarsat-2 images were used to define and estimate microhabitats descriptors. Correlation analysis were performed between the MCIs derived from Earth observation images and field data related to the abundance of mosquito species that can transmit zoonotic vector-borne diseases especially West Nile virus and Eastern equine encephalitis virus that are known to be endemic in Quebec.

MCIs and microhabitats descriptors demonstrate a high spatial and temporal variability of local environmental conditions associated with areas of the occurrence of MBD' vectors. This variability appears to be correlated with the local variation in time and space of mosquito densities. These results will support surveillance of local population dynamic and local outbreak forecasting of MBD.

KEYWORDS: Mosquito-borne diseases, Earth observation, Microclimatic indicators, Microhabitats descriptors

ASTRONAUT PHOTOGRAPHY OF THE EARTH: A LONG-TERM DATASET FOR EARTH SYSTEM RESEARCH, APPLICATIONS, AND EDUCATION

Stefanov W[1]
[1]*NASA Johnson Space Center*

The NASA Earth observations dataset obtained by humans in orbit using handheld film and digital cameras is freely accessible to the global community through the online searchable database at https://eol.jsc.nasa.gov, and offers a useful compliment to traditional ground-commanded sensor data. The dataset includes imagery from the NASA Mercury (1961) through present-day International Space Station (ISS) programs, and currently totals over 2.6 million individual frames. Geographic coverage of the dataset includes land and oceans areas between approximately 52 degrees North and South latitudes, but is spatially and temporally discontinuous.

The photographic dataset includes some significant impediments for immediate research, applied, and educational use: commercial RGB films and camera systems with overlapping bandpasses; use of different focal length lenses, unconstrained look angles, and variable spacecraft altitudes; and no native geolocation information. Such factors led to this dataset being underutilized by the community but recent advances in automated and semi-automated image geolocation, image feature classification, and web-based services are adding new value to the astronaut-acquired imagery.

A coupled ground software and on-orbit hardware system for the ISS is in development for planned deployment in mid-2017; this system will capture camera pose information for each astronaut photograph to allow automated, full georegistration of the data. The ground system component of the system is currently in use to fully georeference imagery collected in response to International Disaster Charter activations, and the autoregistration procedures are being applied to the extensive historical database of imagery to add value for research and educational purposes. In parallel, machine learning techniques are being applied to automate feature identification and classification throughout the dataset, in order to build descriptive metadata that will improve search capabilities. It is expected that these value additions will increase interest and use of the dataset by the global community.

KEYWORDS: Photography, International Space Station,Time Series, Georeferencing, Classification

AUTOMATED EXTRACTION OF SUBSURFACE RADAR LAYERS ON EARTH AND MARS BY USING CONTINUOUS WAVELET ANALYSIS

Xiong S[1], Muller J[1]
[1]*University College London*

Subsurface layering is a phenomenon resulting from stratigraphy of yearly deposits. Subsurface penetrating radars are effective instruments to detect these layering features. The application of ground penetrating radar is very mature, while the airborne and orbiting versions of this instrument have only been successfully applied to image subsurface features in the polar regions of Earth and Mars, to date. The Multi-channel Coherent Radar Depth Sounder (MCoRDS) employed in the NASA's Operation IceBridge (OIB) mission is a low frequency (193.9 MHz) penetrating radar, which extends studies of ice sheets from surface to to deep internal ice sheets in Greenland and Antarctica. It imaged an ice fold near the North Greenland Eemian ice drilling (NEEM) station. On the other hand, the Shallow Radar (SHARAD) is a subsurface penetrating radar (20 MHz) on the NASA Mars Reconnaissance Orbiter, which was launched in June 2005, has imaged the internal stratigraphy of the North Polar Layered Deposits (NPLD), which results largely from different fractions of dust and ice. Automated extraction of subsurface features is a necessary pre-processing step before relating them to geologic process and climate change. There are few studies on extracting layering features from MCoRDS and SHARAD data. However, they fail to consider the intrinsic characteristics of the radar signal when penetrating down to a subsurface area and neglect the interference from the strong background noise. In this study, we analyse and compare radar signals observed by the two above-mentioned instruments from Greenland and the Martian NPLD, and show results from an automated processing chain based on continuous wavelet transform to extract the subsurface radar layering from these two types of subsurface penetration radar. Validation of these layers depths is also considered including the use of their surface expression on Mars and ice cores on the Earth.

AUTOMATED GLOBAL LAND COVER MAPPING - FROM GLC VERSION 2

Gong P[1], Li C[2], Yu L[1], Wang J[3], Ji L[1], Huang H[3], Clinton N[5], Bai Y[1], Biging G[2], Zhu Z[4]
[1]Tsinghua University, [2]University of California, [3]Institute of Remote Sensing and Digital Earth, [4]United State Geological Survey, [5]Google Inc.

The first 30 m resolution global land cover dataset was created in 2011 named FROM-GLC. Several major techological breakthroughs were made to achieve FROM-GLC. They include a unique and cross-walkable two level (29 classes) land cover classification system, a sampling method called Global Mapper that ensures the sampling efficiency by integrating high resolution Google Earth images with Landsat images in the global sampling system, a contextual use of spatial-temporally close samples to classify every individual Landsat image, and use of high-performance computer to realize image classification for the entire world in one day with high computation demanding algorithms like support vector machine. FROM-GLC has subsequently been updated for several times to improve the mosaicing effect caused by the use of a single date Landsat image for any location. Improvement has also been made to cropland and water layers. In 2014, a new global mapping strategy has been developed that is driven by multi-seasonal training data collection with an objective to make the sample data globally applicable in any season. This strategy has been tested in Africa and substantially improved results have been achieved. In this presentation we will present a new version of global land cover mapping results for each month in 2015 to demonstate the effect of the new mapping strategy. In the new global land cover data layers we will integrate world vegetation height data to increase separability of forest and shrubs. Lastly, the capability of using the new multi-seasonal training sample to classify Landsat images in any other years will be demonstrated.

BEYOND CROP PRODUCTION ESTIMATES: INTEGRATED CLIMATE, BIOPHYSICAL AND REMOTE SENSING APPROACHES

Potgieter A[1], Hammer G[1], Brider J[2]
[1]University Of Queensland, [2]Deaprtment of Agriculture and Fisheries

Since early settlement, the existence of most rural communities in Australia has depended on agriculture. Dry land cropping has been one of the main activities contributing towards the long-term viability and sustainability of these communities. This is still true today. However, the operating environment of producers has become more challenging. Food producers are increasingly exposed to variability and change in world markets, commodity prices and climate, thus increasing their vulnerability and threatening their livelihoods. Every few years a major drought occurs across the continent with huge impacts on the Australian economy: for example, the 2002 drought has decreased the gross domestic product (GDP) by 0.75 percentage points. In addition, during almost all drought years net farm incomes are reduced to well below the long-term average. Advance knowledge of the associated risk in crop production, however, can mitigate some of the impacts of such factors. Hence, easily accessible, near real-time, objective and accurate crop production information is becoming increasingly valuable in decision-making for agricultural industry and government agencies. To date, industry and crop forecasters have had a good idea of the potential crop yield for a specific season, but advance, accurate, timely and objective information on crop area for a shire or region has been mostly unavailable. Here we present a holistic framework in linking the islands of knowledge and thus bridging the information gap resulting in generating production (volume, area, yield) estimates across large regions. Finally, this presentation will showcase the Australian experience and likely application of such technologies to assist crop monitoring and food security to enhancing decision-making at a global scale.

BILGE DUMPING DISCRIMINATION FEATURES FROM SAR IMAGERY IN SOUTHERN AFRICA OCEANS

Mdakane L[1], Kleynhans W[1]
[1]*Meraka, Council for Scientific and Industrial Research*

Marine oil pollution is a major threat to sea ecosystem and affects many countries in the world. Bilge dumping (ships dumping wastewater that contains oil during cleaning operations) is reported as the highest contributor of marine pollution. Southern Africa oceans are among the busiest in the world. Several major international shipping routes pass through these regions connecting the Atlantic to the Indian ocean. Heavy ship traffic and oil transportation activities increase the risk of marine pollution significantly. The law requires all ships to dump bilge waste at a port, however, accidental or deliberate dumping can occur at sea. Regular surveillance in these regions is an important component in ensuring marine legislation compliance and the general protection of coastal environments. The approach is to detect the presence of bilge dumping from Synthetic Aperture Radar (SAR) imagery, including the ship that may have illegally dumped the bilge waste. Bilge wastewater appears as darker linear feature in SAR imagery due to the dampening effect of oil on water-surface capillary waves under low to moderate wind conditions. Bilge dump events in SAR imagery can be visually detected by a trained human interpreter or by using image processing techniques to automate the process. The automated detection process can be broadly divided into three steps, dark-spots detection, feature extraction and discrimination. This paper focuses on the feature extraction step where the authors investigate which SAR based features can be successfully used to discriminate bilge dumps from look-alikes (natural occurring phenomena with dampening effects similar to bilge wastewater). The paper reviews and compares bilge dump discrimination features (shape, size, dB-value, texture, current, etc) for Southern Africa's ocean SAR data. This step is critical for an effective discrimination of the Southern African's bilge dump events.

KEYWORDS: Bilge dumping, Maritime Domain Awareness, Synthetic Aperture Radar, Image processing.

BUILDING A ROBUST AND COMPREHENSIVE IN SITU ENVIRONMENTAL OBSERVATION NETWORK IN A DEVELOPING ECONOMY: CHALLENGES AND CRITICAL SUCCESS FACTORS

Pauw J[1]
[1]South African Environmental Observation Network (SAEON), [2]International Long-Term Ecological Research Network (ILTER)

Fourteen years of persistent organic growth by SAEON, a public-funded variant of the Long-Term Ecological Research (LTER) approach, is one significant success story from within the National System of Innovation of South Africa. The paper will provide an overview of the systemic challenges that SAEON had to overcome by adopting sound strategies suitable to the developing state of South Africa's economy. The key success factors of the past, including factors such as community consultation, organisational design, opportunism, open access, collaboration, risk taking and social relevance, will be explored and explained. Going forward, many challenges remain and new challenges have arisen. Ironically, one of those new challenges is how to cope with a drastic increase in the available budget for research infrastructure.

KEYWORDS: South Africa, LTER, SAEON, research infrastructure, institutionalised network

BUILDING OUTLINE DETECTION OF INFORMAL DWELLINGS IN SUB-METRE OPTICAL IMAGERY USING ATTRIBUTE MORPHOLOGY AND SHAPE DESCRIPTORS

van den Bergh F[1]
[1]*CSIR*

Rapid urban growth, particularly in the form of informal settlements, requires regular and effective monitoring to inform infrastructure planning in support of basic service delivery such as water provisioning and waste removal.

Characterizing informal settlements using metrics extracted from optical imagery with a ground sampling distance of 0.5m or more presents unique challenges stemming from the small size of informal dwellings similar to those encountered in the Gauteng province of South Africa. A combination of close inter-structure spacing and poor contrast relative to the surrounding ground surface leads to inconsistent results using standard object-segmentation methods.

Grayscale attribute morphological filtering allows us to process an image down to the lowest possible contrast difference, i.e., where the intensity of an object differs from its surroundings by only one gray level. This is accomplished by successively thresholding the image at each gray level, producing a set of connected components representing potential objects of interest.

Although examples of such attribute morphological filtering applied to imagery over built-up areas exist in the literature, the filtering methods rarely go beyond elementary properties of the connected component shape, such as area, perimeter length or compactness. We propose extending the filtering by introducing full classification of the connected component shape using shape feature descriptors, such as Flusser's moment based invariants, Fourier descriptors and Procrustes analysis.

After component shape identification, shapes are simplified either by combining nested shapes from multiple threshold levels, or non-overlapping but adjacent shapes from multiple threshold levels. This allows the identification of structure outlines composed of multiple simple shape.

Once a set of final compound shapes have been identified, their properties are used to describe the characteristics of an informal settlement. Suitable properties include a distribution of structure sizes, or a histogram of Delaunay triangulation edge lengths to describe the spatial pattern of the structures.

CANOPY GAP ANALYSIS USING LIDAR-DERIVED VARIABLES

Lombard L[1], Poona N[1], Ismail R[2]
[1]*Stellenbosch Univeristy*, [2]*Sappi South Africa*

Gaps in commercial plantations are generally caused by ineffective use of growing space, areas prohibiting plantation such as rocks, and tree mortality. Additionally, natural disturbances such as wind, snowfall, disease, drought, climate change, and fire result in canopy gaps. Remote sensing has revolutionised forest monitoring and management practices and has subsequently been explored as an approach to detect and quantify canopy gaps. In this study, light detection and ranging (LiDAR) data was used to detect and map canopy gaps. A canopy height model (CHM) and an intensity raster, derived from the original LiDAR dataset, was used to delineate and map canopy gaps within a four-year-old Eucalyptus grandis compartment using a rule-based approach. We further tested the utility of the combined CHM and intensity raster to delineate and map canopy gaps. An assessment of positional accuracy of canopy gap classification showed that overall accuracies were all above 90% with Kappa values ranging between 0.8 and 0.9. To test the transferability and robustness of our models, we applied our models to a second E. grandis compartment of similar age (four years). Favourable results were obtained for the test compartment with overall accuracies above 90% and Kappa values ranging between 0.79 and 0.86. Additionally, a comparative area based assessment was undertaken whereby the area of the reference gaps (n = 9) was compared to the area of the derived canopy gaps (n = 9). Promising results were obtained from this assessment with accuracies ranging between 70% and 90% for both compartments.

The results of this study show that using LiDAR-derived variables, i.e. a CHM and intensity raster, is a viable approach to accurately detect and delineate canopy gaps within a commercial forest environment.

CAPACITY BUILDING IN GEOSPATIAL TECHNOLOGY AND ITS APPLICATIONS USING E-LEARNING AND ONLINE INTERACTIVE MODE OF CONTENTS DELIVERY IN INDIA

Karnatak H[1], Srivastav S[1], Tiwari P[1], Kumar A[1]
[1]*Indian Institute of Remote Sensing, Indian Space Research Organisation*

The skill development for effective use of geospatial technologies and its applications in India is critical due to large number of users, distributed geographic locations, multi-lingual environment and multi-disciplinary nature of the domain. Indian Institute of Remote Sensing (IIRS), ISRO has started its outreach programme in year 2007 by connecting 12 universities through satellite based interactive terminals (EDUSAT and INSAT 4CR). The programme has grown with many folds with inclusion of various advanced ICT tools and mode of contents delivery. The IIRS outreach programme was extended to Internet domain to connect more users using interactive Learning Management System (LMS) i.e. A-view platform in year 2012. The online live and interactive classrooms sessions are being conducted for 480+ connected knowledge institutions in the country under this programme where 8000 to 10,000 participants are getting befitted in each live and interactive course. IIRS has also extend its outreach programme through e-learning based online training and education in multi-lingual environment with a concept of 'learning anytime anywhere' by targeting working professionals and researchers. The interactive e-learning contents as a SCROM package for 100+ e-learning hours are created with customized LMS using Moodle. During last eight years, the IIRS outreach programme is quite successful and popular among its users where till date 40,000+ participants are trained from various user departments, ministries and academia in the country. The IIRS outreach programme is being further extended by developing digital contents for different target users at various levels. A web based knowledge repository has been developed for widespread knowledge disseminations using digital workflow and Learning Management System.
This paper presents the experience of IIRS-ISRO in the field of online training and education for geospatial technologies and its applications and also technological implementation and challenges for online mass scale capacity building.

CARBON ASSESSMENT USING EARTH OBSERVATION AND CARBON MODELS IN AFRICA

Le Toan T[1], Bouvet A[1], Mermoz S[1], Dardel C[1]
[1]*CESBIO*

To reduce the uncertainties in the calculations of carbon stocks and fluxes associated to the terrestrial biosphere requires global measurement of forest extent and forest biomass data. In recent years, Earth observation methods have been used to map biomass globally or in specific regions. Among the remote sensing data used, SAR data have been proved the most adapted to measure and monitor changes in forest biomass and extent.

The JAXA L-band SAR (ALOS-PALSAR, ALOS2) has provided yearly global data mosaics, the ESA C-band Sentinel-1 SAR has acquired data globally since 2014, and the forthcoming ESA BIOMASS mission scheduled for 2021 will provide global biomass datasets. A forest monitoring system can be developed based on these systematic and open data. Sentinel-1 will be used for monitoring deforestation, ALOS data for measuring biomass in the low biomass range, and BIOMASS data for biomass covering the high range of biomass in tropical forests.

For carbon flux calculations, carbon models can be used, several of which are embedded in climate models. The comparison of biomass data with model outputs will provide a way of testing the models. Comparison of existing carbon models indeed indicates that they are very dissimilar as regards the absolute magnitude of biomass and its spatial distribution, and that a large part of the misrepresentation of biomass comes from incorrect modeling of functioning processes.

In this paper, we will provide an illustration over Africa for which a biomass map has been generated based on ALOS data and global Land use Land cover datasets. The comparison with outputs from the ORCHDEE carbon model has provided insights in the functioning of various vegetation types and indicated the need for model improvements.

CHALLENGES & OPPORTUNITIES IN IMPLEMENTING THE DATA REVOLUTION FOR SUSTAINABLE DEVELOPMENT

Haigh T

The UN report on the data revolution, A World that Counts: Mobilising The Data Revolution for Sustainable Development. set out some of the challenges that would need to be addressed to achieve the global goals for sustainable development. The report highlights two big global challenges in relation to data: The challenge of invisibility (gaps in what we know from data, and when we find out). The challenge of inequality (gaps between those with and without information, and what they need for decision making). The talk will explore some examples of how these challenges play out in practice and how open data and effective data discovery and access tools can contribute to helping address them.

CHANGE DETECTION USING GLOBALLY AVAILABLE DIGITAL SURFACE MODELS (DSM) – A FEASIBILITY STUDY FROM THE KRUGER NATIONAL PARK, SOUTH AFRICA

Baade J[2], Schmullius C[1]
[1]*University Jena, Department for Earth Observation,* [2]*University Jena, Department for Physical Geography*

The German TanDEM-X mission acquired data for a new and truly global Digital Elevation Model (DEM) from January 2010 to December 2015. Since October 2016, the final DEM is available and first results suggest an accuracy of about 1 m; an order of magnitude higher than the initially targeted benchmark for the linear error (LE90 < 10 m). Being sensible not only to buildings and other infrastructure but as well to canopy cover, the TanDEM-X DEM actually represents a Digital Surface Model (DSM) as compared to a Digital Terrain Model (DTM). In this property the TanDEM-X is similar to the Shuttle Radar Topographic Mission (SRTM) acquired in February 2000. Both data sets are available in 1 and 3 arcsec posting. This inevitably calls for an investigation into the feasibility of a 15 years change detection assessment using these two Synthethic Aperture Radar (SAR) derived data sets. Potential changes detectable include change of canopy and relief due to natural drivers or human impact.

A recent accuracy assessment for Kruger National Park (KNP) has shown that the TanDEM-X DSM provides submeter accuracy for open terrain while the SRTM data set is characterized by an offset in the order of about 5 m and random errors in the order of 5 m. The later limits the surface change detection to about 5 m in height. Another issue to take into account is the different SAR wavelengths used in the acquisitions in 2000 versus 2015.

This contribution provides the results of a feasibility study centred around the Kruger National Park in South Africa. Natural drivers of surface change inside KNP are fire and especially mammal pressure on the vegetation. Outside KNP, forestry, mining and the general development of the communities have impacted the SAR scattering layer.

KEYWORDS: TanDEM-X, SRTM, change detection, Kruger National Park

CHARACTERISTICS OF FROZEN SOIL AND SNOW ESTIMATED USING SMOS LEVEL 3 BRIGHTNESS TEMPERATURES

Lemmetyinen J[1], Schwank M[2,3], Naderpour R[2], Rautiainen K[1], Mätzler C[3], Wiesmann A[3], Wegmüller U[3], Roy A[4], Toose P[5], Derksen C[5], Pulliainen J[1]

[1]Finnish Meteorological Institute, [2]Swiss Federal Research Institute WSL, [3]Gamma Remote Sensing AG, [4]Universite de Sherbrooke, [5]Environment and Climate Change Canada

Dry snow cover is conventionally though to have only a minimal influence on microwave emission at L-band (~1-3 GHz), due to the inherent high penetration depth of microwave energy in dry snow at this frequency range. However, dry snow cover causes a non-negligible influence on L-band emission as detected by means of microwave remote sensing, due to the dual effects of refraction and impedance matching at the soil-snow interface. Dry snow cover thus influences estimates of e.g. soil moisture and soil freeze/thaw state in snow covered areas. The effects also introduce the potential of retrieving characteristics of snow cover (namely, snow density) in dry snow conditions using microwave remote sensing at L-band.

This study describes the retrieval of dry snow density and ground permittivity using SMOS Level 3 gridded brightness temperatures. The inherent coarse scale spatial resolution of SMOS necessitates accounting for forest cover and water bodies to account for mixed-pixel effects in heterogeneous observation scenes. A comprehensive forward modeling environment developed for L-band, exploiting a two-stream approach for transmission of incoherent microwave energy, is applied. Methodologies applying different polarization combinations from the multi-angular, dual polarized SMOS data are introduced, and applied over two dedicated test sites (Sodankylä, Finland, and Saskatoon, Canada). .

CHARACTERIZATION OF AEROSOL EMISSIONS FROM AFRICAN BIOMASS BURNING

Ichoku C[1], Ellison L[1,2], Pan X[1,3]
[1]Nasa Goddard Space Flight Center, [2]Science Systems & Applications, Inc., [3]University of Maryland

A series of studies conducted over the last decade have shown that the total aerosol particulate matter emitted from open biomass burning is directly proportional to the fire radiative energy (FRE), whose instantaneous rate of release or fire radiative power (FRP) is measurable from space. By leveraging this relationship, we developed a global, gridded smoke-aerosol emissions dataset based on FRP and aerosol optical thickness (AOT) measurements from the MODIS sensors aboard the Terra and Aqua satellites. The first version of this Fire Energetics and Emissions Research (FEER.v1) global product is available at http://feer.gsfc.nasa.gov/.

Analysis of FEER.v1 emissions datasets for the period of 2003-2014 shows that biomass burning aerosol emissions have been on the decrease at an estimated rate of -3%/yr in northern sub-Saharan Africa (defined as the region bounded by 0°-20°N, 20°W-55°E). In a region where regular anthropogenic use of fires for agricultural and pastoral purposes is a staple of the rural economies, this decrease over the last decade may have far-reaching environmental and climate implications and their potential impact on the regional communities. We compared FEER.v1 aerosol datasets with several other major global biomass-burning emissions datasets, and ingested each one in NASA's GEOS-5 model, whose outputs were compared with a suite of satellite-derived AOT products and those of the available ground-based AERONET sunphotometers. These comparisons have provided us the opportunity to visualize the relative performances of the different biomass burning emissions inventories, of which FEER.v1 did considerably well overall. Planned upgrades of future versions of the FEER emission products are only expected to improve the product's accuracy, which bodes well for future modeling and analysis efforts, particularly in Africa, where biomass burning is still depended upon for the livelihoods of the rural societies.

KEYWORDS: Africa, Fire Radiative Power, Emissions, Aerosol, FEER

CKAN: OPPORTUNITIES & CAPABILITIES FOR EARTH OBSERVATION DATA

Moleski S

CKAN is a free and open source technology, a data management system providing tools to streamline and reuse data. It is a powerful, well-used and well-tested technology that supports GEOSS in it's mission to share, find and use (open) EO data. It publishes links to the original source and is therefore a scaleable solution for GEOSS as a big data infrastructure. It broadens the reach as data can be published in any format and will be available for users to download. Data hub harvesters are demonstrated technology, prototypes that qualify for integration with each infrastructure to be automatically harvested. This presentation will shed some light on different aspects of the technology and catalog techniques. It will demonstrate functionality with examples and best practices, especially its potential for EO. It will showcase opportunities and challenges that implementing bodies, such as the European Commission and the United Kingdom government, have faced.

CKAN supports collaborative development of innovative solutions that lead to a resilient sustainable world, it stimulates to work across scales and disciplines.

CLAIRE: A CANADIAN SMALL SATELLITE MISSION FOR MEASUREMENT OF GREENHOUSE GASES

Germain S[2], Sloan J[1], McKeever J[3], Latendresse V[4], Durak B[3]
[1]S&A Research, [2]GHGSat Inc., [3]Xiphos Systems Co., [4]MPB Communications Inc.

CLAIRE, a Canadian mission operated by GHGSat Inc. of Montreal, is the world's first satellite designed to measure greenhouse gas emissions from single targeted industrial facilities. Claire was launched in June 2016 into a 500 km polar sun-synchronous orbit selected to provide an acceptable balance between return frequency and spatial resolution. Extensive simulations of oil gas facilities, power plants, hydro reservoirs and even animal feedlots were used to predict the mission performance. The principal goal is to measure the emission rates of carbon dioxide and methane from selected targets with greater precision and lower cost than ground- based alternatives. CLAIRE will measure sources having surface areas less than 10 x 10 km2 with a spatial resolution better than 50 m, thereby providing industrial site operators and government regulators with the information they need to understand, manage and ultimately to reduce greenhouse gas emissions more economically. The sensor is based on a Fabry-Perot interferometer, coupled with a 2D InGaAs focal plane array operating in the short-wave infrared with a spectral resolution of about 0.1 nm. The patented, high etendue, instrument design provides signal to noise ratios that permit quantification of emission rates with accuracies adequate for most regulatory reporting thresholds. The very high spatial resolution of the density maps produced by the CLAIRE mission resolves plume shapes and emitter locations so that advanced dispersion models can derive accurate emission rates of multiple sources within the fi of view. The satellite bus, provided by the University of Toronto's Space Flight Laboratory, is based on the well-characterized NEMO architecture, including hardware that has significant spaceflight heritage.

CLIMATE CHANGE - CENTRAL TO GEO'S ENVIRONMENTAL ACTIVITIES

Obregon A[1]
[1]*Group on Earth Observations (GEO)*

The talk will give an overview on the climate activities within GEO.
Climate change is affecting most, if not all, of GEO's Societal Benefit Areas. GEO engages climate data providers and users to ensure a sustained dialogue around information needs, particularly of those seeking to integrate climate products and services into adaptation processes and decisions. The GEOSS Portal makes information and knowledge available in support of effective policy responses for climate change adaptation and mitigation.

In the GEO Mexico City Declaration, adopted in November 2015, Ministers affirmed that GEO and its Earth observations and information will support the implementation of the United Nations Framework Convention on Climate Change (UNFCCC). The 2016 Global Climate Observing System (GCOS) Implementation Plan outlines the Essential Climate Variables, indicators and actions that are needed to support the Paris Agreement. There is a clear policy need for research, systematic observations and scientific data emerging from this treaty. Responding to the Paris Agreement, the GEO Carbon and Greenhouse Gas (GHG) Initiative aims to ensure overall coherence of existing efforts, to provide comprehensive knowledge on changes in the global carbon cycle and GHG emissions, and to support decision makers with timely, policy-relevant information.

With respect to the important role of the cryosphere in the Earth's climate, GEO has launched a Cold Regions Initiatives (GEOCRI) to provide coordinated Earth observations and information services across a range of stakeholders. GEOCRI facilitates well-informed decisions and supports the sustainable development of the cold regions globally.

In terms of climate services, GEO is working to build appropriate linkages with the priority areas of the Global Framework for Climate Services (GFCS). Operational climate services are already contributing to the implementation of GEOSS, such as the Copernicus Climate Change Service (C3S) that is providing authoritative, quality-assured data and services to inform decision makers.

CLIMATE CHANGE AND LAND SURFACE PHENOLOGY IN THE NORTHEAST CHINA

Zhao J[1]
[1]*Northeast Normal University*

Northeast China is located at high northern latitudes and is a typical region of relatively high sensitivity to global climate change. Studies of the land surface phenology in Northeast China and its response to climate change are important for understanding global climate change. In this study, the land surface phenology parameters from the Global Inventory Modeling and Mapping Studies GIMMS 3g dataset that was collected from 1982 to 2013 were estimated to analyze the variations of the land surface phenology in Northeast China at different scales and to discuss the internal relationships between phenology and climate change. We examined the phonological changes of all ecoregions. The average start of the growing season (SOS) did not exhibit a significant trend throughout the study area; however, the end of the growing season (EOS) was significantly delayed by 4.1 days or 0.13 days/year ($p < 0.05$) over the past 32 years. The SOS for the Hulunbuir Plain, Greater Khingan Mountains and Lesser Khingan Mountains was earlier, and the SOS for the Sanjing, Songnen and Liaohe Plains was later. In addition, the EOS of the Greater Khingan Mountains, Lesser Khingan Mountains and Changbai Mountains was later than the EOS of the Liaohe Plain. The spring temperature had the greatest impact on the SOS. Precipitation had an insignificant impact on forest SOS and a relatively large impact on grassland SOS. The EOS was affected by both temperature and precipitation. Furthermore, although temperature had a lag effect on the EOS, no significant lag effect was observed for the SOS.

KEY WORDS: land surface phenology; GIMMS 3g; NDVI; climate change; northeast of China

CLIMATE CHANGE EFFECTS ON URBAN LEVEL: CITIZEN HEALTH AND BUILDING ENERGY DEMAND

San Jose R[1], Pérez-Camanyo J[1], Pérez L[1], Gonzalez-Barras R[2]
[1]*Technical University Of Madrid (upm), Computer Sciences School,* [2]*Complutense University of Madrid (UCM)*

The future impacts of climate change on citizen health and building energy demand have been researched considering two possible IPCC global climate scenarios: RCP 4.5 (stabilization emission scenario) and RCP 8.5 (little effort to reduce emissions). The climate scenarios have been dynamically downscaled from 1º to 50 meters of spatial resolution over three European cities: Madrid, Milan and London. Air quality has also been simulated up to streets levels. Climate and air pollution information are used as input to the health impact and building energy demand assessment tools. The impacts are calculated as future (2030, 2050 and 2100) minus present (2011. The short term health impact assessment includes mortality and morbidity related with changes in the temperature and air pollution concentrations. Also the cost of health impacts is calculated. Concentration-response coefficients were taken from the recent environmental epidemiological literature. The morbidity and mortality costs arising from climate change are then evaluated for each health outcome separately by multiplication of the number of cases with the respective cost estimates. The larger increase of costs of mortality and morbidity was noted in the increasing scenario (RCP8.5) for year 2100, because RCP 8.5 is characterized by temperature increments. Maps of the spatial distribution of the costs of the climate change have showed Building energy demand simulations have been achieved with the EnergyPlus model using specific prototype buildings based on ASHRAE 90.1 Prototype Building Modeling Specifications and urban climate information by each building. .The results show an increase in cooling demand with RCP 8.5 because future will be cooler that the present. The conclusions show that climate change will have a large effect in the building energy demand and citizen health and the used methodology can be used to design strategies to reduce the effects of climate change.

KEYWORDS: climate, urban, air quality, health, energy

CLOSE RANGE REMOTE SENSING (DRONES) TO SUPPORT NATIONAL HERITAGE AUTHORITIES IN CAMPECHE, MEXICO

De Maeyer P[1], Hernandez M[2], Benavides A[3]
[1]Ghent University, Department of Geography, [2]ISPRS, [3]INAH Campeche

In general for archaeologists going to a heritage site and taking measurements is a cumbersome task resulting in 2D plans that are often not accurate enough. In order to assist the heritage authorities of the Institute of Anthropology and History of Campeche (INAH-Campeche), the University of Ghent, Department of Geography has undertaken a series of field campaigns to the Maya Archaeological site of Edznà (located 55 km SE of the city of Campeche). The heritage site is spread in an area of approximately 25 square-kilometers. INAH is interested in using digital 3D models in order to further undertake archaeological research as well as for the promotion of the site through virtual tours. Drones using panchromatic as well as infrared cameras have been flown over selected buildings of the large heritage site of Edznà. In addition, a large number of topographical measurements and GNSS (Global Navigation Satellite System) observations to create a highly accurate metrical context were completed. The thousands of photos and the accurate field survey allow in an integrated approach to process the data to derive highly accurate 3D models.

The elaboration of a virtual archaeological site will be the next step, using an approach with different levels of detail (entire site - separate structures - objects of interest on these structures); data set that will be made available on an online environment. In addition, the models will be part of a 3D-GIS model.
The work will be completed with satellite images (SENTINEL) and if possible with some TERRASAR images in order to detect if one of the main monuments has some signs of incidence.

Since this work is based on the expertise obtained through the research work done, using satellite images to support a 4D Heritage Information System for INAH-Campeche, we will also illustrate such a research project.

CLOUD BASED CROPWATCH：GLOBAL REMOTE SENSING MONITORING ONLINE SYSTEM

Zeng H[1], Wu B[1], Zhang X[1], Zhang M[1]
[1]*Institute of Remote Sensing and Digital Earth, Chinese Academy of Sciences*

The cloud technology is an effective tool for people to produce, to browse and analysis the information anywhere. To better produce and generate the bulletin and provide an alternative way to access agricultural monitoring indicators and results in CropWatch, a new simple all-in-one solution platform named CropWatch online system based on the cloud techniques has been developed.

CropWatch online system provide four main components: CropWatch Processing, CropWatch Explore, CropWatch Analysis and CropWatch Bulletin. CropWatch Processing provides an online interface for people producing CropWatch indicators at anytime and anywhere from county to global scales. By taking advantages of Alibaba cloud flexible system with parallel computing's architecture, the data production efficiency for each bulletin increased by 50 times and no consumption in the idle time.

CropWatch Explore provides a visualization web service for our users to smoothly explore CropWatch data including agro-climatic indicators, agronomic indicators related with crop area, crop yield and crop production. Information or indicators are visualized using vector or raster according to their features. Vector mode provides the statistic value for all the indicators over each monitoring units which allows users to compare current situation with historical values (average, maximum, etc.). Raster mode provides pixel based anomaly of CropWatch indicators globally. In this mode, users can zoom in to the regions where the notable anomaly was identified from statistic values in vector mode.

CropWatch Analysis is cloud collaboration documentation tool for the CropWatch teams or invited people from over the world analyzing their CropWatch indicators anywhere. It provides create document, allocate and manage tasks, monitor schedule and publish the document online functions which let people over the world finish their documents together on the cloud platform.

CropWatch Bulletin publishes the CropWatch bulletin for each season, CropWatch updates for each month and global crop information on http://www.cropwatch.com.cn.

CLOUD CONTAMINATION IN MODIS COLLECTION 6 SURFACE REFLECTANCES

Kharbouche S[1], Muller J[1], Danne O[2]
[1]*Mullard Space Science Laboratory, University College London,* [2]*Brockmann Consult*

In the www.QA4ECV.eu project, surface reflectance products are used as inputs to an inversion scheme to derive BRDF, which in their turn are used as inputs to produce LAI and fAPAR. Hence, the evaluation of the quality of surface reflectances in terms of atmospheric correction and cloud detection plays a key role in the accuracy of the final product. However, contrary to the uncertainty of atmospheric correction, whose value can be modelled and propagated through derived products, the uncertainty associated with cloud detection is much harder to model, mainly due to its binary state (cloudy, cloud-free). Furthermore, poor cloud detection in surface reflectance does not only lead to wrong BRDF values but also to misleading BRDF quality indicators, because these depend mainly on the number and angular distribution of surface reflectance samples which are assumed to be cloud-free.

In QA4ECV, we ingest surface reflectances from the entire time series of MERIS, VEGETATION and MODIS (collection6). Our first results revealed that MODIS appears to be more cloud-contaminated than both MERIS and VGT, therefore an in-depth investigation has been carried out, also making use of the Brockmann-Consult 'IdePix' cloud masking tool. The dataset was scanned looking for pairs (MODIS vs MERIS and MODIS vs VGT) that are similar in terms of geometry angles and acquisition dates. Then the extracted samples were processed in two steps: 1) considering only samples having very different values in overlapping bands in both sensors, and 2) considering samples whose spectral response is close to those of cloud. The results confirmed, mainly over vegetated-land, the high cloud contamination of MODIS. In this presentation, we will show and discuss the derived results and their implications for future use of MODIS data.

This work was supported by QA4ECV, a EU-FP7/2007-2013 project (no. 607405).

KEYWORDS: Cloud Contamination, Surface-Reluctances, MODIS, MERIS, VGT.

CLOUD-BASED AGRICULTURAL SOLUTION: A CASE STUDY OF NEAR REAL-TIME REGIONAL AGRICULTURAL CROP GROWTH INFORMATION OVER DELMAS (MPUMALANGA, SOUTH AFRICA)

Hiestermann J[1], Ferreira S[1], Thompson M[1]
[1]*Geoterraimage*

Recent advances in cloud-based technology has led to the rapid increase of geospatial web-based applications. The combination of GIS and cloud-based solutions is revolutionizing product development in the geospatial industry and is facilitating accessibility to a wider range of users, planners and decision makers. Accessible through an internet browser, web applications are an effective and convenient method to disseminate information in multiple formats, and they provide an interface offering interactive access to geospatial data, real-time integration and data processing, and application specific analysis tools. An example of such a web application is GeoTerraImage's monthly crop monitoring tool called GeoFarmer. This tool uses climatic data and satellite imagery processed through a complex model to determine monthly climatic averages and anomalies, and most importantly the field specific crop status (i.e. is the field fallow, or is the crop emerging, or if the field has been harvested). Delmas (Mpumalanga, South Africa) and its surroundings have been part of an ongoing monthly crop monitoring program run by GeoFarmer. Monthly field verification has formed a part of calibrating the growth classification outputs to further improve the accuracy of its monthly agricultural reporting. The goal of this application is to provide timely data to decision makers to assist them in field-level and regional crop growth monitoring, crop production and management, financial risk assessment and insurance, and food security applications. This web application has the unique advantage of being highly transportable to other regions, since it has been designed so it can easily be adapted to other seasonal growth response patterns, and up-scaled to regional or national coverages for operational use.

KEYWORDS: crop, monitoring, growth, regional, web-application

CNES ACTIVITIES IN TELE-EPIDEMIOLOGY

Vignolles C[1]
[1]Cnes

Emerging/re-emerging infectious diseases with high epidemiological potential risks, lead public health managers to adapt their policies. Adaptation includes early knowledge of risks requiring new tools to prevent re-emerging risks.

Key factors, involved in these diseases, can be environmental, climatic, demographic, socio-economic and/or behavioral. Some can be identified from space, requiring the development of effective methods to use remote sensing for risk factor characterization, mapping and monitoring. Data from Earth observation satellites do not directly concern the pathogens, but their environment – they will therefore be used to measure these favorable factors.

The French Spatial Agency (CNES) with its partners has developed a conceptual approach called tele-epidemiology which consists in studying the links between the environment, ecosystems and etiological agents responsible for these diseases, based on space products truly adapted to the needs of health actors. This multidisciplinary approach is based upon the study of the key mechanisms favoring the surge of those diseases. Analysis of processes is a key step in the development of new and original risk mapping using EO satellite data. The mission is to provide to health actors additional tools helping them in diseases surveillance and in the implementation of strategies to diseases control.

This concept has been applied with success for the Rift Valley Fever in Senegal and the Dengue in La Martinique. Objectives were to provide dynamic entomological risk maps, and then to study adaptation processes for controlling management. For those examples, results will be presented. The effectiveness of risk prevention could be improved by providing health authorities with these maps predicting “when and where” there will be an entomological risk . If regularly updated, risk mapping could provide useful information to optimize vector control measures.

KEYWORDS: tele-epidemiology, public health, infectious diseases, environmental factors, remote sensing, risk mapping, spatio-temporal dynamics

CNES' EARTH OBSERVATION PROGRAMME: CONTRIBUTING TO THE GLOBAL EARTH OBSERVING SYSTEM OF SYSTEMS

Ultré-Guérard P[1]
[1]*CNES*

The principal characteristics of CNES's (Centre National d'Etudes Spatiales, French Space Agency) Earth observation programme are presented here, including key facts and figures from past missions, those currently being operated and those under development. Information will be provided on many missions, from those with a science focus, observing ocean surface topography (Jason series, Altika), clouds and aerosols (Calipso, Parasol), ocean salinity and soil moisture (SMOS), tropical atmospheric water (Megha Tropiques) and Earth's magnetic field, to operational missions such as IASI for atmospheric Infra Red sounding (temperature, water and composition of the atmosphere) on board Metop A and B, and SPOT, Pléiades 1-A and 1-B for optical high resolution imagery. Some examples of significant science results are shown in the fields of oceanography, physics and composition of the atmosphere, solid Earth and land surfaces.

Among the missions currently under development, Jason 3 for ocean surface topography, Venus for multispectral imagery, CFOSAT for sea waves, IASI-NG on board Metop-SG, SWOT wide swath altimetry for sea and inland water levels and Merlin and Microcarb for atmospheric methaneand carbon dioxide respectively.

The complementarity with other Earth observation programmes in particular the European Space Agency's (ESA) Earth Explorer programme and the European Union's Copernicus programme will be highlighted. CNES's contribution to coordinating international multilateral action through the Committee on Earth Observation Satellites (CEOS), will be presented highlighting this committee's role in coordinating the space component of the GEOSS (Global Earth Observation System of Systems). Finally, some perspectives on the future science missions prioritized with the French science community during the French "scientific prospective seminar" held in La Rochelle 2014 are provided.

CO2 EMISSIONS FROM FOREST FIRES USING GOSAT DATA

Guo M[1]
[1]*School of Geographical Sciences, Northeast Normal University*

Wildfire has been thought as a major disturbance in Russian forests as approximately 95% of Russian forests are boreal forests, and 71% of those boreal forests are dominated by coniferous forest with higher fire risk. In the summer of 2010, intense wildfires occurred in western Russia, near Moscow, due to high temperatures and the absence of precipitation. CO2 emissions from Russia forest fire 2010 were estimated using GOSAT (Greenhouse gases Observing SATellite) data. We used GOSAT CAI (Cloud and Aerosol Imager) to identify the forest fire smoke plumes and then used GOSAT SWIR (Short-Wavelength InfraRed) L2 data to monitor the CO2 concentration increase (ΔXCO2). Then we calculated the amount of CO2 released from forest fire in Russia 2010 and found that 255.76 Tg CO2 was released. We also compared our result with the Biomass Burning Model and emission ratios methods and found the similar results. This paper proposed a new method to evaluate CO2 emissions from wildfires using remote sensing data, which could be used to improve knowledge of biomass burning at regional or continental scales.

KEYWORDS: Wildfire; CO2 Emissions; GOSAT Data; Biomass Burning Model (BBM)

CODE-DE - A COPERNICUS DATA AND EXPLOITATION PLATFORM FOR GERMANY

Müller A[1], Keuck V[2], Schreier G[1], Reck C[1]
[1]DLR - German Aerospace Center - German Remote Sensing Data Center, [2]DLR - German Aerospace Center - Space Administration

The Copernicus Data and Exploitation Platform – Deutschland (CODE-DE) is the German entry point to the EU Copernicus Sentinel Satellite Systems and their data products. The Sentinel-Satellites form the core of the European Earth Observation Program Copernicus.

In Germany the Ministry of Transport and digital Infrastructure (BMVI) exploits the potential of Copernicus to generate a modern powerful infrastructure to foster Earth Observation technologies and to generate new chances for services in the geo-information market (www.d-copernicus.de).

CODE-DE is particularly dedicated to grant German entities - administrative bodies and civil services, research organizations or private companies, or even members of the public fast, secure and easy to use access to data of all operational Sentinel satellites as well as all derived information generated by the Copernicus services.
A continuously updated data catalogue allows access to Sentinel-Data filtered by time, location and additional meta-data information such as cloud coverage. All data can be downloaded from Online-server-platforms.

Currently up to date data from Sentinel 1a and b, Sentinel 2a as well as Sentinel 3a are available online. In the future (starting Q42017) the service platform will feature the possibility to process data online. Selected user groups will be able to use the online processing power von CODE-DE to generate tailored information products even without own computing infrastructure. Both the use of pre-installed Earth observation tool boxes provided by CODE-DE as well as the deployment of customer developed external processors will be supported.

In the presentation the technical concept, the hard- and software realisation of the system and results of first performance measurements will be addressed.

COFFEE CROP 'HEALTH' MONITORING USING SATELLITE DATA – CAN IT WORK?

Hausknecht P[1], Hawkins O[1]
[1]*Earth-i Ltd.*

Coffee crop monitoring using satellite data poses a challenge, since the fruits usually cannot be monitored directly from space and some crop stressors do not cause directly detectable alteration of the foliage. However very high spatial resolution data, combined with other data sources, offers actionable indicators of the plantation condition. Integrated with time series information from high resolution satellites such as Sentinel2 for seasonal monitoring, may then facilitate much improved crop assessments. In addition one can potentially utilise proxies derived from other information sources, to overcome cloudy periods which can be quite common in coffee growing regions of the globe. When combined with mapping and supply chain information derived from satellite imagery, such insights provide valuable guidance for farmers that can significantly improve crop yields.

This paper will look at some examples and compare the information content from the different sensors. Demonstrating the DMC3/TripleSat high resolution data capability we show some image examples, amongst others from central Africa.

COMBINING RED SUN-INDUCED CHLOROPHYLL FLUORESCENCE AND VEGETATION INDICES IMPROVES DIURNAL ESTIMATION OF GPP: MODELING AND VALIDATION OVER THE WINTER WHEAT CANOPIES

Guan L[1,2], Liu L[1], Liu X[1], Hu J[1,2], Du S[1,2]
[1]Institute of Remote Sensing and Digital Earth, Chinese Academy of Sciences, [2]University of Chinese Academy of Sciences

Accurate estimation of gross primary production (GPP) is of great importance to global change research and also food and fuel security. Previous studies have demonstrated that remote sensing of sun-induced chlorophyll fluorescence (SIF) is a novel optical tool for providing a direct measure of ecosystem GPP, and research works have focused almost exclusively on the application in the far-red bands. However, red SIF signals that are mainly contributed by Photosystem II activity have been suggested to be more related to photochemical processes. Since red SIF data are difficult to be accurately obtained with existing instrumentation and they suffer the influence of reabsorption by canopy, only few studies have successfully retrieved red SIF, nor have they investigated the relationship between red SIF and canopy GPP measurements. In this context, we present a combined modeling and vegetation indices-based approach to predicting the GPP of winter wheat from the red SIF data while minimizing their reabsorption effects. The coupled fluorescence-photosynthesis model SCOPE was used to simulate GPP and SIF at the diurnal scale across a range of LAI values and leaf chlorophyll contents (Cab). The results showed that the relationship between simulated GPP and red SIF was highly sensitive to LAI and Cab variations, and integrating VIs can develop a unique GPP-SIF688 model for winter wheat canopy with various levels of LAI and Cab. The combined method was then validated using ground truth measurements at different growth stages, and it revealed that VIs were able to enhance the power of red SIF for GPP inversion.

KEYWORDS: Photosynthesis, GPP, Chlorophyll fluorescence, SCOPE, Vegetation Indices

COMMUNITY CLIMATE-BASED MAPPING AND MODELLING OF INVASIVE SPECIES

Ouko E[1]
[1]*Rcmrd*

Invasion of alien species plants into African savannas poses great threat to the native biodiversity and changes ecosystem functioning. Invasive or alien – is a non-native species that has been introduced either accidentally or intentionally, of which the intentional introductions have been attributed to economic, environmental and social considerations. In the forest sector, for instance Acacia species are important sources of fuel-wood, yet at the same time they have increased strain on water resources and shrunken forage spaces for both livestock and wildlife. In recently infested regions, invasive species can progress through the stages of introduction, establishment and dispersal to a full range. Currently there is much worldwide interest in predicting distributions of invasive species, and several organizations will be faced with questions of whether and how to tackle such environmental challenges, or how to interpret predictions from the science community. Conservation practitioners require mapped estimates of where species could persist in a given region, and this is associated to information about the biotope – i.e. the geographic location of the species' niche. The process of collecting species distribution data for identifying the potential distribution of the invasive species in the invaded ranges has become a challenge both in terms of resource and time allocation. This study highlights application of innovative approaches in crowdsourcing validation data in mapping and modelling invasive (Acacia reficiens and Prickly Pear) through involvement of the local communities.

KEYWORDS: Invasive species, Mapping, Modelling, Prediction, Crowdsourcing, Climate

COMPARATIVE 3D EVALUATION OF FINE-SCALE GLOBAL AND LOCAL NORMALISED DEMS: THE TSHWANE CASE STUDY

Breytenbach A[1]
[1]*Csir*

To properly inform decision-making in the urban planning and engineering domains beyond the normal 2D mapping perspective, topographic surface data has become standard input in processes geared towards the sufficient representation of a real world situation at a particular location. With the progressive advancement of modern remote sensing technology and software in this regard, quality ortho-optic products are generated nowadays with high levels of detail and accuracy at various scales. It could include both the seamless digital surface and terrain model that, through simple subtraction, offer users a normalised DEM or height model of the area in question. This study used three different normalised DEMs (nDEMs) at three spatial resolutions, namely 2 m, 4 m and 13 m, that was derived from stereo aerial image pairs, tri-stereo Pléiades images and Tandem-X InSAR data, respectively. Covering an area over the capital City of Tshwane, South Africa, these nDEMs were compared volumetrically within nine land cover classes against a LiDAR derived reference surface at matching spatial resolutions. The results indicated that the higher the spatial resolution of the sample DEM, the more accurate the fundamental volumes per land cover class became, particularly in built-up areas, but less so in dense vegetation.

KEYWORDS: Accuracy, DEM, land cover, topography

COMPARATIVE ASSESSMENT OF THE COPERNICUS LAND MONITORING PAN-EUROPEAN HIGH RESOLUTION LAYER ON IMPERVIOUSNESS DEGREE AND THE EUROPEAN SETTLEMENT LAYER

Sannier C[1]
[1]*Sirs Sas*

The validation of a dataset such as the Copernicus Pan-European imperviousness degree high resolution or European Setllement layer requires considerable effort. A stratified systematic sampling approach was developed based on the LUCAS sampling frame focusing on a 2 stage stratification approach. A two-stage stratified sample of 20,164 1ha square primary sampling units (PSU) was selected over EEA39 based on countries or groups of countries which area was greater than 90,000km^2 and a series of omission and commission strata. In each PSU, a grid of 5 x 5 Secondary Sample units (SSUs) with a 20 m step was applied. These points were photo-interpreted on orthophotos with a resolution better than 2.5m.

Initial results for the Imperviousness degree layer based on the binary conversion of the map by applying the 30% threshold indicate a level of omission and commission errors substantially greater than the required maximum level of 15% set in the product specifications . However, this assumes that complete information is available for each PSU which is not the case. An alternative procedure was applied to the quantitative continuous data considering the sampling error due to the SSUs selection which is expected to exhibit a more realistic assessment of the amount of omission and commission.

As expected, the European Settlement Layer provides lower values than the Imperviousness degree layer and non-built-up (roads and other artificial surfaces) impervious areas appear to be excluded with building appearing more clearly. In addition, there seems to be a relatively strong relationship between the two layers.

KEYWORDS: urban sprawl, urban extent, accuracy assessment, sampling design, stratification, Sentinel-2, Landsat, Dense time series, Time-series analysis, Cloud and cloud shadow masking

COMPARATIVE STUDY OF TREND RUN AND PCA MODEL FOR PERFORMING THE TREND CALCULATION ON SAWS (SOUTH AFRICAN WEATHER SERVICE) ATMOSPHERIC DATA

Krishnannair S[1], Venkataramanan [2]
[1]*University Of Zululand,* [2]*University of Kwazulu Natal*

The trend and variability of surface temperature from South African Weather Service (SWAS) at Cape Point station were investigated using a principal component analysis (PCA) model and its performance was compared with the TREND RUN model. The contribution of the different atmospheric forces such as annual , semi- annual Quasi-Biennial oscillations, El-Nino Southern Oscillation (ENSO), and the 11-years solar cycle (SSN) and Indian Ocean Dipole (IOD) were determined by using PCA and Trend Run. We used the rainfall, surface minimum and maximum temperature datasets collected from 1980 to 2004 from Cape Point station to calculate the decadal trend. It is shown that the most significant temporal and spatial portion of different atmospheric forces are characterised by the first few principal components in the PCA model simultaneously. The results show that PCA model is more effective than Trend RUN model.

COMPARING IMERG AND TMPA PRECIPITATION DATA IN AFRICA

Dezfuli A[1], Ichoku C[2], Huffman G[2]
[1]NASA/GSFC & USRA, [2]NASA/GSFC

The ground-based precipitation data have a relatively poor spatio-temporal coverage in Africa, and that has made it difficult to study the drivers of rainfall variability in various parts of this continent. Satellite-based data, in particular the TRMM Multi-Satellite Precipitation Analysis (TMPA)-3B42 version 7 have been successfully used as an alternative to fill this void. The TMPA incorporates precipitation estimates from several satellites and is calibrated with gauge records from the GPCC. The recently-launched Global Precipitation Measurement (GPM) mission continues the legacy of TMPA in monitoring space-based precipitation observations through the "Integrated Multi-satellitE Retrievals for GPM" (IMERG) project. The IMERG incorporates more satellite microwave precipitation estimates than the TMPA and provides significantly improved spatial and temporal resolution. This data set will replace the TMPA products by early 2018 when its retrospective processing over the TRMM era is expected to be complete. It is therefore essential to evaluate the performance of the "final run" of IMERG and compare it with the TMPA. Here, we have addressed this by examining the spatial patterns of various evaluation measures including the correlation coefficient (CC), mean absolute difference (MAD), multiplicative bias, probability of detection (POD), false alarm ratio (FAR), frequency bias (FBS), and Heidke skill score (HSS). The results show general agreement between the two products, though the consistency can vary by region. For example, the correlation coefficients over parts of the Ethiopian Highlands and Morocco are quite low. That is similarly reflected in other evaluation indices, using a 1 mm/day threshold to identify rainy days. Some systematic differences, primarily in POD, CSI and HSS are also apparent over the western part of the Gulf of Guinea coastal area and East Africa. One important finding was the strong sensitivity of the IMERG rainfall data to inclusion of the GPCC gauges used for calibration during transition seasons.

COMPARISON OF LITHOLOGICAL MAPPING RESULTS FROM AIRBORNE HYPERSPECTRAL VNIR-SWIR, LWIR AND COMBINED DATA

Rivard B[2], **Rogge D[1]**, Feng J[2]
[1]German Remote Sensing Data Center, [2]University of Alberta

This study investigates using AISA SWIR and Sebass LWIR data independently and in combination to produce detailed lithologic maps in support of mineral exploration in a the Cape Smith Belt, Canada, where mafic, ultramafic and sedimentary rocks are exposed in the presence of lichen coatings. Continuous wavelet analysis (CWA) is used to improve the radiometric quality of the imagery through the minimization of random noise and the enhancement of spectral features, the minimization of residual errors in the ISAC radiometric correction and target temperature estimation in the case of the LWIR data, the minimization of line to line residual calibration effects, and the reduction in variability of the spectral continuum introduced by variable illumination and topography. CWA also provides a platform to directly combine the wavelet scale spectral profiles of the SWIR and LWIR after applying a scalar correction factor to the LWIR such that the dynamic range of two data sets have equal weight. Lithologic maps are generated using an iterative spectral unmixing approach with image spectral endmembers extracted from the SWIR and LWIR imagery. Results demonstrate clear benefits to using the CWA combined imagery. SWIR and LWIR imagery highlight similar regions and spatial distributions for the three ultramafic units (dunite, peridotite, pyroxenite). However, significant differences are observed for quartz-rich sediments, with the SWIR overestimating the distribution of these rocks whereas the LWIR provided more consistent results compared with existing maps. Both SWIR and LWIR imagery were impacted by the pervasive lichen coatings on the mafic rocks (basalts and gabbros) although the SWIR provided better results than the LWIR. Limitations observed for the independent data sets were removed using the combined data resulting in all geologically meaningful units mapped correctly. In combining the SWIR and LWIR data, we achieved an overall improved mapping capability for all geologically meaningful units.

COMPLEXITIES OF MULTI-TEMPORAL REMOTE SENSING IN STRONGLY SEASONAL ENVIRONMENTS AND THE NEED FOR SENSITIVITY ANALYSIS OF TIMES SERIES DEPTH AND IMAGE SPACING

Crews K[1]
[1]*University Of Texas*

Time series analysis of remotely sensed data represents one of the most critical environmental monitoring and management tools available today. In between the more traditional high spatial resolution Landsat-based change detections and the more signal processing-oriented analysis of hundreds of MODIS images is a middle numbers problem: how to deconvolve time signals into cyclic and longer-term components when the community has now amassed a greater inventory of high resolution imagery but not to the level of using techniques such as Fast Fourier Transform. A protocol was developed based on a project under the NASA Safari Y2K program combining harmonic regression and waveform analysis using a series of 85 Landsat-derived vegetation indices over 25 years in the Okavango Delta, Botswana – a strongly seasonal environment listed as both a Ramsar Wetland of Importance and a UNESCO World Heritage site. However, the luxury of such a time-series of high radiometric quality is unusual, and thus a sensitivity analysis was performed on the input time series depth and spacing. By assessing both systematic and random (100 iterations) image dropouts, the impact of such changes was assessed. Of particular surprise and concern was noting that at certain significant drops of images, accuracy actually increased – artificially – over that resulting from inputting the complete time series. These results suggest that in strongly seasonal environments, a inadequately populated image time series may artificially result in higher accuracies of temporal trends, likely due to the dithering of seasonal compared to longer-term signals. This work implies the need in time-series work to include sensitivity analysis of time-series depth and image / date placement as standard protocol in strongly seasonal environments and also suggests the need to rerun earlier studies that may not have taken this phenomenon into account.

CONNECTINGEO GAP ANALYSIS METHODOLOGY AND RESULTS IN THE GEOSS IN-SITU DATASETS

Maso J[1], Serral I[1], Plag H[2]
[1]*CREAF*, [2]*TIWAH*

The EC H2020 project ConnectinGEO is conducting a gap analysis process in the GEOSS in-situ Earth Observation datasets. GEOSS has decided to include a Foundational task in the work program that what to regularly repeat the exercise. This communication informs about the characteristics of the gap analysis conducted by ConnectinGEO. The methodology defines a common model to describe a gap. This model includes also fields that need to be populated to assess the priority such as: remedy, cost, feasibility, impact and timeframe. These extra fields will be used to numerically determine a priority factor. The gaps collected are a combination of top-down and bottom up approaches where goals and consultation are use respectively. The gap analysis is done in loops that allow for external review and feedback by means of exposing the gaps in a queriable table on the Internet. With this methodology, ConnectinGEO has collected around 2 hundred gaps that are summarized in a set of high priority list. The result of these gaps, not only inform GEOSS, but also the funding agencies that can finance the process of filling these gaps by releasing calls for projects or tenders. The gap analysis methodology has been co-designed with the EC H2020 project GAIA-Clim

CONTRIBUTION OF HIGH SPATIAL RESOLUTION SATELLITE IMAGERY FOR ACCURATE ECOLOGICAL MODELING OF THE VECTORS OF ZOONOSES: PRESENTATION OF RODENT STUDIES IN THE INDIAN OCEAN

Herbreteau V[1], Révillion C[2], Biscornet L[3], Tortosa P[3]
[1]IRD, UMR ESPACE-DEV (IRD, UAG, UM, UR), [2]Univ. Réunion, UMR ESPACE-DEV (IRD, UAG, UM, UR), [3]Univ. Réunion, UMR PIMIT (INSERM, CNRS, IRD)

The spatio-temporal dynamics of vector-borne diseases is strongly linked to environmental factors that condition the distribution and dynamics of their vectors, which are often insects or mammals. The overall distribution areas of species are generally related to climatic and environmental factors that have a barrier role. Locally, within these areas, species occupy a smaller ecological niche, depending on favorable environmental conditions, but also on other factors that may constrain their presence, such as predators. High spatial resolution satellite data is fundamental to describe the ecology of vectors. Together with meteorological and climatic data, they are the rare data available at metric scales and necessary for such studies.

We present here how we calculated landscape information from SPOT 5 images to characterize the ecology of rodents, vectors of several zoonoses. We used object-oriented classification (eCognition® software) to integrate textural and topographic variables and build an accurate land use map. Then we calculated several landscape metrics to define the environment of the animals (vectors) that were trapped for these studies. Beyond each animal, these ecological variables made it possible to characterize rodent species and the pathogens that they convey.

We finally discuss the use of high spatial resolution data in ecological modeling and the need for data integration in order to model species temporal dynamics.

KEYWORDS: Health, Ecology, Remote sensing, Rodent-borne diseases, leptospirosis

CONTRIBUTION OF REMOTE SENSING FOR MONITORING AND INVESTIGATING VECTOR-BORNE DISEASES IN CROSS-BORDER CONTEXTS - APPLICATION TO THE INTERNATIONAL BORDERS OF THE BRAZILIAN AMAZON

Roux E[1], Catry T[1], Dessay N[1]
[1]*Institut de Recherche pour le Développement (IRD), UMR ESPACE-DEV (IRD, UM, UR, UA, UG)*

While Malaria transmission has been decreasing worldwide in the last decade, international borders have been identified as major obstacles for malaria elimination. In fact, several cross-border related factors – cross-border human (and thus parasite) mobilities, health care access and patient follow-up difficulties, informal treatment seeking behaviors from populations in illegal situation, etc. – participate to the persistence of malaria transmission foci and the development of treatment resistances, and facilitate parasite re-introduction.

In such contexts, exploiting remotely sensed data to provide key cross-border information on environment, human settlement distribution, urban typologies, and demography, appears particularly relevant. In that sense, remote sensing can significantly contribute to build cross-border observatories at adequate spatial and temporal scales and resolutions for: 1) providing a shared vision of the epidemiological situation; 2) identifying hotspots of transmission; 3) relating malaria transmission with territorial, environmental and climatic factors; 4) targeting actions of control and evaluating their impacts; 5) early identifying disease reemergence.

Research activities carried out in the cross-border area between French Guiana and Brazil for the last five years illustrate the contribution of remote sensing to spatial epidemiology studies in cross-border contexts. Some significant results will be presented. They show: 1) how remote sensing can contribute to the mapping of the transmission risk, through the estimation of individual and populational vector exposure risks. This includes vector habitat modeling and spatializing, the characterization of the human settlements and activities, and the quantification of the role of the landscape in the human-vector encounter risk; 2) the contributions of very high spatial resolution remote sensing for estimating population denominators, required for incidence rate computation and thus for malaria surveillance and understanding.

Eventually, the reproducibility of the proposed approaches, to other cross-border contexts and vector-borne diseases (Aedes-borne diseases) will be discussed.

KEYWORDS: Remote sensing, Cross-border malaria, Vector-borne diseases, Observatories.

CONTRIBUTION OF REMOTE SENSING TO THE BRAZILIAN CLIMATE AND HEALTH OBSERVATORY

Barcellos C[1]
[1]*Oswaldo Cruz Foundation*

Global environmental and climate changes may produce impacts on human health in various ways and intensities. Some of these changes are evident and may be directly observed and monitored from remote-sensing data. Other processes involving climate and health are indirect and may be manifested in different dimensions such as changes to ecosystems, their biodiversity and biogeochemical cycles, societal organization and social inequalities. Unlike other fields of GIS and RS applications, health data is not obtained by remote means. Data on population´s health conditions should be actively sought through surveys and censuses, or passively through epidemiological surveillance systems. Therefore, modeling the relations between climate and health conditions is composed by many methodological and technological challenges. In this work, the potential usage of remote sensing-derived data is critically examined, focusing the proposals, potentials and challenges of environmental health observatories, with emphasis in climate change processes. The implementation of the Brazilian Climate and Health Observatory is described and two stages was defined: the requirement analysis stage which had to identify the national and regional institutional players in their roles as data producers/users; the negotiation stage, resulting from thematic workshops, dealing with water-related diseases, vector-borne diseases, disasters and climate extreme events and health problems derived from droughts and forest fires. The Climate and Health Observatory is an example of making information on climate and health are available through an internet site where data from different origins can be accessed on a common platform. Complex queries are made by users and can be executed over multiple sites, geographically distributed, with all technical details hidden from the end user. At this stage of the prototype, some remote sensing-derived data are made available, such as land use mapping, reanalysis of meteorological variables and climate forecasts, soil moisture and vegetation indices.

COPERNICUS MARINE ENVIRONMENT MONITORING SERVICE, THE UNIQUE EU CAPACITY AT THE SERVICE OF BLUE GROWTH AND SUSTAINABLE OCEANS: USE CASES AND SUCCESS STORIES

Thomas-Courcoux C[1]
[1]*Mercator Ocean*

The Copernicus programme places a world of insight about our planet at the disposal of citizens, public authorities and policy makers, scientists, entrepreneurs and businesses on a full, free and open basis. It consists of a complex set of systems which collect data from multiple sources: earth observation satellites and in situ sensors such as ground stations, airborne and sea-borne sensors. Its processes these data and provides users with reliable and up-to-date information through a set of services related to environmental and security issues: marine, atmosphere, land, climate change, and emergency and security services.

The Copernicus Marine Service covers 4 areas of benefits: Marine Resources, Maritime Safety, Marine and Coastal Environment and Climate/Seasonal Forecasting and counts about 8000 subscribers worldwide (Oct2016). It is designed to serve many public, commercial and scientific purposes including major EU policies such as the Marine Strategy Framework Directive, combating pollution, and protection of marine species, maritime safety and routing, sustainable exploitation of ocean resources, marine energy resources, climate monitoring and hurricane forecasting. Copernicus Marine Service also strives to reach out to the General Public. A pedagogical approach is essential to address strong appetite of citizens towards knowledge about oceans, about ocean protection and fight against pollution and about climate change.

COPERNICUS SENTINELS EO-DATA FOR URBAN LANDFILLS DETECTION AND MONITORING

Cadau E[1], Vingione G[1], Palumbo G[1], Laneve G[2]
[1]Serco SPA, [2]SIA - University of Roma "La Sapienza", Rome

Over the last three decades there has been increasing global concern over the public health impacts attributed to environmental pollution. The WHO estimates that about a quarter of the diseases facing mankind today occur due to prolonged exposure to environmental pollution. Improper management of solid waste is one of the main causes of environmental pollution and degradation in many cities and suburban areas, especially in developing countries. Many cities lack solid waste regulations and proper disposal facilities, including for harmful waste management. Municipal waste dumping sites are designated places set aside for waste disposal. Poor waste management poses a great challenge to the well-being of city residents, particularly those living adjacent to the dumpsites endangered by pollution of water and food resources through the contamination of the aquifers, land, air and vegetation.

Satellite remote sensing data exploitation, has been demonstrated in previous works to be an important means in providing valid and accurate information to the local stakeholders for the identification and monitoring of contaminated sites using non-invasive and costs efficient methods. Some national Italian precursor studies have assessed the suitability of EO data, both Optical and SAR, as a powerful and proficient tool to ensure prompt acquisition of useful information over wide vulnerable areas with high repetition frequency. Remote sensing can be hence fruitfully adopted in environmental impact studies in order to promptly detect and evaluate any potential critical contamination due to the landfills malfunctioning.

This paper aims at describing the major results that can be achieved with the exploitation of the Copernicus Sentinels (S1 and S2) constellation image data detailing the algorithms adopted and describing the possible output products. In particular two urban landfill test cases have been selected: the Bisasar Landfill in the Durban province of South Africa and the Wellington Landfill in the Western Cape province.

CRAFT: LINKING SPATIAL DATA WITH SIMULATION MODELS AND CLIMATE FORECASTS FOR CROP YIELD FORECASTING

Hoogenboom G[1], Shelia V[1], Hansen J[2]
[1]University of Florida, [2]Columbia University

The CCAFS Regional Agricultural Forecasting Toolbox (CRAFT) is a framework for running multiple crop simulation models under a unified user interface and for spatial aggregation of the simulated results into interactive thematic maps. CRAFT consists of three main components including a user-friendly client application, a spatial database that contains all input and output data required for the models, including crop management, soil, weather, and climate data, and an integrated GIS object, which is used for the visualization of gridded results using thematic maps. CRAFT is designed to use spatial data schemes through the use of either 5 or 30 arc minute resolution grids. Schematization at three different spatial scales, including country, state/province and district levels, can be implemented. CRAFT has been integrated with external engines including one for crop modeling for spatial crop simulations and one for seasonal climate forecasts using the Climate Predictability Tool (CPT) developed by the International Research Institute for Climate and Society (IRI). The crop modeling engine provides the interface for the support of multi-crop model capabilities using the crop model translation tools that have been developed by the Agricultural Model Intercomparison and Improvement Project (AgMIP). In the current version of CRAFT the Cropping System Model (CSM) of Decision Support System for Agrotechnology Transfer (DSSAT), APSIM, and SARRA-H have been implemented. CRAFT simulates yield for each individual grid cell using predefined inputs, and based on statistical forecasting and the seasonal predictors yield is adjusted. CRAFT includes options for hind-cast analysis, de-trending, and post-simulation calibration of model predictions from historical agricultural statistics. The system was recently evaluated for rice and wheat yield prediction in Nepal under the auspices of the Nepal Food Security Monitoring System (NeKSAP) through a collaboration between CCAFS and the World Food Programme. CRAFT is available from the Google Drive upon request.

CROP MONITORING FOR FOOD SECURITY EARLY WARNING USING FEWS NET PROCESSES, TOOLS AND EARTH OBSERVATION DATASETS

Magadzire T[1], Verdin J[2]
[1]*FEWS NET*, [2]*USGS*

The USAID FEWS NET activity regularly produces eight-month outlook reports detailing the current and expected future state of food security in more than thirty-five countries around the world, based on an analysis of market and trade, agroclimatology, livelihood and nutrition-related factors. FEWS NET science partners process and disseminate earth observation and agroclimatology model datasets to assist food security experts analyze the climate-related component of the food security equation. Web-based and offline tools have been developed to help analysts interrogate these data within a food security context. On a monthly basis, food security analysts use available information to draw up preliminary assumptions on agroclimatological conditions that can affect the food security outlook. FEWS NET climate experts in Africa, Latin America, and the U.S. then review these assumptions through an interactive monthly process that examines best-available evidence in the form of earth observation datasets, field reports and seasonal climate forecasts. The assumptions are then confirmed or revised, as appropriate, and they are factored into the FEWS NET eight-month food security outlook update issued each month at www.fews.net. The monthly agroclimatology review also provides the basis for FEWS NET contributions to the monthly development of the GEOGLAM Early Warning Crop Monitor report by partners in international, regional, and national organizations. This multi-agency process to classify crop conditions provides a consensus, evidence-based, early warning of adverse crop conditions that can potentially lead to reduced food availability and ultimately, food security crises in the absence of early action.

KEYWORDS: food security, crop monitoring, agroclimatology, early warning

CURRENT STATUS OF HYPERSPECTRAL IMAGER SUITE (HISUI) ONBOARD INTERNATIONAL SPACE STATION (ISS)

Matsunaga T, Nakamura R[3], Iwasaki A[2], Tsuchida S[3], Mouri K[4], Iwao K[3], Yamamoto H[3], Kato S[3], Osamu Kashimura J[4], Tachikawa T[4]

[1]National Institute for Environmental Studies, [2]University of Tokyo, [3]National Institute of Advanced Industrial Science and Technology, [4]Japan Space Systems

Hyperspectral Imager Suite (HISUI) is a future spaceborne hyperspectral Earth imaging system being developed by Japanese Ministry of Economy, Trade, and Industry (METI). It will be deployed on Japan Experiment Module / Exposed Facility in International Space Station (ISS) in 2018-2019.

HISUI will observe the Earth's surface with 185 spectral bands ranging from 0.4 to 2.5 µm with 20 x 30 m spatial resolution and 20 km swath from ISS's nominal altitude of 400 km. It consists of a reflective telescope and two spectrometers which cover the visible and near infrared region (VNIR) and the shortwave infrared region (SWIR). Each spectrometer consists of a grating and a two dimensional detector. SWIR spectrometer has a Stirling cooler for the SWIR detector.

The objective of HISUI is to obtain data necessary to start a full-scale practical application development through manufacturing and in-flight performance verification of a hyperspectral imager with high spectral resolution using ISS by FY2021. More specifically, through the deployment of the manufactured instrument on ISS, its calibration and mission planning, and data acquisition/analysis, we will verify its usefulness in various applications such as oil resource exploration, evaluate the potential of the manufactured instrument, and acquire knowhow to put hyperspectral imagers into practice.

DATA SHARING PRINCIPLES AND THE DEVELOPMENT OF AFRICAN DATA STRATEGIES

Hodson S[1]
[1]*CODATA, Committee on Data of the International Council for Science*

The conduct of research, the opportunities for government and development and many aspects of civil society engagement with policy are being transformed by the digital revolution and the availability of open data of various sorts. These transformations are particularly important in the application of Earth observation data and in their potential for assisting the processes of sustainable development.

Essential if the promise of the data revolution is to be realised are the adoption of open data policies that are appropriate to African countries, the enhancement of capacity for the use of data (covering both processes, skills and infrastructure) and - to coordinate this - effective regional and national data strategies.

The GEO Data Sharing Principles and Data Management Principles can play a valuable role in increasing access to and the use of Earth observation data. The Science International Accord on Open Data in a Big Data World presents an inclusive vision of the need for and the benefits of Open Data for science internationally, and in particular for Lower and Middle Income Countries.

CODATA and ASSAf are collaborating on a DST-funded initiative for an African Open Science Platform which will engage a number of African countries. The development of an open science and innovation platform depends not only on the physical infrastructure for acquiring, curating and disseminating data and information, but also on protocols, policies and procedures in the science system that provide the structure and support to ensure that science objectives are achieved. An open science platform is conceived as an integrated set of arrangements that provides a policy, capacity-building and infrastructural framework for enhanced accessibility and impact.

This presentation will explore the opportunities and challenges relating to the development of African data strategies, with particular reference to Earth observation data and the Sustainable Development Goals.

DEALING WITH AND EXPLOITING BIG EARTH OBSERVATION DATA

Sawyer G[1]
[1]*Earsc*

The world of satellite services is changing. New digital technology targeted on big data, many new satellites being launched offering large volumes of imaging data and the diversity of operational data generated by the Copernicus system are all creating new opportunities for agile companies selling geospatial services. Nevertheless, the market for these services remains rather obscure. Big opportunities seem to be there but connecting users to service providers and data we consider to be the next key challenge. This is particularly so for small companies which are creating new ideas and innovating new products but which struggle to reach a global market.

For this reason, EARSC has launched an initiative on behalf of the European EO service providers, to create a Marketplace for EO services. This will offer new market development opportunities for service providers as well as serving new and innovative enterprises.

So MAEOS, as the initiative is known, will offer companies a full choice of participation and/or co-operation. They can benefit from the services that EARSC and MAEOS will offer or not. They can also decide to enter fully into the Marketplace which in that case would require a new legal entity and a commitment to sharing an integrated front-end and a possible new back-end transaction system.

All the efforts are being driven by the opening-up of Copernicus data and information through a new service to become established by 2018. Joining up the data and encouraging new products and services will be a priority. Our goal is to try to ensure that the European geospatial services industry is fully able to reap the benefit of the European level actions. This paper will explain how.

KEYWORDS: Marketplace, Geospatial, services, Copernicus

DECADAL CYCLIC CHANGES IN MANGROVES IN AUSTRALIA'S ICONIC KAKADU NATIONAL PARK INFORM FUTURE IMPACTS OF CLIMATE CHANGE

Finlayson M[1], Lucas R[2], Asbridge E[3], Mitchell A[4], Woodroffe C[5], Rogers K[6], Bartolo R[7], Hacker J[8]

[1]Institute for Land, Water and Society, Charles Sturt University, [2]Centre for Ecosystem Sciences, Biological, Earth and Environmental Sciences, the University of New South Wales, [3]Centre for Ecosystem Sciences, Biological, Earth and Environmental Sciences, the University of New South Wales, [4]Centre for Ecosystem Sciences, Biological, Earth and Environmental Sciences, the University of New South Wales, [5]School of Earth and Environmental Sciences, Faculty of Science, Medicine and Health, University of Wollongong, [6]School of Earth and Environmental Sciences, Faculty of Science, Medicine and Health, University of Wollongong, [7]Environmental Research Institute of the Supervising Scientist (ERISS), Supervising Science Branch, Science Division, Department of the Environment, [8]Airborne Research Australia / Flinders University

The recent extensive dieback of mangroves along much of the northern Australian coastline has raised concerns about the impact of climate-related phenomena (fluctuations in sea level, rising temperatures, increased storm intensity and frequency) on the long-term integrity and viability of this ecosystem. Such changes were initially detected at sites along the coastline of the Gulf of Carpentaria in late 2015/early 2016 but were subsequently observed in Kakadu National Park (NP) during a September 2016 airborne campaign. The extent of dieback is particularly worrying because all mangroves in Australia have a high degree of protection and those in the north have not been cleared or significantly degraded by humans. Hence, the causes of the change can be attributed to an adverse deviation in prevailing environmental conditions and may be a consequence of human-induced climate change.

A comparison of true colour aerial photography, Compact Airborne Spectral Imager (CASI) and LIDAR acquired in 1991, 2002 and 2011 respectively over Kakadu NP and including the main rivers and islands of the Alligator Rivers Region (West, East and South Alligator, Wildman, Field and Barron Islands) suggested that, whilst the total area occupied by mangroves had remained relatively similar over this period, significant redistribution had occurred as a consequence of storm damage (losses) and colonisation of accreted sediments and upper reaches of creeks (gains) as a consequence of changing hydrological conditions and sea level rise. As with the Gulf, the dieback event observed in Kakadu NP defied the trends of mangrove expansion that had been observed in previous decades. However, whilst there has been considerable and justified alarm about the dieback of mangroves, we present evidence that such changes might have occurred in past decades, albeit not to the same severity, extent and rapidity.

DEEP LEARNING ARCHITECTURE COMPARISON FOR SHIP DISCRIMINATION IN SYNTHETIC APERTURE RADAR

Schwegmann C[1,2], Kleynhans W[1,2], Mdakane L[1,2], Meyer R[1,2]
[1]*Council For Scientific And Industrial Research,* [2]*University of Pretoria*

Maritime Domain Awareness is an important concept for any sea-bordering country and continually improving ways to monitor large ocean areas is an active field of research. Synthetic Aperture Radar provides a mechanism whereby these large maritime environments can be monitored independent of weather or time of observation. One main usage of Synthetic Aperture Radar is the detection of ships at sea when conventional ship transponder systems such as the Automatic Identification System fail. Early work in Synthetic Aperture Radar ship detection combined detection and discrimination but the ability to detect ships and then separate them from false positives can be better accomplished by separating these processes and applying different techniques to each stage. Ship detection can be accomplished using a low threshold Constant False Alarm Rate prescreening algorithm. This highlights all ships brighter than their background as well as possible look-alikes. Following this, machine learning can be used to discriminate amongst all the detections to extract actual ship targets. Previous work has indicated that Deep Learning Highway Networks provide sufficient accuracy and efficiency for ship discrimination across a large Synthetic Aperture Radar dataset. New research indicates that ship discrimination performance can be improved by applying a Deep Learning architecture known as Convolutional Neural Networks to the ship discrimination stage. Convolutional Neural Networks have been designed specifically for object recognition in images as opposed to Highway Networks which are a general extension to conventional Neural Networks. This paper will compare these two Deep Learning architectures and provide results and recommendations for which architecture provides the best current ship discrimination architecture.

KEYWORDS: Machine Learning, Ship Discrimination, Image Processing, Synthetic Aperture Radar, Maritime Domain Awareness

DEFORMATION EFFECTS OF DAMS ON COASTAL REGIONS USING SENTINEL-1 IW TOPS TIME SERIES: THE WEST LESVOS, GREECE CASE

Karamvasis K[1], Vassilia K[1]
[1]*Laboratory of Remote Sensing, National Technical University Of Athens*

Coast zones are vulnerable to erosion and loss by level sea rise. Subsidence caused by the reduction of fluvial sediments in coastal zones found close to dams, is another important deformation factor. Quantification of the deformation rate of coastal region is essential for natural and anthropogenic activities. The study utilizes Interferometric SAR (Synthetic Aperture Radar) techniques and exploits the archive of Sentinel-1 TOPS data for the full period of their availability (2014-2016). The freely available, wide ground coverage (250x250km) and small temporal resolution Sentinel-1 TOPS datasets are promising for coastal applications. Persistent Scatterer Interferometry (PSI) methodologies are considered state-of-the-art remote sensing approaches for land deformation monitoring. The selected PSI method is the Small Baseline Subset (SBAS) multitemporal InSAR technique. The study area of this study is the coastal zone of west region of Lesvos Island, Greece. The main characteristic of the area of interest is the reduction of the fluvial sediment supply from the coastal drainage basins due to construction of dams and the abstraction of riverine sediments. The study demonstrates the potentials of the SBAS method for measuring and mapping the dynamic changes in coastal topography in terms of subsidence rates and discusses its advantages and limitations. The results show that natural and rural environments appear to have diverse ground deformation patterns of the subsidence. A relationship between the observed deformation and the natural (sea level rise) and anthropogenic (existence of dam) factors, which are responsible for the ongoing subsidence can be established.

KEYWORDS: Sentinel-1, PSI, land subsidence, coastal environments, dams

DEIMOS IMAGING: CURRENT MEDIUM AND HIGH RESOLUTION SATELLITES FOR AGRICULTURE AND ENVIRONMENTAL MAPPING AND MONITORING

Chivato R[1]
[1]*Deimos Imaging*

Deimos Imaging has played a key role in the design, implementation and operational consolidation of highly relevant services in precision agriculture in different parts of the world. DEIMOS-1 has demonstrated large-scale coverage capacity since 2011 by providing the US Department of Agriculture (USDA) with the satellite imagery to retrieve in-season crop statistics and crop monitoring for the 48 contiguous states. This is done in cooperation with its twin satellite UK-DMC2, owned by Airbus.
Apart from USDA, major precision agriculture services in the world use Deimos Imaging as data source. The main contribution is the continuous optimization in the provision of DEIMOS-1 and UK-DMC2 data to empower the precision agriculture users at different stages: (i) Planning; (ii) in-season; and (iii) yield.
During the planning stage, the data provided by Deimos Imaging is being used in:

- The definition of uniform management zones and yield potential based on the analysis of the historical performance of the crops
- Planning of soil sample
- The definition of maps for variable rate of seeding and fertilizer

During the in-season frequent monitoring, the data is being used in:

- Prescriptions for variable rates: fertilizer and plant protection products (pesticides, herbicides, etc.) In function of the growth anomalies
- Prescriptions for irrigation support systems.

After harvesting, yield values are compared to the management zones maps and yield potential maps. Satellite archive record is reanalyzed for a better understanding of crop performance and for readjusting the management zones for the next season.
This presentation we will cover the results of various campaigns in agriculture, carried out with DEIMOS-1 and UK-DMC2 since 2011. Finally, Deimos Imaging will introduce "UrtheDaily", a new constellation of satellites for a worldwide daily monitoring of entire landmass of Earth.

KEY WORDS: Precision agriculture, advanced geoanalytics services, multi-disciplinary applications, long time data series, crop performance analysis

DEMO-WETLANDS: COPERNICUS-BASED DETECTION AND MONITORING OF TROPICAL WETLANDS IN RWANDA

Amler E[1], Strauch A[1], Röse M[1], Kirmaier S[2], Cornish N[2], Franke J[2], Greve K[1], Hentze K
[1]*University Of Bonn, Remote Sensing Research Group,* [2]*Remote Sensing Solutions (RSS) GmbH*

Wetlands protection and restoration is an important goal of the newly set up Sustainable Development Goals within the post 2015 development framework. For the implementation, international initiatives like the Ramsar convention on the wise use of wetlands need spatial data on wetlands and their changing surfaces. The objectives of the DeMo-Wetlands project are to develop improved methods and tools for remote sensing based detection and monitoring of tropical wetlands and to demonstrate and implement these on the national scale in Rwanda. The project results and developed methods and products shall directly feed into a national wetland information system of Rwanda and support international organizations and initiatives aiming towards the wise use of wetlands. To reach these goals, a good cooperation with academic and agency partners in Rwanda and with international partnering conventions and initiatives is established. The operational approach of DeMo-Wetlands is concerned with collection and analysis of Copernicus-related sensor system datasets and the automation and implementation of results on the national demonstrator Rwanda. TanDEM-X and Sentinel-1 data are used to identify flooded areas and potential wetland locations based on geomorphologic conditions, followed by a verification of wetland sites based on a high spatial and temporal resolution optical dataset (Sentinel-2). This detection of wetlands is augmented by an ongoing monitoring component which considers inundation and land cover/ land use of wetlands. The multisensoral working approach combined with a multitemporal analysis and continuous automation to enable a sustainable use of the demonstrator will be presented besides the general objectives of DeMo-Wetlands.

KEYWORDS: Wetlands mapping and monitoring, multisensor remote sensing, Copernicus, automation, Rwanda

DETECTING AND ANALYZING GAS FLARES USING SENTINEL 3 AND FIREBIRD

Rücker G[1], Caseiro A[2], Kaiser J[2], Lorenz E[3], Leimbach D[1]
[1]Zebris Gbr, [2]Max Planck Institute for Chemistry, [3]DLR

According to recent research, black carbon has the second strongest effect on the climate system after carbon dioxide. In high Northern latitudes, industrial gas flares are an important source of black carbon, especially in winter. This fact is particularly relevant for the relatively fast observed climate change in the Arctic since deposition of black carbon changes the albedo of snow and ice, thus leading to a positive feedback cycle.
Here we present results of developing algorithms for detection and characterization of industrial gas flares based on nighttime observations of the Sea and Land Surface Temperature Radiometer (SLSTR) on board of the Sentinel 3 satellite.

Routinely obtained gas flare fire radiative power (FRP) from Sentinel 3 shall be used to estimate flaring fluxes of black carbon and trace gases globally. The methods developed shall be integrated into the Copernicus Atmosphere Monitoring Service.

Using nighttime thresholding in the Shortwave Infrared (SWIR) and signal persistence, gas flare locations were identified from Sentinel 3 SLSTR data and verified using high resolution data and SWIR fire detections from Sentinel 2 and Landsat.

Using nighttime radiances from SLSTR bands from shortwave to thermal infrared and curve fitting techniques, flare parameters (flare temperature, flare area and fire radiative power) were estimated based on the Planck function. Additionally, FRP only was estimate from a subset of data using the midwave infrared (MWIR) radiance elevation for day and nighttime observations. MWIR radiance based FRP estimates from Sentinel 3 SLSTR and the FireBird TET-1 satellite obtained near simultaneously over gas flares in the Russian Arctic showed good correlations but lower FRP estimates of TET-1. Planck curve fitting seemed to overestimate fire temperatures, which could probably be corrected by additionally using night time observations of visible and near infrared channels.

KEYWORDS: Gas flaring emissions, fire radiative power, Sentinel 3, FireBird

DETERMINATION OF THE PRESENT VEGETATION STATE OF A WETLAND WITH UAV IMAGERY

Boon M[1], Tesfamichael S[2]
[1]Department of Zoology University of Johannesburg, [2]Department of Geography, Environmental Management and Energy Studies University of Johannesburg

The compositional and structural characteristics of wetland vegetation play a vital role in the services that a wetland supplies. Apart from being important habitats, wetland vegetation also provide services such as flood attenuation and nutrient retention. South Africa is known to be a water scarce country. The protection and continuous monitoring of wetland ecosystems is therefore important. Factors such as site transformation and disturbance may completely change the vegetation of a wetland and the use of Unmanned Aerial Vehicle (UAV) imagery can play a valuable role in high-resolution monitoring and mapping. This study assessed if the use of UAV imagery can enhance the determination of the present vegetation state of a wetland. The WET-Health level two (detailed on-site evaluation) methodology was followed for the vegetation assessment, where wetland health is a measure of the deviation of a wetland's structure and function from its natural reference condition. The mapping of the disturbances classes was then undertaken using ultra-high resolution orthophotos, point clouds and digital surface models (DSM). The WET-Health vegetation module completed with the aid of the UAV products still indicates that the vegetation of the wetland is largely modified ("D" PES Category) and that the vegetation of the wetland will further deteriorate (change score). These results are the same as determined in the baseline study. However a higher impact (activities taking place within the wetland) score were determined. The assessment of various of the WET-Health vegetation indicators were significantly enhanced using the UAV imagery and derived products. The UAV products provided an accurate vantage point over the wetland and surroundings, and assisted to easily refine the assessment of the disturbance classes and disturbance units.

KEYWORDS: wetland vegetation, aerial photography, UAV, 3D point clouds, DSM

DEVELOPING AND TESTING A RANGELAND MONITORING SYSTEM WITH NAMIBIAN FARMERS

Van der Waal C[1]
[1]*Agri-ecological Services*

Namibia is the most arid sub-Saharan country in Africa. Due to the aridity, extensive animal production is the most important form of land use outside protected areas. The productivity of the rangelands fluctuates widely from year-to-year, in response to the variable climate, and subsequently creates a serious management challenge to balance animal forage requirements with the forage produced. This challenge is further compounded by a long dry season, when herbaceous plants are mostly dormant. Poor management decisions may result in animal mortalities, production losses and missed marketing opportunities, as well as rangeland degradation manifesting in the loss of desirable, perennial grazing plants, bush thickening and accelerated soil erosion.

To support decision making at a ranch or grazing area level, near real-time NDVI anomaly and Vegetation Condition Index information are disseminated to rangeland managers as the season develops. The information is made available through a website and an email service and farm boundaries superimposed on the map products to allow individual land users to track the status of their properties as the season develops. More than 70% of respondents in a service satisfaction survey indicated that they found the products useful. Farmer supplied long-term data showed good correspondence between herbaceous biomass production (end of the dry season) and livestock production (kg produced per ha per season) with aggregated satellite-derived vegetation indices. Users also indicated that estimates of forage production at the end of the growing season would be useful for dry season fodder flow planning. Research is underway to predict end-of-growing season herbaceous biomass and cover. If successful, this will allow rangeland managers to set more realistic stocking rates than is currently the case.

ISRSE37
TSHWANE 2017

DEVELOPING REMOTE SENSING GOOD PRACTICE GUIDANCE DOCUMENTS FOR SUB-INDICATORS OF SDG TARGET 15.3.1: LAND COVER, LAND PRODUCTIVITY AND CARBON STOCKS ABOVE AND BELOW GROUND

Sims N[1], Newnham G[1], England J[1], Held A[1]
[1]*CSIRO*

In November 2016, CSIRO was engaged to coordinate the production of good practice guidance documents (GPGs) for three sub-indicators of United Nations Sustainable Development Goal (SDG) 15.3.1 (the proportion of land that is degraded over total land area). These are 1. Land cover and land cover change, 2. Land productivity and 3. Carbon stocks above and below ground. In addition, we were asked to produce a GPG describing methods for bringing the sub-indicators together to report at the SDG level. Our approach was to link GPGs for the land productivity and carbon stocks sub-indicators to levels of detail available in landcover descriptions. Using default mean estimates of productivity and carbon stocks within broad land cover units might suffice at when few data are available. Where sufficient data exist, more advanced approaches might include the use of measured correlates of NPP (such as NDVI and green cover fraction) and carbon stocks (such as vegetation productivity and structure from radar) to inform and refine those estimates within more defined landcover classes. Recent improvements in the quality and availability of freely available satellite data will likely improve the ability to measure and monitor changes in these sub-indicators for many countries. This presentation will present the GPGs and describe available data and monitoring options at different levels of assessment.

KEYWORDS: SDG 15.3.1, landcover, land productivity, carbon stocks, environmental monitoring

DEVELOPMENT OF REGIONAL SEA SURFACE TEMPERATURE AND OCEAN COLOUR MODIS PRODUCTS FOR ECOSYSTEM MONITORING OF THE SOUTH AFRICAN WEST AND SOUTH COAST SHELF SEAS

Whittle C[1], Bernard S[1]
[1]*Council For Scientific And Industrial Research*

The South African west and south coast shelf seas constitute the southern-most part of the Benguela Current Large Marine Ecosystem. It hosts a complex marine region that is influenced by coastal upwelling processes, typical of the Benguela Upwelling System, and also dynamic upwelling resulting from the meandering of the Agulhas Current. High primary production is observed at preferred locations along the coast and along the 100m isobath of the eastern Agulhas Bank. The Agulhas Bank provides both a spawning ground and nursery area for economically important pelagic fish and is linked to the highly productive St Helena Bay nursery region via a coastal jet current. Small pelagic fish abundance and distribution are closely linked to the variability of the upwelling systems in which they exist. Routine monitoring cruises and in situ sampling at key embayments within the region do not provide the temporal and spatial time series data that adequately resolve the frequency and extent of production events over the connected ecosystem. Earth observation data provide wide coverage at a high spatial resolution with frequent repeat sampling that readily resolve the mesoscale surface signatures associated with high productivity events. Satellite derived sea surface temperature (SST) and ocean colour indices serve as important indicators of overall ecosystem health and variability for the assessment of climate change impacts. Previous studies of the southern BCLME using standard NASA 4km satellite products inadequately represent the spatial and temporal event scale of upwelling occurrence. Regionally developed 1km SST and Ocean Colour products, derived from the MODIS instrument aboard the Aqua and Terra satellites, provide significantly improved spatial and temporal data retrieval. These new products present the opportunity to not only investigate the seasonal and interannual variability, but also the event-scale variability of upwelling and bloom events in St Helena Bay and on the Agulhas Bank.

DIAS : DATA INTEGRATION AND ANALYSIS SYSTEM

Mukaida A[1], Inoue J[1]
[1]*Remote Sensing Technology Center Of Japan*

The goals of the Data Integration and Analysis System (DIAS), phase I of which launched in 2006, are to collect and store earth observation data; to analyze such data in combination with socio-economic data and convert it into information useful for crisis management with respect to global-scale environmental issues, natural disasters, and other threats; and to make this information available within Japan and overseas. We aim to help resolve global issues through policy-making assistance; the development of applications and tools through cooperative planning and production with the industrial world; and the creation and social implementation of new public benefits.

In order to archive, analyze, and simulate ultra-high volumes of data, DIAS has an massive data storage/analytical space totaling 25 Pb and an analysis cluster. It is also connected to the National Institute of Informatics' Science Information Network (SINET) for high-speed data transfer with data centers and supercomputers at remote organizations.

Various information is necessary to solve environmental issues, including global-scale observation data; reanalysis data; local observation data for down-scaling and the construction of advanced models and so on. The CMIP3 and CMIP5 are also essential for long-term forecasts. DIAS archives such kinds of data according to the priority of the associated project, and provides an integrated support portal for metadata creation, quality management, and data input.

To date, the application has mainly been applied to the field of climate and hydrology in relation to climate change measures. Moving forward, applications and tools using data from the field of biological and ecological systems, as well as multidisciplinary fields such as health, agriculture, and urban planning are planned to be added in a virtuous cycle, eventually leading to integrated problem resolution and the creation of new research and business fields. We will present the current status and future direction of this project.

DIGITALGLOBE - CHANGING THE WORLD WITH CLOUD, CROWD AND GEOSPATIAL BIG DATA

Fortescue A[1]
[1] *DigitalGlobe*

This paper will demonstrate how DigitalGlobe's entire archive of over 90 Petabytes is made available online via cloud services. It will include examples of how this online content has been deployed with crowd applications in disaster response, baseline mapping and macro economic trend analyses. It will include an introduction to DigitalGlobes Geospatial Big data platform and availability for developers.

KEYWORDS: Cloud, crowd, Geospatial big data, disaster monitoring,

DISCUSSION: FEDERATION OF EARTH OBSERVATION CLOUD PLATFORMS

Wagner W[1], Loekken S[2]
[1]Vienna University of Technology (TU Wien), [2]European Space Agency

An increasing number of earth observation cloud platforms has been put into operations in recent years, and even more are under development. They all differ in terms of their data holdings, processing capabilities and thematic orientation, but obviously there are also overlaps. In this discussion round we would like to raise the questions, when federation makes sense and what are possible solution for such a cooperation on a technical and organisational level?

DO WE HAVE ADEQUATE EARTH OBSERVATION SENSORS FOR DETECTING AND MONITORING THE SPATIAL EXTENT OF FRESHWATER ECOSYSTEMS?

Van Deventer H[1], Cho M[1], Naidoo L[1], Mathieu R[1], Nel J[2]
[1]CSIR, [2]Nelson Mandela Metropolitan University (NMMU)

The tremendous loss and decline of freshwater ecosystems since the 1900 warrants methods for the mapping and monitoring of the structure, function and phenology freshwater ecosystems over time. The aim of this paper is to evaluate the current capabilities of existing Earth Observation sensors (for example Sentinel 1A and 2A, RapidEye and WorldView) in detecting the spatial extent of different freshwater ecosystem types. Fine-scale wetland datasets from a number of biomes (grassland/savannah, fynbos and arid) in South Africa were used to ascertain the adequacy of the sensors in detecting the spatial extent of seven natural and two artificial wetland types. Previous studies showed that large artificial wetlands will be easier to detect with space-borne multispectral sensors owing to the presence of open water throughout a phenological cycle. In contrast, small artificial wetlands, seeps and wetland flats are small in extent with little open water across a seasonal cycle. The variation of herbaceous aboveground biomass across wetland ecosystem types and adjacent herbaceous terrestrial ecosystems was furthermore compared to assess whether they differ significantly. The Leaf Area Index (LAI) developed from the Moderate Resolution Image Spectrometer (MODIS) was used to estimate aboveground biomass for the wetland types in each biome. It is expected that most of the wetland types in the grassland and fynbos biomes will have aboveground biomass, whereas the arid biome will have no significant biomass to detect or monitor. The aboveground biomass of terrestrial and wetland ecosystems are expected to be similar. The ability of existing space-borne multispectral sensors to detect and monitor these ecosystems will be crucial for attaining certain Aichi goals by 2020.

DYNAMICS OF GLOBAL WETLANDS REFLECTED IN THE DAILY TIME SERIES PRODUCT GLOBAL WATER PACK

Klein I[1], Gessner U[1], Dech S[1], Kuenzer C[1]
[1]*DLR*

Wetlands are rich ecosystems and essential for human life. They provide wildlife habitats, are crucial to agriculture and fishery, purify and store water, and maintain its availability during dry periods. Key functions and services of wetlands are closely related to the temporal variations of water levels and inundation cycles. Daily time series of optical medium resolution earth observation data with global coverage are now available for more than 15 years, and open the possibility for long-term monitoring of wetland ecosystems.

The Global Water Pack is a global and daily 250 m time series product developed at DLR that captures inland waterbody extents, their seasonal dynamics, and their multi-annual alterations. The main input data for the Global Water Pack is daily spectral information of MODIS (Moderate Resolution Spectroradiometer), complemented by auxiliary data including a global DEM and selected thematic MODIS products. Water is first classified on a daily basis using a dynamic thresholding approach. In a subsequent time series analysis, cloud gaps within the resulting daily water masks are reduced. The dataset has been validated based on more than 300 Landsat-based reference classifications revealing an accuracy of more than 90% for pixels of more than 75% sub-pixel water coverage.

The Global Water Pack provides crucial information for global wetland ecosystems reflecting the dynamics of open water surfaces. Using examples of major wetlands in Africa, Asia and America, the potential of the Global Water Pack to capture relevant parameters such as inundation frequency and duration, timing of flooding and water retreat, as well as freezing and thawing cycles, is presented. Results are compared with high resolution spatial reference data and in-situ measurements, and the application potential of the Global Water Pack time series for monitoring and assessing wetland ecosystem services and changes are discussed.

EARTH OBSERVATION ACTIVITIES AND FUTURE PERSPECTIVES IN EGYPT

Abou El-magd I[1]
[1]*National Authority for Remote Sensing and Space Sciences*

Egypt was one of the first developing countries in Africa that used earth observation and remote sensing in various applications since 1970s. It has grown up in the last decades to build its own capacity in space science and technology that ended up by launching earth observation satellites. At the same time Egypt continued to develop the capacity in EO applications and contribute to the national development plans. In this domain NARSS, the governmental research institute that lead the EO and space applications has completed many research and development projects in EO applications in mineral resources exploration, coastal and marine resources, air quality, water resources management, food security, etc. This was via operational projects with the stakeholders and users to ensure sustainability and operation of the services. For example, NARSS has developed an operational system to monitor the national crop rice using EO information that capable to provide the actual land planted with rice and predict the yield. The system has enabled to provide recommendations for other plots of land that suitable for rice plantation. In the area of environmental hazards, many projects on the flash floods and the vulnerability to flash flood hazards were developed providing decision makers with vulnerability maps and Atlases on national level. Further details on the EO activities and future plans at NARSS, Egypt will be presented in this paper.

EARTH OBSERVATION ANALYSIS WITH SAP HANA

Gildhoff H[1]
[1]*SAP SE*

SAP works closely together with ESA and provides since end of 2016 a new HANA cloud service for transparently processing earth observation data. The goal is to close the gap between the business and earth observation community and provide a simple REST API for developers to extract business relevant information and integrate it into standard business and analytics tools.

EARTH OBSERVATION ASSISTED DECISION SUPPORT TOOLS FOR EVIDENCE BASED DECISION MAKING

Vagen T[1], Neely C[1], Winowiecki L[1], Bourne M[1]
[1]*World Agroforestry Centre (ICRAF)*

Earth Observation (EO) has the potential to provide decision makers with tools offering evidence on land health and land degradation at a range of different spatial scales. Through systematic assessments of land health using the Land Degradation Surveillance Framework (LDSF), it is now possible to produce predictive maps based on remote sensing data for a range of metrics related to soil and land health, as well as land degradation risk. The level of accuracy that is achievable with these approaches has been improving steadily with an increasing database of ground observations from the global tropics, coupled with advanced spatial models. Hence, accuracy is now at a level where this information can be applied in decision making, as well as for targeting of land management interventions and monitoring of progress over time. Through a combination of stakeholder engagement processes and tools, EO-based evidence has significant potential to alter the way decision making is being made by embedding evidence into the various stages of decision making processes. We demonstrate this approach, including tools, with case studies from Kenya.

EARTH OBSERVATION FOR SUPPORTING AND TRACKING PROGRESS OF SUSTAINABLE DEVELOPMENT GOALS: BEST PRACTICE EXAMPLE FROM THE AUSTRALIAN TERRESTRIAL ECOSYSTEM RESEARCH NETWORK (TERN)

Metternicht G[1], Held A[2], Phinn S[3], Christensen B[2], Kerblat F[3], Sims N[3], Guershman J[3]
[1]School of Biological, Earth and Environmental Science, UNSW Australia, [2]CSIRO, [3]University of Queensland

In September 2015, the United Nations adopted the 2030 agenda for sustainable development, a plan of action for people, the planet and prosperity. The 17 Sustainable Development Goals with 169 associated targets integrate the economic, social and environmental dimensions of sustainable development. The goals address a variety of wicked problems related to human-environment interactions –environmental degradation, climate change, sustainable management of natural resources, fresh water scarcity, loss of biodiversity - whose achievement will require integrated solutions and collaborative work across disciplines. The goals are universal, though countries are responsible for selecting national targets, and determining their own priorities and level of ambition in terms of the changes pursued.

The UN resolution states that goals and targets will be followed-up and reviewed using a set of global indicators, and it recognizes the contribution to be made by earth observation and geo-spatial information in supporting implementation and tracking progress. Accordingly, this presentation highlights the potential of EO for supporting countries in their own target setting process, as well as planning and implementation of related strategies, and tracking progress towards the goals (e.g creating baselines and input for indicator monitoring frameworks). We use efforts by the Committee on Earth Observation Satellites (CEOS) to identify goals, targets and indicators that can profit from earth observation and geospatial information, and the extent of the benefit (high, moderate, low). A case study of an Australian initiative, the Terrestrial Ecosystem Research Network (TERN) AusCover is used to illustrate how EO could quantitatively and economically guide the implementation of target-based SDG strategies (particularly goals of 15,14,13,and 11), given its extensive operational role in the country's current environmental monitoring and management. TERN's -EO facility is a partnership of over 12 government and academic institutions that collects data about ecosystems using satellites, airborne sensors, and onground systems.

EARTH OBSERVATION IN SUPPORT OF SUSTAINABLE WATER RESOURCE MANAGEMENT, THE TIGER INITIATIVE – LOOKING AFTER WATER FOR AGRICULTURE IN AFRICA

Koetz B[1]
[1]*European Space Agency*

Reliable access to water, managing the spatial and temporal variability of water availability, ensuring the quality of freshwater and responding to climatological changes in the hydrological cycle are prerequisites for the development of countries in Africa. Water being an essential input for biomass growth and for renewable energy production plays an integral part in ensuring food and energy security. Water, as a source of safe drinking water, is furthermore the basis for ensuring the health of citizens and plays an important role in urban sanitation. The recently announced Sustainable Development Goals (SDG) of the United Nations include for the first time a dedicated goal (6th SDG) on "Ensure water availability and sustainable management of water for all". Earth Observation can support the assessment and monitoring of several targets and indicators asscociated to the 6th SDG.

The concept of Integrated Water Resource Management (IWRM) is seen as an opportunity to help manage water variability and the wide spread water scarcity in Africa. One key component missing from IWRM in Africa is the limited knowledge of the available extent and quality of water resources at basin level. ESA's TIGER initiative aims at enabling African water authorities to fill this information gap by monitoring water resources at adequate temporal and spatial scales based on Earth Observation (EO) technology.

This contribution will present the use of the Sentinels for monitoring water resources in the context of the agriculture sector. TIGER currently supports 10 African-European research projects addressing different aspects regarding inventorying, modelling, monitoring and planning the (re)allocation and use of water resources in the agricultural sector.

EARTH OBSERVATION METHODS FOR MONITORING SALT ACCUMULATION IN IRRIGATED AREAS

Van Niekerk A[1], Muller J[1], Vermeulen D[1]
[1]*Stellenbosch University*

Salt accumulation is of major concern in many parts of the world due to its negative environmental impact on agricultural areas, which often leads to a reduction in crop yields. Salt accumulation can also lead to lower property values, eutrophication of rivers, damage to infrastructure, increased soil erosion and engineering difficulties. Conventional methods for monitoring salt accumulation within irrigation schemes involve regular field visits to collect soil samples for laboratory analysis. Remote sensing has been proposed as a less time-consuming, cost-effective alternative as it provides imagery covering large areas throughout the year. This paper overviews a range of experiments that were carried out to evaluate the efficacy of optical very high resolution (VHR) imagery to map areas affected by salt accumulation. The experiments involved the direct (targeting exposed soil) and indirect (crop condition monitoring and terrain analyses) approaches. A large set of soil samples were collected in nine irrigation schemes throughout South Africa. The samples were analysed in a laboratory to determine the electric conductivity (EC) of soils. The EC data was used in various statistical and machine learning techniques to differentiate affected and non-affected areas. The results showed that, although machine learning produced relatively high accuracies, the resulting maps varied considerably and that none of the models were robust (transferable). The variations in tolerances of crops to saline conditions, as well as the temporal variations within irrigation schemes, were found to be the main obstacle to developing robust models. A geographical object-based image analysis (GEOBIA) approach that circumvents inter-field conditions is proposed and demonstrated as a solution.

EARTH OBSERVATION MISSIONS AND APPLICATIONS IN BRAZIL

Fonseca L[1], Ferreira H[1]
[1]*National Institute For Space Research (inpe)*

Earth Observation data are used to enable better understanding and improved management of the Earth and its environment and resources. As they gather information continuously across both space and time, they provide valuable information for research in different applications such as biodiversity, public health, water resources, agriculture and forest management. The objective of this presentation is to highlight major achievements in EO applications and activities in Brazil, and present an outlook of future EO program.

EARTH OBSERVATION TECHNOLOGY FOR MONITORING LAND DEGRADATION IN SOUTH AFRICA: A SYNTHESIS FOR THE DEPARTMENT OF ENVIRONMENTAL AFFAIRS UNCCD FRAMEWORK

Chirima G[1]
[1]*Agricultural Research Council*

Land degradation is highly variable, discontinuous, arising from different causes, and affect people differentially depending economic, social, and political circumstances. South Africa faces threats to sustainable food production due to climate change linked meteorological hazards, and loss of productive land because of land degradation processes. Soil degradation is severe and increasing in most communal crop and grazing lands. The UNCCD acknowledges that about 30 % of land area of the world is affected by desertification, and about 73 % of drylands in Africa. In worst forms, land degradation leads to irreversible biophysical change and permanent loss of service provision by the land.

The current paper covers a UNCCD research on desertification, land degradation, drought and land cover indicators mapping for South Africa. The aim was to determine the 2009-2013 status of land degradation and the rate of change from that reference year. Because, of the extensive areas needed to be covered within the 1-year time span of the project, use of satellite data and technology for these purposes was investigated.

The study showed the approach to be very successful and practical. Most notably, land degradation 'hotspots' in the country were identified. The extent of drought affected lands, and desertified lands and rates of land degradation can be identified. In line with the UNCCD convention, and other international obligations, it recommended that DEA regularly conduct land degradation, desertification, erosion, and drought assessments using these approaches in order to inform policies, rehabilitation programs and strategies.

EARTH OBSERVATION-SUPPORTED SERVICE PLATFORM FOR THE DEVELOPMENT AND PROVISION OF THEMATIC INFORMATION ON THE BUILT ENVIRONMENT – THE TEP-URBAN PROJECT

Esch T[1], Balhar J[2], Boettcher M[3], Boissier E[4], Kuchar S[5], Marconcini M[1], Mathot E[4], Metz A[1], Permana H[3], Soukup T[2], Üreyen S[1], Svaton V[5], Zeidler J[1]

[1]German Aerospace Center (DLR), Earth Observation Center (EOC), [2]GISAT s.r.o., [3]Brockmann Consult GmbH, [4]Terradue Srl, [5]IT4Innovations

The upcoming suite of Sentinel satellites in combination with their free and open access data policy will open new perspectives for establishing a spatially and temporally detailed monitoring of the Earth's surface. The Sentinels will provide a so-far unique coverage with Earth observation (EO) data and new possibilities with respect to the implementation of innovative methodologies, techniques and geo-information products and services. However, the capability to effectively and efficiently access, process, analyze and distribute the mass data streams from the Sentinels and high-level information products derived from them poses a key challenge. This is also true with respect to the necessity of flexibly adapting the processing and analysis procedures to new or changing user requirements and technical developments. Hence, the implementation of operational, modular and highly automated processing chains, embedded in powerful hard- and software environments and linked with effective distribution functionalities, is of central importance. This paper introduces concepts for the utilization of modern information technology functionalities and services to bridge the gap between the technology-driven EO sector and the information needs of environmental science, planning, and policy. Key components of such systems are currently developed in the project Urban Thematic Exploitation Platform (U-TEP). This includes the implementation of an open, web-based platform employing distributed high-level computing infrastructures (Platform as a Service – PaaS) providing key functionalities for i) high-performance access to thematic data (Information as a Service – InaaS), ii) modular and generic state-of-the-art pre-processing, analysis, and visualization (Software as a Service – SaaS), iii) customized development and dissemination of algorithms, products and services, and iv) networking and communication. These services and functionalities are supposed to enable any interested user to easily exploit and generate thematic information on the status and development of the environment based on EO data and technologies.

KEYWORDS: Service Platform, Urban, Monitoring, Earth Observation

EFFECTIVE AND SCALABLE COMPUTATION OF TEMPORAL COLOUR COMPOSITES FROM EARTH OBSERVATION BIG DATA

Karmas A[1], Tzotsos A[1], Karantzalos K[1]
[1]*National Technical University Of Athens*

Earth Observation data are increasing with tremendous rate in volume, variety and complexity. As a consequence new challenges are emerging in regard to their access, archiving, processing and analysis. Towards this direction, an EO big data platform capable of efficiently analysing big data for geospatial applications has been developed and evaluated. The developed platform, in order to achieve the goals of efficient analysis, storage and handling of big data, integrates a number of tools, libraries and processing models in a distributed computer cluster environment. In particular, the core functionality consists of Geotrellis, a geospatial data processing engine for high performance applications, responsible for the storage, handling and processing of the data as well as the Apache Spark framework for distributing computations and data across the cluster. Various algorithms were implemented in Scala programming language both for applying ETL (Extract, Transform, Load) procedures on the raw data and for the access and distributed analysis of multispectral satellite data. The developed system in its current form covers partially the Greek territory with multispectral satellite data which are acquired by Landsat 8 and Sentinel-2 satellites and that are stored and processed in an automated way in the infrastructure which is in our disposal for demonstration purposes. The developed data processing queries are focused mainly in agricultural applications and create artificial color composites, based on well defined indices, that provide valuable information regarding the crops' state over time per pixel and per year as well as value-added products that map seasonality from which it is possible to extract patterns for the life cycle of the monitored crops.

ENABLING ANNUAL LAND COVER MAPPING ACROSS EAST AFRICA

Oduor P[1], Ouko E[1], Healey S[2], Cohen W[2], Zhiqiang Y[3], Wilson S[4], Gorelick N[5]
[1]RCMRD, [2]US Forest Service, [3]Oregon State University, [4]US Geological Survey, [5]Google Corporation

Storage of atmospheric carbon is an important terrestrial ecosystem service because it mitigates the effects of anthropogenic greenhouse gas emissions. International reporting of this service places a premium on the specificity and precision of monitoring data used to estimate carbon storage or emission. An inventory of land cover change is a critical component of most national-level accounting systems, and the Landsat series of satellites is a uniquely positioned to provide this land cover change "activity data." Such maps are also useful in food security and water demand assessments. In Eastern Africa, there are already high-quality Landsat-based cover maps for 2 or 3 points in time. However, these maps do not provide the annual land cover change information needed for many applications, and land cover changes inferred from independent maps at different dates cannot easily be assigned a level of uncertainty.

We are developing innovative, annually produced land cover maps to meet needs of national partners within the East Africa region. Several different change detection algorithms will be run in parallel on the full series of archived Landsat images since 2000, and locally developed reference data will be used to calibrate a model that optimally integrates the output maps from each separate algorithm. This strategy was developed under an inter-agency, national change mapping project in the United States: the Landscape Change Monitoring System (LCMS). LCMS pilot research strongly suggests the value of integrating an ensemble of different approaches to map cover change. Cloud computing has only recently made this data- and processing-intensive approach feasible, and the fact that images no longer have to be downloaded and stored locally circumvents traditional access barriers in bandwidth-limited locations. Our methods are globally scalable, and moving forward, they have the potential to support land cover change monitoring across multiple sectors and geographies.

EO4URBAN: MULTITEMPORAL SENTINEL-1A SAR AND SENTINEL-2A MSI DATA FOR GLOBAL URBAN SERVICES

Ban Y[1]
[1]*KTH Royal Institute Of Technology*

With more than half of the world population now living in cities, and 2.5 billion more people expected to move into cities by 2050, urban areas pose significant challenges on the environment. Although only a small percentage of global land cover, urban areas significantly alter climate, biogeochemistry, and hydrology at local, regional, and global scales. Thus, accurate and timely information on urban land cover and their changing patterns is of critical importance to support sustainable urban development.

The objective of this research is to evaluate multitemporal Sentinel-1A SAR and Sentinel-2A MSI data for global urban services using innovative methods and algorithms. Ten cities around the world in different environmental conditions are selected as study areas. Multitemporal multi-resolution Sentinel-1A IW SAR and Sentinel-2A MSI data were acquired during the vegetation seasons in 2015 and/or 2016 to maximize the difference between urban and rural areas. Historical ENVISAT ASAR and ERS-1/2 SAR data were also collected for urbanization monitoring. The methodology in this research involves multi-scale analysis of multitemporal Sentinel-1A SAR, Sentinel-2A MSI and historical SAR data including image processing, urban extent extraction, object-based image classification, change detection and accuracy assessment.

The results demonstrate that multitemporal Sentinel-1A SAR and Sentinel-2A MSI data are very promising for urban extent mapping at global scale. Compared to the urban extraction results from ENVISAT ASAR or ERS SAR data in 1995 and 2005, the urbanization patterns and trends were analyzed. For urban land cover classification, both Sentinel-2A MSI data and fusion of SAR and MSI data produced over 80% accuracy. Fusion of S-1A SAR and S-2A MSI data could reduce the confusions among several classes. For mapping urban green structure and their changes, Senitnle-2A MSI and Fusion of Sentinel-1A SAR and Sentinel-2A MSI data are suitable.

KEYWORDS: EO4Urban, Sentinel-1A SAR, Sentinel-2A MSI, Data Fusion, Global Urban Services

ESA'S EARTH EXPLORER MISSIONS – AN OVERVIEW OF THE RESEARCH SATELLITES OF ESA'S "LIVING PLANET PROGRAMME"

Rast M[1]
[1]*ESA ESRIN*

ESA's Earth Explorer missions are the core element of ESA's Living Planet Programme. They focus on the atmosphere, biosphere, hydrosphere, cryosphere and the Earth's interior and provide an important contribution to further our understanding of Earth.

To date, eight missions have been selected for implementation:

GOCE
The Gravity field and steady-state Ocean Circulation Explorer (GOCE), launched on 17 March 2009 and ended on 11 November 2013 provided high resolution gravity-gradient data to improve global and regional models of Earth's gravity field and geoid.
SMOS
The Soil Moisture and Ocean Salinity (SMOS) mission, launched on 2 November 2009, with its interferometer provides information on soil moisture and salinity in the surface layers of the oceans.
CryoSat
CryoSat, launched on 8 April 2010, measures with a SAR/interferometric radar altimeter provides fluctuations in the thickness of ice on land and sea to determine how Earth's ice is changing.
Swarm
Swarm, launched on 22 November 2013, provides a survey of the geomagnetic field and its temporal evolution. The resulting geomagnetic models will provide new insights into Earth's interior.
ADM-Aeolus
The aim of the Atmospheric Dynamics Mission is to demonstrate measurements of vertical wind profiles from space. The mission due for launch in 2018 employs a high-performance Doppler wind lidar.
EarthCARE
Earth Clouds Aerosols and Radiation Explorer (EarthCARE), due for launch in 2019, will improve the representation and understanding of Earth's radiative balance in climate and numerical forecast models.
Biomass
The Biomass mission will, with the first P-band SAR in space, provide measurements to determine the amount of biomass and carbon stored in forests, thereby improving our understanding of the carbon cycle.
FLEX
The Fluorescence Explorer will map vegetation fluorescence to improve our understanding of the way carbon moves between plants and the atmosphere and how photosynthesis affects the carbon and water cycles.

ESA'S FLUORESCENCE EXPLORER (FLEX): MISSION CONCEPT AND SCIENTIFIC GOALS

<u>Rascher U</u>, Moreno J[1]
[1]*University Of Valencia*

Mapping at a global scale the actual photosynthesis of terrestrial vegetation is of particular interest for the improvement in the predictive capability of global models through new parameterizations for canopy photosynthesis and the corresponding exchange processes of energy, water and carbon between the surface and the atmosphere. Current remote sensing techniques can only provide an estimate of the "potential" photosynthesis, rather than "actual", but sun-induced chlorophyll fluorescence is a sensitive indicator of the actual photosynthesis in both healthy and physiologically stressed vegetation, which can be used as a powerful non-invasive marker to track the status, resilience, and recovery of photochemical processes.

The ESA's Earth Explorer FLEX (Fluorescence EXplorer) mission is the first space mission focused on the estimation of fluorescence emission by terrestrial vegetation on a global scale with high spatial resolution (300 m) and resolving the spectral shape of fluorescence emission. The FLEX mission also includes explicit measurement of photochemical changes in reflectance, canopy temperature measurements and all the relevant variables (chlorophyll content, Leaf Area Index, etc.) needed to assess the actual physiological status of vegetation and to provide quantitative estimates of photosynthetic rates and vegetation stress conditions.
The FLEX mission concept consists in a single platform that carries the Fluorescence Imaging Spectrometer (FLORIS), which has been designed and optimised for discrimination of the fluorescence signal in terrestrial vegetation. FLEX will fly in formation with Copernicus Sentinel-3 in order to further enhance the spectral coverage from measurements made by the Sentinel-3 instruments OLCI and SLSTR, exploiting the synergy between their data and helping in the proper characterization of the atmospheric state and cloud screening, essential for a reliable retrieval of fluorescence emission.

In this paper, we provide the relevant scientific background and an overview of the FLEX mission concept, measurement methods and scientific challenges, describing current status and perspectives.

ESSENTIAL BIODIVERSITY VARIABLES MEASURED BY IMAGING SPECTROMETERS – EXAMPLES FROM THE KRUGER PARK AREA

Skidmore A[1]
[1]*ITC, University Twente*

Directly measuring EBVs using Remote Sensing, and placing these biodiversity measurements in the context of Kruger National Park and its immediate surrounding areas is discussed. Very encouraging results have been obtained using image spectroscopy mostly obtained from aircraft platforms. For example, biomass and LAI can be directly estimated, as can plant traits such as specific leaf area and foliar nitrogen can also be retrieved from hyperspectral remote sensing. More complex EBVs can be generated, for example from the spectral variation hypothesis (SVH) that proposes that the variability in satellite remote sensing reflectance is a proxy for environmental heterogeneity and thus an indicator of biodiversity. Remote sensing can be complemented with the in situ sensor networks in the Kruger Park as well as other ground-based data for the modeling and validation of species data and other essential biodiversity variables obtainable from field records, museum data and citizen science.

ESTIMATING FORAGE QUALITY ACROSS DIFFERENT GRASS COMMUNITIES IN TELPERION GAME RESERVE USING HIGH RESOLUTION REMOTE SENSING DATA

Chabalala Y, Mashamba T, Adam E, Oumar Z
[1]*University of the Witwatersrand,* [4]*KZN Department of Agriculture & Rural Development*

Grass quality, as determined by the concentration of nitrogen, phosphorus, potassium, calcium, and sodium is an important factor influencing the distribution of grazing mammals. Mapping grass quality is important to understand ecosystem dynamics, functions, wildlife and livestock feeding as well as distribution patterns. Understanding the spatial distribution of grass quality provides essential information for sustainable planning and management by identifying crucial areas for conservation and restoration. A number of studies have shown that grass quality at a large scale is influenced by climatic and environmental variables and grass diversity. Techniques that can estimate grass quality on such scale are therefore critical in understanding and explaining wildlife and livestock foraging and migration patterns.

In this study, we aimed at examining the use of new generation's multispectral sensor such as Sentinel-2 and WorldView-2 in mapping grass communities and estimating canopy nitrogen (N), phosphorus (P) and sodium (Na) across different common grass communities in Telperion. A machine learning-based approached was used on this study for classification and regression.

The study demonstrated the possibility to accurately map the common grass communities and grass quality at a large scale using multispectral sensors with the red edge waveband with a high spatial resolution.

KEYWORDS: Sentinel 2, Nitrogen; Grass Species; Wildlife

ESTIMATION OF CANOPY HEIGHT IN A COMPLEX VEGETATION ENVIRONMENT USING SENTINEL-1 SAR

Booyzen A[1], Tesfamichael S[1]
[1]*Department of Geography, Environmental Management & Energy Studies; University of Johannesburg*

Synthetic Aperture Radar (SAR) data has attracted great interest for estimating various vegetation characteristics. Much of the focus in this regard has been on large spatial areas with distinct vegetation characteristics. There is a need to explore the performance of SAR in localized areas that have complex vegetation composition. This study investigates the performance of a high-spatial resolution Sentinel-1 C-band dual-polarisation (vertical-vertical (VV) and vertical-horizontal (VH)) SAR image for the estimation of canopy height in a localised and complex vegetation community. Tree height was collected in 51 sample plots using Vertex Hypsometer. A number of statistical metrics were extracted from SAR backscatter per plot and per polarisation (VV and VH). A generalised linear regression model was used to relate the SAR metrics as independent variables with field observation as dependent variable in R-project. The results returned a better correlation for VV than VH polarisation. In conclusion, SAR data is a promising tool for the estimation of canopy height even in complex vegetation environments. Further investigation is needed to improve the accuracy of SAR by increasing the sample size and fusing the data with higher spatial resolution optical data.

KEYWORDS: SAR, Sentinel, Canopy height, Crown cover, Vegetation characteristics

ESTIMATION OF CHLOROPHYLL-A IN EUTROPHIC INLAND WATERS USING SENTINEL 3 OCEAN AND LAND COLOR INSTRUMENT

Kravitz J[1], Matthews M[2], Griffith D[3], Faniso Z[3]
[1]Department of Oceanography, University Of Cape Town, [2]CyanoLakes (Pty) Ltd, [3]Council for Scientific and Industrial Research

The quality of South African freshwater systems has become of great concern as the prevalence of eutrophication and potentially toxic algal blooms have increased due to a growing population and a changing climate. The highly successful Medium Resolution Imaging Spectrometer (MERIS) provided unprecedented capabilities for the monitoring of inland water trophic status and toxic algal species identification until its unfortunate termination in 2012. The European Space Agency hopes to continue that success with the recently launched Ocean and Land Color Instrument (OLCI) aboard the Sentinel 3 satellite. The success of the mission will depend on extensive validation efforts for the development of accurate and robust in-water algorithms. One of the biggest hurdles to algorithm validation for inland water bodies is the difficulty of performing an accurate atmospheric correction. High turbidity and close proximity to land pixels make traditional correction schemes unsuitable for inland waters. The objective of this study is to assess the potential of OLCI for the monitoring of trophic status of three small inland water bodies within South Africa. Radiometric errors associated with OLCI Top of Atmosphere (TOA) radiances will be assessed by validating with in-situ measurements of water color radiometry modelled to Top of Atmosphere. A selection of various atmospheric correction procedures based on the method of aerosol retrieval and ease of implementation will be evaluated with in situ water color radiometry measured simultaneously with imagery acquired from OLCI. The uncertainties and limitations of using a full or partial atmospheric correction procedure will be evaluated in terms of their radiometric accuracy and comparison of resulting chl-a concentrations from a paired bio-optical model with in situ bio-geophysical measurements. Initial results illustrate the difficulty of performing a successful full atmospheric correction over small inland water targets.

ESTIMATION OF GRASS NITROGEN CONCENTRATION USING THE INVERSION OF PROSAIL RADIATIVE TRANSFER MODEL – IN THE HETEROGENEOUS SAVANNA ECOSYSTEM – TOWARDS SEASON INDEPENDENT AND REGIONAL APPROACH

Ramoelo A[1], Cho M[1], Masemola C[1], Dziba L[1]
[1]*Council for Scientific and Industrial Research , Pretoria, South Africa*

Empirical modelling has been successfully used to estimate leaf nitrogen concentration (N) in various ecosystems, including the savanna as an indicator of rangeland quality. Most of this series of models are based on the use of vegetation indices, including the red edge band or position in hyperspectral and commercial satellite sensors such as RapidEye, WorldView and now freely available Sentinel-2 data. Empirical model are known to be data, site and season specific in nature, grass nitrogen has often achieved during peak productivity. Alternatively, the physical based models including PROSAIL are known to be robust and could be useful in the estimation of grass N independent of season and are rarely explored. The objective of this study is to investigate the plausibility of inverting the PROSAIL model to estimate leaf nitrogen using Sentinel-2 data, through the relationship between leaf N and chlorophyll content – Kruger National park area, South Africa. The input parameters of the model shall also be constrained using field and other ancillary data sets, to avoid ill-posed problem. The inversion process shall be achieved by using spectral indices, lookup-table and machine learning tools. The results indicate that the estimation capability of the leaf N using inversion process yielded comparable results to those of empirical modelling. Though in most cases the physical based modelling results yields lower prediction capability of leaf biochemical, but their main advantage is the robustness and transferability of the models. Information on leaf N is critical for understanding the movement and feeding patterns of the herbivores and could be useful for planning and managing grazing systems.

KEYWORDS: Leaf nitrogen, PROSAIL, radiative transfer model, rangeland, quality, savanna

ESTIMATION OF ICE THICKNESS ON LARGE NORTHERN LAKES FROM JASON-2 RADAR ALTIMETER AND PASSIVE MICROWAVE DATA

Duguay C[1,2], Zakharova E[3,4], Kouraev A[3,4,5], Kheyrollah Pour H[1,6], Kang K[1,2]
[1]University of Waterloo, [2]H2O Geomatics, [3]LEGOS, Université de Toulouse, CNES, CNRS, IRD, UPS, [4]State Oceanography Institute, St. Petersburg Branch, [5]Tomsk State University, [6]Environment and Climate Change Canada

Lake ice grows steadily between the end of the freeze-up and onset of the break-up periods as a result of the thermodynamics of freezing water as well as dynamic ice motion on the surface. In thermodynamic thickening, the conductive heat flow controls the ice growth rate and the ice thickness, and the ice thickens downward as a result of heat loss at the top of the ice cover (or snow if present on the ice surface). The value of passive microwave and radar altimeter data from satellite altimetry missions has been shown for the determination of ice dates (freeze-up and break-up), but it has yet to be evaluated for the estimation of ice thickness. Data acquired from current satellite missions (e.g. Jason-2/3, Sentinel-3) provide an opportunity for investigation of the development of lake ice thickness retrieval algorithms.

The objective of this study is to evaluate the potential of radar altimeter and passive microwave radiometer data from the Jason-2 satellite mission (launched on 20 June 2008 with a 10-day repeat cycle) for the estimation of ice thickness on large northern lakes. Backscatter and brightness temperature (Tb) measurements from its nadir-pointing radar altimeter (Ku-/C-band 13.575/5.3 GHz) and passive microwave radiometer (18.7 and 34 GHz) are compared against ice thickness estimates obtained with a thermodynamic lake ice model and in situ measurements collected on three lakes in northern Canada (Baker Lake, Great Bear Lake and Great Slave Lake) over the course of eight ice seasons (2008-2009 to 2015-2016). The temporal evolution of backscatter and Tb is then explored to estimate ice thickness. Results show that both passive microwave and radar altimeter data acquired in the 5-19 GHz frequency range offer a promising means for estimating ice thickness on high-latitude lakes.

KEYWORDS: Lake, ice thickness, altimeter, radiometer, Jason-2

ESTIMATION OF TREE DISTRIBUTION AND CANOPY HEIGHTS IN IFAKARA, TANZANIA USING UNMANNED AERIAL SYSTEM (UAS) STEREO IMAGERY

Asige I[1], Odera P[2]
[1]RCMRD, [2]JKUAT

Tree height estimation is fundamental in forestry inventory especially in the estimation of biomass. Multiple return LiDAR capabilities offer convenient solutions for height estimation though at equally increased costs. This study seeks to provide an assessment of the accuracy of UAS stereo imagery in establishing tree distribution and canopy heights in open forests as an inexpensive alternative. This study attempts to: generate accurate 3 dimensional surface and bare earth models from UAS data and using these products, establish tree distribution and estimate canopy heights using data filters; and validate these accuracies using ground methods.

A Sirius UAV fitted with a 16 MP camera and flying at an average height of 371m AGL was used to image approximately 2km^2 capturing 400 images per flight. An image overlap of up to 85% was sufficient for stereo generation at a GSD of 10cm for a flight period of 40 minutes. The stereo imagery captured were processed into photogrammetric point clouds with an average point density of 23 points per square meter. Point cloud segmentation revealed tree distribution patterns in the area, with the Near Infrared band proving useful in filtering out trees from non-vegetated areas. From the tree height estimations and with validation information from 30 trees from 2 sample plots yielded r^2=0.85.

The study highlights a cost-effective approach for generation of accurate 3D models from stereo UAS data. With a survey grade GPS/IMU for direct georeferencing, limited controls were required which reduces the cost of the project. With the ease of varying the size of imagery overlap and flying height, imagery with improved radiometry can be obtained hence improving the determination of tree distribution, and with multi-view image matching algorithms processing of UAS imagery is made accurate and inexpensive.

KEYWORDS: UAS, photogrammetric point clouds, GPS/IMU, Sirius UAV

EUROPEAN NETWORK OF EARTH OBSERVATION NETWORKS FOR IN-SITU COORDINATION IN EUROPE

Maso J[1], Serral I[1], McCallum I[2], Plag H[3]
[1]CREAF, [2]IIASA, [3]TIWAH

The European Network of EO Networks (ENEON) is a network of in-situ Earth observation networks to provide integrated and harmonized perspective of observations, helping to reduce redundancies and detect gaps in the European EO. In the context of GEO, no global coordination group exists for the in-situ observation community so ENEON aims to function as the European in-situ node in GEOSS. ENEON has conducted a study of the current EO networks in Europe and their relations, released as a dynamic graph that can be found at www.eneon.net/graph. A user feedback system has been integrated to accept input on how to enhance or correct the dynamic graph. The graph shows that the number of in situ networks is vast and they are very heterogeneous. Some of them do not interact with GEOSS or are not even aware of GEO. ENEON is also creating a "commons" to support the networks by providing an open and inclusive computer-based collaboration environment that facilitates the discovery, execution, reuse of information. It will serve In-situ networks , Scientists and Policy makers in different ways. The later will be able to formulate questions and get interpretations and options from data providers working with their experts. In practice it will be an adaptation of an environment developed by 52North in another EC H2020 (WaterInnEU). A pilot of the ENEON commons has been conducted as a proof of concept that offers a comprehensive description for three selected networks with their offerings in terms of observation variables. It includes a connection to the GEOSS Common Infrastructure. ENEON is currently having a very light structure that is supported by the ConnectinGEO partners and in the future will continue in the EC H2020 ERA-Planet program.

EVALUATING THE INFLUENCE OF AGROMETEOROLOGICAL PARAMETERS FOR WINTER WHEAT YIELD FORECASTING

Mashaba Z[1,3], Chirima G[1,3], Botai J[2,3], Combrinck L[3,4]
[1]Agricultural Research Council ISCW, [2]South African Weather Services, [3]University of Pretoria, [4]Hartebeesthoek Radio Astronomy Observatory

Crop yield forecasting is a crucial process for ensuring food security in a country. However, current crop yield prediction techniques used in South Africa rely on manual field surveys which is not timely for decision making and costly. Therefore, there is a need for using remotely sensed data which is freely available for crop yield forecasting. In this study, the Normalized Difference Vegetation Index (NDVI) was related to meteorological parameters derived from satellite imagery for wheat yield forecasting. Ten years of wheat yield data, NDVI, soil moisture, evapotranspiration and surface temperature were used to calibrate and validate a multi-linear regression forecasting model. This model was tested using five years of independent data. Agrometeorological parameter were evaluated for wheat yield forecasting using a correlation matrix and principal component analysis. The calibrated model had a coefficient of determination (R^2) of 0.88 and a p-value of 0.0141. The Root Mean Square Error (RMSE) was close to zero indicating a good level of accuracy and the Mean Bias Errors (MBE) gave an indication that the predicted yield was similar to the observed yield. Percentage relative errors were ±10% for the model testing data with exception to 2010 and 2013 which indicated that the level of accuracy was reasonable. Parameters identified as important for wheat yield forecasting were the NDVI (r=0.88) and evapotranspiration (r=0.58). This study proved that remote sensing can be used at high levels of accuracy for forecasting wheat yield to aid timely decision making regarding imports and exports.

KEYWORDS: wheat yield, agrometeorological parameter, food security, remote sensing

EVALUATING THE PERFORMANCE OF SUPPORT VECTOR MACHINE AT PIXEL AND OBJECT-BASED IMAGE ANALYSIS PLATFORMS IN DELINEATING AREAS OF FRAGMENTED SMALLHOLDER SUGARCANE GROWERS USING LANDSAT-8 OLI IMAGERY

Maake R[1]
[1]*Agricultural Research Council*

The delineation of smallholder farm's areal extent in fragmented landscapes can be a challenge due to their complex spatial configuration (. i.e. patchy field sizes) and un-uniform planting and harvesting dates. However, delineating and area estimation of such farming systems is essential in crop yield estimation for food security. Moreover, estimating the areal extent of fragmented smallholder farms can provide an insight of their natural resource use as well as their contribution to carbon pool. Exploring different classification platforms could aid in finding the optimal classification to overcome the challenges of delineating fragmented smallholder farms. This study sought to evaluate the performance of support vector machines (SVM) at object-based image analysis (OBIA) and pixel-based image analysis (PBIA) platforms in delineating the areal extent of fragmented smallholder farms using Landsat-8 OLI imagery. The SVM classifier at both classification platforms was evaluated based overall accuracy and statistical significance. Findings of this study showed overall accuracies of 79.79%, and 77.97% for OBIA-SVM and PBIA-SVM respectively. Results from the McNemar's test indicated an insignificant statistical difference ($p > 0.05$) in overall accuracies obtained by the SVM between the two classification platforms. This study demonstrates that SVM can be utilised to improve the accuracy in delineating fragmented smallholder farms while reducing the accuracy gab between OBIA and PBIA.

EVALUATING THE SENTINEL-2 MOSAICKING PROCESSOR FOR FEATURE EXTRACTION IN CLOUD-FREE COMPOSITES

De Lemos H[1]
[1]*South African National Space Agency*

In this study, features extracted from Sentinel-2 composites computed from the mosaic processor will be evaluated. The mosaic processor also known as the mosaicking tool is included on the open-source Sentinel Application Platform (SNAP) software package, which is the common architecture for all Sentinel Toolboxes and SMOS Toolbox. The mosaic tool is able to compute cloud-free Sentinel-2 composites from Level-2A products and their associated cloud, cloud shadow and snow masks. Feature extraction has a number of applications in earth observation, which include mapping and change detection analysis. Cloud, shadow and snow are some of the major limiting factors encountered during feature extraction from optical remote sensing data. Thus, a methodology is proposed which makes use of multiple Sentinel-2 images to produce cloud, shadow and snow –free composites for feature extraction.

For the purposes of this study, a one month multi-temporal stack of Sentinel-2 satellite images acquired over Johannesburg will be used. A combination of urban and rural features will be extracted using an Object Based Image Analysis (OBIA) approach on the cloud-free Sentinel-2 composite and SPOT6/7 imagery. The SPOT6/7 imagery will serve in visually validating the feature extractions, however ANOVA (Analysis of variance) and F-test statistics will be used to compare the means and variations of the feature extractions respectively.

KEYWORDS: Sentinel-2, mosaic, feature extraction, cloud-free, composite, object based image analysis.

EVALUATION OF FIRE DANGER AND FIRE POTENTIAL INDICES FOR SOUTH AFRICA

Burgess M[1], Frost P[1], Coetzee S[2], McFerren G[1]
[1]*CSIR*, [2]*University of Pretoria*

Wildfires can be harmful to the environment and can cause a loss of life if not properly managed. Not all wildfires should be viewed as negative as some ecosystems require regular burning to maintain ecosystem health. Information should be provided to fire managers to do planning and mitigation to decrease the negative effects some fires may have. Fire potential indices can be used to provide the necessary information to fire managers. Fire potential indices are based on a number of variables that have an influence on fire ignition and spread. Some of the variables include fuel moisture, elevation, slope, aspect, temperature, wind and relative humidity. Fire potential indices have been developed, but their suitability in South Africa has not been determined. Fire potential indices were implemented and evaluated for two provinces in South Africa namely the Western Cape and Mpumalanga. The evaluation included fire potential indices and fire danger indices. Active fire events, detected via satellite, were used in the statistical analysis of the indices. The indices performed differently in each province. The overall performance was not very good. Testing index suitability is challenging because high fire potential will not be accompanied by a fire event if an ignition source is not present.

KEYWORDS: Wildfires, Wildfire Management, Fire Potential Indices

EVALUATION OF QA4ECV LAND ECVS GENERATED USING AN OPTIMAL ESTIMATION FRAMEWORK

Muller J[1], Kharbouche S[1], Lewis P[2], van Leeuwen M[2], Danne O[3], Blessing S[4], Giering R[4], Gobron N[5], Marioni M[5], Zunz V[6], Govaerts Y[6], Schulz J[7], Doutriaux Boucher M[7], Lattanzio A[7]

[1]University College London - Mullard Space Science Laboratory, [2]University College London - Department of Geography/NCEO, [3]Brockmann Consult GmbH, [4]FastOpt GmbH, [5]European Commission – Joint Research Centre, Knowledge for Sustainable Development & Food Security, [6]Rayference, [7]EUMETSAT

Building on the ESA-DUE GlobAlbedo project (http://www.GlobAlbedo.org), a 35 year record of land surface albedo, fapar and LAI was generated as part of the EU-FP7 QA4ECV project on the STFC CEMS cloud computers from the US AVHRR & GOES and European AVHRR2, METEOSAT, VEGETATION, MERIS and Proba-V sensors using optimal estimation. This is based on 3 broadbands (0.4-0.7, 0.7-3, 0.4-3μm) and 7 MODIS spectral wavelengths along with one at MISR blue and one in the UV, which are useful for NO2 and HCHO retrievals respectively. We report on the use of a modified GlobAlbedo retrieval scheme and spectral mapping using a MODIS prior created using MOD09 from Collection 6 of the MODIS BRDF product to generate daily, 1km as well as monthly 0.05º and 0.5º broadband and spectral BRDF/albedos. For polar regions, NASA MISR BRFs and MODIS sea-ice have been employed to generate instantaneous daily, weekly and monthly 1km BRDF/albedos for 16 years.

Employing the Two Inversion Package method surface albedos are employed to retrieve a set of FAPAR/LAI. In parallel in order to continue with the series of products generated using optimal estimation at JRC, AVHRR data from the NASA LTDR have been employed to generate a 35 year record of 10-daily and monthly FAPAR.

We describe their production and validation using in situ data from the fair and open subset of FLUXNET tower measurements and their upscaling using NASA CAR data.

* QA4ECV has received funding from the European Union's Seventh Framework Programme (FP7/2007-2013) under grant agreement no. 607405; for CEMS diskspace and computing resources from the NERC Big Data grant and from NCEO/NERC.

EVALUATION OF RELATIONSHIPS BETWEEN SAR PARAMETERS AND VINEYARD STRUCTURAL PARAMETERS

Barratt K[1]
[1]*Stellenbosch University*

South Africa is one of the world's leading wine producers and recently there is growing concern regarding the annual loss of approximately 5% of existing vineyards. The wine producing industry is increasingly focused on methods that allow for the acquisition of accurate and timely information pertaining to vineyard condition and vineyard spatial properties in order to facilitate the streamlining of vineyard management practises and ensure the highest possible grape quality and yield. Although SAR has been widely utilised in the monitoring of grass crops there has been little research into viticulture applications; though its sensitivity to the geometric configuration of crops and changes in vegetation indicate that SAR shows high potential for vineyard monitoring and mapping and further research into these applications is a necessity.

The aim of this project is to determine if quad-polarised SAR data is able discriminate between different trellis types and differing vineyard row directions. The demand for this data stems from the need for decision support regarding vineyard development and re-planting procedures. Understanding the manner in which the trellis structure and row direction of vineyards may influence SAR observables also allows for the possible mitigation of such effects in future research.

The project is a multi-date approach involving the use of 4 RADARSAT-2 images, captured over the course of 2016. An object based classification method will be assessed in terms of its ability to differentiate between row directions and trellis types. A number of SAR parameters will be extracted, including polarimetric and textural features; and the use of feature selection for dimensionality reduction will be considered. Results will be presented as classification accuracies and feature importance rankings.

ISRSE37
TSHWANE 2017

EVALUATION OF TIP FAPAR AND LAI FROM A MULTI-SENSOR BHR DERIVED IN THE QA4ECV PROJECT

Blessing S[1], Giering R[1], Danne O[2], Kharbouche S[3], Muller J[3]
[1]Fastopt GmbH, [2]Brockmann Consult GmbH, [3]University College London - Mullard Space Science Laboratory

The Two-Stream-Inversion package (TIP) is an efficient method to estimate the partitioning of fluxes from top-of-canopy reflectances in a 1-D canopy model. This has successfully been used to retrieve effective LAI (Leaf AreaIndex) and FAPAR (Fraction of Absorbed Photosynthetically Active Radiation) from satellite retrieved top-of-canopy bi-hemispherical reflectances (BHR) at visible (VIS, 300--700 nm) and near-infrared (NIR 700--300 nm).

This study compares the more direct approach of using BHRs at equivalent MODIS (Moderate-resolution Imaging Spectroradiometer) channels 1 (620-670 nm) and 2 (841-876 nm) at 1 km resolution to the results to those retrieved from VIS and NIR BHRs of the same source. The BHRs stem from the multi-sensor optimal estimation of the EU-FP7 project QA4ECV (Quality Assurance for Essential Climate Variables) product (Muller et al., this conference).

As the processing is ongoing, quantitative results will not be available before a few months after the writing of this abstract. General characteristics of TIP FAPAR and LAI, and evaluation against comparable retrievals will be presented

* QA4ECV has received funding from the European Union's Seventh Framework Programme (FP7/2007-2013) under grant agreement no. 607405; for CEMS diskspace and computing resources from the NERC Big Data grant and from NCEO/NERC

EVIDENZ: EARTH OBSERVATION BASED INFORMATION PRODUCTS FOR DROUGHT RISK REDUCTION AT THE NATIONAL LEVEL

Dubovyk O[1], Graw V[1], Jordan A[2], Kussul N[3], Post J[4], Szarzynski J[5], Wallz Y[5]

[1]Center for Remote Sensing of Land Surfaces (ZFL), University of Bonn, [2]Disaster Management Training and Education Centre for Africa, University of the Free State, [3]National Space Research Institute of Ukraine, [4]United Nations Platform for Space-based Information for Disaster Management and Emergency Response (UNSPIDER), [5] Institute fir Environment and Human Security, United Nations University (UNU-EHS)

In the last decades, occurrence of drought events, their duration and intensity have generally increased threatening agricultural production and, eventually, food security. The EvIDENz project (Earth Observation based information products for drought risk reduction at the national level) focuses on supporting the SFDRR (Sendai Framework for Disaster Risk Reduction) for the cased of drought risk reduction. Specifically, we aim at providing new approaches and information products to allow the Earth Observation (EO) based outputs meeting the demands of SFDRR goals, targets and indicators as well as needs by national level mandated stakeholders.

Using South Africa and Ukraine as case study countries which have recently been severely affected by agricultural drought events, the EvIDENz project focuses on: (i) providing risk knowledge through EO-based drought hazard monitoring, exposure and vulnerability of the agricultural sector, (ii) calculating the specific SFDRR indicators such as number of people affected by droughts and (iii) calculating direct economic losses due to drought events. Drought hazard assessment on the national level is carried out using a combination of different sensors and highlighting the use of data from new Sentinel missions. Lastly, to contribute to national monitoring systems in the field of drought risk and agriculture, EO-based information products will be generated by applying the developed in EvIDENz methods and approaches.

EXPECTATIONS OF THE GEOSS INNOVATIVE INFRASTRUCTURE

Ochiai O[1]
[1]*Group On Earth Observations Secretariat*

The GEO Strategic Plan 2016-2025: Implementing GEOSS states that, "GEOSS is a set of coordinated, independent Earth observation, information and processing systems that interact and provide access to diverse information for a broad range of users in both public and private sectors. GEOSS links these systems to strengthen the monitoring of the state of the Earth. It facilitates the sharing of environmental data and information collected from the large array of observing systems contributed by countries and organizations within GEO." In line with the big picture above, the session will discuss on how the innovative infrastructures in various IT technologies and Tools support the delivering of Earth Observation solutions to the scientific community and to decision makers, with particular focus on needs of developing countries. The presentation will state expectations on the GEOSS innovation infrastructure and architecture in order to meet the needs of developing countries.

EXPLOITATION PLATFORMS IN THE EUROPEAN EARTH OBSERVATION PAYLOAD DATA GROUND SEGMENT

Loekken S[1], Laur H[1], Pinto S[1]
[1]*European Space Agency*

The operations concept for the European Space Agency's Earth Observation Payload Data Ground Segment has evolved over three decades in support of the data exploitation scenarios of ERS, ENVISAT, Earth Explorers, and ESA's Third Party Missions. Previously the focus has been primarily on disseminating satellite data to single users, but opportunities and challenges stemming from evolution of technology, shifts in user expectations, and steep increase in data volumes and complexity have driven the emergence iin recent years of a complementary operations concept, based on collaborative Exploitation Platforms (EP), and with ESA and the EC as coordinators of multiple, synergistic European efforts.

An EP is a complete virtual workspace, providing the user community with access to large volumes of data and all the resources required to exploit them, including massive computing resources, toolboxes, collaborative functionalities, development environments etc. The discriminating characteristic of this operations concept is therefore bringing the users to the data (and to the resources required to exploit them), rather than just bringing data to the users.

The concept is complementary to the still dominant traditional service of data dissemination, but has been widely adopted over recent years, to the point where existing and planned EPs now constitute perhaps the principal capability of user-facing ground segments, and consequently a main pillar of the European ground segments evolution strategy.

This presentation will focus on the on-going seminal ESA Thematic Exploitation Platform R&D projects – addressing the Coastal, Food Security, Forestry, Geohazards, Hydrology, Polar, and Urban earth science fields and user communities – as well as the natural extension of these activities within the context of 'EO Innovation Europe', a ground segment evolution strategy for Europe.

EXPLORING THE JOINT POTENTIAL OF SENTINEL 2 AND 3 IN MONITORING WATER USE AND STRESS IN AFRICAN SAVANNA ECOSYSTEMS: A CASE OF THE KRUGER NATIONAL PARK, SOUTH AFRICA

Andreu A[1], Dube T[2], Nieto H[3], Mudau A[4], Guzinski R[5], Shoko C[6], Mutanga O

[1]UNU-FLORES, [2]Department of Geography & Environmental Studies-University of Limpopo, [3]Institute for Sustainable Agriculture – IAS, [4]Global Change and Ecosystem Dynamics Research Group, Natural Resources and Environment, Council for Scientific and Industrial Research - CSIR, [5]ESRIN D/EOP-SEP, European Space Agency, [6]School of Agriculture, Earth & Environmental Science, University of KwaZulu Natal

Drought periods and erratic rainfall patterns across large parts of Africa result in water-limited environments like savannas, highly sensitive to changes in land management practices and climate. However, savannas are also a key productive landscape, covering half of the continent and supporting livestock, crops and livelihoods. Monitoring water use and natural vegetation stress in these ecosystems can support management, shaping the actions needed to maintain productivity. To face this challenge, TIGER project 401 explores the joint potential of Sentinel 2 (S2) and 3 (S3). High spatial (10m) and temporal (~5 days) resolution of VIS/NIR S2 images allow for a continuous monitoring of vegetation cover and unstressed evapotranspiration (ET-using Kc-FAO56). S2 provides the required resolution for farm scale applications, tracking separately the seasonal variations of each canopy layer (grass-tree), with different spectral, phenological and physiological characteristics. Meanwhile, lower spatial resolution (1km) S3 thermal data allow for derivation of a regional water stress index (ratio between actual ET, using Two Source Energy Balance-TSEB, and potential ET), supporting the detection of vulnerable areas and drought prediction. SPOT 4/5 and AATSR/MODIS, with similar characteristics to S2 and S3 respectively, are used to validate the procedure (2010-2012) over two experimental savanna areas with different ecotones (Skukuza and Malopeni). The model framework is later applied with S2 and MODIS (2016) and with S2 and both S3/MODIS (2017) over Kruger Park. The meteorological data required as inputs and the flux data needed for validation are obtained from eddy covariance towers and weather stations (gaps filled with ERA Interim data). By combining these robust approaches, Kc-FAO56 (Allen et al., 1998) and TSEB (Kustas and Norman, 1999), and taking advantage of the different conceptual and operational capabilities of Sentinel-2 & 3 missions, savanna water fluxes could be reliably monitored and strategies for sustainable management developed. Savanna, Sentinel 2-3, waterstress

EXPLORING THE POTENTIAL ROLE OF THE UPCOMING EO-SAT1 SATELLITE IN MARITIME AND COASTAL APPLICATIONS

Sibandze P[1], **Mhangara P[1]**
[1]*South African National Space Agency (SANSA)*

The spatial, spectral, temporal and radiometric resolutions of the upcoming South African EO-SAT1 satellite are well configured to support the maritime and coastal applications through the provision of suitable geographic information at appropriate timescales and spatial resolution sufficient for mapping ocean-related infrastructure and spatial planning. The technical configurations for EO-SAT1 are ideal for ship detection, offshore oil and gas exploration, mapping coastal mega infrastructure, monitoring fishing activities and to support marine protection priorities that are identified under South Africa's Operation Phakisa's Ocean Economy Programme. Operation Phakisa is a government programme that aims to fully exploit ocean and coastal resources for economic benefit. EO-SAT1's high (2.5m and 10m) and medium spatial resolution sensors (15m and 60m) will provide high spatial resolution imagery and medium resolution imagery required to support the South Africa's National Marine Spatial Planning Framework with timely spatial information on the status and dynamics of the coastal and ocean environment. The spatial and spectral information provided by EO-SAT1 has considerable potential to contribute towards monitoring developments of built-up areas and land degradation along coastal zones. The objective of this paper is to demonstrate how the spatial, spectral and temporal configurations of the upcoming EO-SAT1 could be exploited for coastal and maritime applications in support of the Operation Phakisa Programme.

KEYWORDS: EO-SAT1, Ocean Economy, Operation Phakisa, Coastal Mapping, Maritime Monitoring

EXPLORING THE RELATIONSHIP BETWEEN SPECIES DIVERSITY AND TREE COVER IN THE SAVANNAH WOODLAND OF SOUTHERN AFRICA

Madonsela S[1,2], Cho M[1,2], Ramoelo A[1,3], Mutanga O[2], Naidoo L[1]

[1]Earth Observation Research Group, Natural Resources and Environment, Council For Scientific And Industrial Research, [2]School of Agriculture, Earth and Environmental sciences, University of KwaZulu-Natal, [3]Risk and Vulnerability Assessment Centre, University of Limpopo

Tree species in the southern African savannah provides multiple benefits to the ecosystem by maintaining nutrients in the system, providing breeding sites for birds, supporting large faunal species and also acts as a safety net against poverty for the neighbouring communities. The loss of tree species diversity leads to a decline in ecosystem productivity and in southern African savannah, trees are heavily impacted upon by elephants and human activities. Understanding the spatio-temporal variation of tree species diversity is essential for management of biodiversity resources. The aim of this study was to explore the relationship between tree species diversity and tree cover in the savannah woodland of southern Africa. A total of 49 plots of 90m X 90m were randomly distributed inside and outside Kruger National Park and tree species diversity within each plot was calculated using Shannon index. Tree cover in each plot was extracted from existing fractional woody cover data produced with Synthetic Aperture Radar (SAR) data. Simple linear regression was applied to test the relationship between species diversity and tree cover. Our results indicates a significant relationship between the two variables ($r2 = 0.26$, $p < 0.001$). The observed linearity in the relationship indicates that tree cover could be a proxy for estimating tree species diversity in southern African savannah. This is the first study to establish an empirical relationship between tree species diversity and tree cover and could have significant implications for large scale monitoring of biodiversity using remote sensing devices.

KEYWORDS: savannah, tree cover, species diversity, Shannon index, remote sensing

EXTRACTION OF COASTAL OCEAN WAVE CHARACTERISTICS USING REMOTE SENSING AND COMPUTER VISION TECHNOLOGIES

Johnson M[1], Lück-Vogel M[1,2]
[1]Stellenbosch University, [2]Council for Scientific and Industrial Research (CSIR)

The coastal zone occurs at the interface of three major natural systems. These systems include the atmosphere, the ocean and the land surface. Ocean waves are among the most important forces shaping the world's coastlines. They drive environmental processes and human activity that occur within the coastal zone as well as on the open ocean. The assessment of wave characteristics, such as wave direction, wavelength, wave period and wave velocity, is critical to understanding coastal processes as a baseline for better coastal management. However, monitoring and assessment of wave characteristics is challenging, given the high complexity of the ocean dynamics and large spatial extent.

Traditionally, wave observation instruments such as wave buoys, wave poles, pressure transducers, inverted echo-sounders and current meters have been used to record ocean wave characteristics. Although delivering very accurate measurements, they only record punctual data.

The work presented here is assessing whether optical imagery from the RapidEye satellite can be used to extract ocean wave characteristics such as wave direction, wavelength, wave period and wave velocity. If successful, the advantage of the proposed remote sensing-based approach would be the spatially continuous provision of wave characteristics for large areas, including the near-shore in a very cost-effective way. As ground truth data for validating open ocean wave conditions are sparse, a lab test with simulated, controlled wave conditions was conducted to assess various approaches for the extraction of wave characteristics from remote sensing imagery. The techniques identified and developed under lab conditions are to be tested using RapidEye imagery on two study areas on the South African coast. Results are expected to significantly contribute to a more comprehensive understanding and monitoring of waves dynamics for better coastal planning.

KEYWORDS: RapidEye, Coastal ocean waves, Wave Characteristics, South Africa

FINITE MIXTURE MODELS FOR SUB-PIXEL COASTAL LAND COVER CLASSIFICATION

Ritchie M[1,2], Lück-Vogel M[4], Debba P[3], Goodall V[2]
[1]Modelling and Digital Science, Council For Scientific And Industrial Research, [2]Department of Statistics, Nelson Mandela Metropolitan University, [3]Built Environment, Council For Scientific And Industrial Research, [4]Natural Resources and the Environment, Council for Scientific and Industrial Research

Medium spatial resolution sensors (10-30 m pixel size) have been used for land cover classification and monitoring for decades. However, these sensors do not have the required resolution to detect coastal specific land cover classes and boundaries thereof as the spatial extent of the target features frequently is too small (e.g. bands of dune vegetation or the water line). Higher resolution satellite imagery which would be more suitable is, however, frequently too costly for operational coastal monitoring and management. A solution for this problem might be the spectral unmixing classification approach on medium resolution imagery (e.g. Landsat 8; Sentinel-2) which have no acquisition cost and are therefore affordable for operational use. Finite mixture models have been used to generate sub-pixel land cover classifications, however, traditionally this makes use of mixtures of normal distributions. However, these models fail to represent many land cover classes accurately, as these are usually not normally-distributed. A potential improvement could be to use models using other distributions which are more robust to non-normally distributed feature classes, such as the student-t distribution.

This presentation aims to show the results of the fitting of various finite mixture models to land cover class signatures derived from radiometrically corrected WorldView-2 imagery of the Strand region of Cape Town. We aim to determine which finite mixture model best fits the signatures for this region. WorldView-2 imagery is used as it allows for the extraction of pixels with "pure" spectral signatures, that is pixels containing one land cover class. However, the long-term goal of this project is to apply finite mixture models for the monitoring of land cover using medium resolution imagery (Landsat 8; Sentinel-2). If successful, this will provide a more robust land cover classification algorithm which is affordable, for routine monitoring land cover in coastal environments.

KEYWORDS: Sub-pixel Classification; Coastal Monitoring

FLUID LENSING & APPLICATIONS TO REMOTE SENSING OF AQUATIC ENVIRONMENTS

Chirayath V[1]
[1]*Nasa Ames Research Center*

The use of fluid lensing technology on UAVs is presented as a novel means for 3D imaging of aquatic ecosystems from above the water's surface at the centimeter scale. Preliminary results are presented from airborne fluid lensing campaigns conducted over the coral reefs of Ofu Island, American Samoa (2013) and the stromatolite reefs of Shark Bay, Western Australia (2014), covering a combined area of 15km2. These reef ecosystems were revealed with centimetre-scale 2D resolution, and an accompanying 3D bathymetry model was derived using fluid lensing, Structure from Motion and UAV position data. Data products were validated from in-situ survey methods including underwater calibration targets, depth measurements and millimetre-scale high-dynamic-range gigapixel photogrammetry. Fluid lensing is an experimental technology that uses water-transmitting wavelengths to passively image underwater objects at high-resolution by exploiting time-varying optical lensing events caused by surface waves. Fluid lensing data are captured from low-altitude, cost-effective electric UAVs to achieve multispectral imagery and bathymetry models at the centimetre scale over regional areas. As a passive system, fluid lensing is presently limited by signal-to-noise ratio and water column inherent optical properties to ~10 m depth over visible wavelengths in clear waters. The datasets derived from fluid lensing present the first centimetre-scale images of a reef acquired from above the ocean surface, without wave distortion. The 3D multispectral data distinguish coral, fish and invertebrates in American Samoa, and reveal previously undocumented, morphologically distinct, stromatolite structures in Shark Bay. These findings suggest fluid lensing and multirotor electric drones represent a promising advance in the remote sensing of aquatic environments at the centimetre scale, or 'reef scale' relevant to the conservation of reef ecosystems. Pending further development and validation of fluid lensing methods, these technologies present a solution for large-scale 3D surveys of shallow aquatic habitats with centimetre-scale spatial resolution and hourly temporal sampling.

FOG COMPUTING PERSPECTIVES IN CONNECTION WITH THE CURRENT GEOSPATIAL STANDARDS

Panidi E[1]
[1]*Saint Petersburg State University*

Cloud computing technologies and Cloud-based Geographical Information Systems have became widely used in recent decades. However, the complexity and size of geospatial datasets remains growing and sometimes become going out of the Cloud infrastructure limitations. Some geospatial data types (or data sources) acquired too complex characteristics and became simultaneously the big (in the meaning of Big Data paradigm), distributed, multisource and real time. Good example of such data are the Volunteered Geospatial Data (commonly used term is Volunteered Geographic Information) and data collected by Smart Sensor Grids. The hyperdataset term can characterize these data appropriately. The hyperdatasets become too big and too distributed to manage them in the common Cloud data center, and multilevel management techniques are demanded supported with the capabilities of horizontal (Client/Client) data flows in addition to vertical (Cloud/Client) data flows. Similar tendencies in information technologies in general have led to the appearance of Fog Computing paradigm that extend the Cloud infrastructure with the capabilities of computational facilities of client devices and implements client-side data storage, management and interchange.

Currently we are summarizing and discussing these tendencies in connection with lineup of available Open Geospatial Consortium standards. We are covering the standards (first of all, Web Map/Feature/Coverage/Processing Service and Sensor Web Enablement stacks of standards) which can be recognized as the first steps on the way of Fog Computing implementation into geospatial domain, and analyzing their strong and weak features from the Fog Computing compliance point of view. The analysis is built upon our experience of implementation of the geospatial Web services test projects, which are the Web Processing Service compatible technique of client-side geoprocessing, and portable Web Coverage Service compatible software servers. In conclusions, we are summarizing the research directions needed for geospatial Fog Computing implementation.

FORECASTING THE EXCESS OF MALARIA CASES, MALI 2015

Griffiths K[1], Sagara I[2], Sissoko M[2], Kone D[3], Cisse B[3], Kayentao K[2], Coulibaly D[2], Doumbo O[2], Gaudart J[1]
[1]Aix Marseille University (SESSTIM), [2]Malaria Research and Training Center - USTTB, [3]National Malaria Control Program - Ministry of Health

Introduction
In 2015, Mali reported an increase of 201845 malaria cases compared to the same time period in the previous year. To understand this excess we analysed malaria cases from 2010 to 2015, forecasting malaria in 2015, taking into account environmental factors.
Methods
The weekly number of suspected malaria cases were collected from nine administrative regions. The populations were calculated according to the national census in 2009 and the World Bank growth rates. Daily cumulative precipitations (TRMM3B42RT) and daily maximum surface temperatures (AIRX3STD.006) were issued from remote sensing datasets. As the Niger river level is correlated with precipitation from its source, daily cumulative precipitations in the Guinean highlands were similarly assessed.
A general additive model (GAM) was estimated by using the number of cases 2010-2014, adjusted on precipitations (shifted in time according to cross-correlation lag), temperature, political events and seasonal variations. A negative binomial distribution was used taking into account overdispersion in case counts. Populations were used as a log-offset, in order to estimate incidence ratios. This model was used to forecast case counts for 2015 in each region.
Results
The final statistical model included time lags between malaria cases and precipitations (4-17 weeks) and temperature (11-20 weeks). The statistical model demonstrated the significance of precipitations, temperature and political events in 6/9, 5/9 and 7/9 regions respectively ($p<0.05$). Real excess of cases (above the predicted 95% confidence interval) were only identified in two regions, Bamako and Gao, which had mean average errors of 1827 and 1047 respectively.
Conclusion
Meteorological data estimated from remote satellite sources can be successfully used to forecast the increased number of malaria cases reported in Mali in 2015. This indicates meteorological forecasts can be used in the future to provide information to guide prevention and elimination plans.

KEYWORDS: Malaria, Forecasting, Remote sensing

FOREST AND FOREST LOSS MAPPING WITH SAR FROM THE ENVISAT/ALOS ERA TO THE SENTINEL-1/ALOS-2 PRESENT IN THE MAI-NDOMBE DISTRICT, DRC

Haarpaintner J[1], Mazinga A[2], Mane L[2]
[1]*Norut - Northern Research Institute,* [2]*OSFAC - Observatoire Satellital des Forêts d'Afrique Centrale*

The Mai-Ndombe district in the Democratic Republic of Congo (DRC) is the demonstration area of the DRC Emission Reductions Program Idea Note (ER-PIN) as well as of the ESA DUE Innovator III project "SAR for REDD". The aim of "SAR for REDD" is to complement optical remote sensing capacities in African REDD countries with synthetic aperture radar (SAR) capabilities. "SAR for REDD" is also participating in the R&D component of the Global Forest Observation Initiative (GFOI).

The full data archive from C-band ENVISAT ASAR (2003-2012) and L-band ALOS-PALSAR Fine-Beam-Dual (2007-2010) were acquired and used for forest and forest change mapping in comparison with current data from Sentinel-1 CSAR and ALOS-2 PALSAR-2 FBD, both operational since 2014.

The maximum forest cover from the ALOS-PALSAR 2007-2010 period served as a baseline for forest loss detection for yearly forest losses as well as for the approximate 5-year forest loss detected between the ENVISAT/ALOS era and the SENTINEL-1 /ALOS-2 era. Forest loss is detected by thresholding the decrease in backscatter in cross polarization in forested areas.

20cm-resolution aerial image mosaics collected with a remotely piloted quadrocopter during a field campaign in September 2016 in the Kwamouth region (in the south-west of the Mai-Ndombe district) confirmed detected forest loss areas between ALOS and ALOS-2 and revealed intense deforestation and agricultural activities like "slash and burn". GPS positions of forest/non-forest land cover transition and ground photography collected along the road to Kwamouth are also used to validate the forest land cover products, showing that the forest boarders and different non-forest land covers (savannah, dry and wet grassland) are well detected in the 30m-resolution remote sensing products. The Mai-Ndombe-wide forest loss detected from SAR data is finally compared to Landsat-based yearly loss data from the Global Forest Change 2000–2014 data set (Hansen et al., 2013).

FOREST-OBSERVATION-SYSTEM.NET – TOWARDS GLOBAL REFERENCE DATABASE FOR FOREST BIOMASS

Schepaschenko D[1], Chave J[2], Phillips O[3], Davies S[4], Fritz S[1], Perger C[1], Dresel C[1], Lewis S[3], Scipal K[5]
[1]IIASA, [2]CNRS, [3]University of Leeds, [4]STRI, [5]ESA

The Forest Observation System (FOS) is an international cooperation implemented at the initiative of the ESA BIOMASS mission, to establish and maintain a global in-situ forest biomass database to support earth observation and to encourage investment in relevant field-based science. It is designed to be able eventually fulfill the ground data requirements for algorithm training and validation of spaceborne forest related missions, i.e. ESA BIOMASS and SAOCOM-CS; NASA GEDI and NISAR; JAXA PALSAR, etc.

The FOS serves as an interface between the remote sensing and ecological communities. Data sharing nowadays is one of the biggest problem despite of the fact that everyone can benefit. Ecologists sometimes do not realize how important their data are for calibration/validation of remote sensing products. Remote sensing community can provide additional rationality and arguments for investments in measurements on sample plots.

The implementation of the FOS is guided by following principles. The ground data should be of high quality and collected on permanent plots from 0.25 ha upwards by size. Sites should cover a broad range of geographical and environmental conditions, so as to maximize the robustness of models. The procedures for ground data acquisition and database compilation should be transparent and proofed extensively.

Project web portal (http://forest-observation-system.net/) presents besides several base maps, two types of data: (1) metadata: where and what were measured on permanent sample plots; (2) sample plot data for subset of plots where authors agreed to share the data: aggregated to 50x50 m live biomass, tree height, wood density and tree composition. In the proof-of-concept phrase, FOS includes the Center for Tropical Forest Science (CTFS-ForestGEO), RAINFOR, AfriTRON and IIASA networks. FOS is an open initiative and we expect more networks to join for common benefits, incl. joint publications and application for funding of fieldwork.

FRAMEWORK AND ARCHITECTURE FOR NEXT GENERATION DIGITAL EARTH

Fan X[1], zhan Q[1], du x[1], liu j[1]
[1]*Key Laboratory of Digital Earth Science, Institute Of Remote Sensing And Digital Earth, Cas*

The advent of big data and data-intensive sciences is speeding up the development process of Digital Earth and hence there is a great need for the next-generation Digital Earth. This paper proposes a framework and architecture for the next-generation Digital earth, in order to provide a new solution for the construction of Digital Earth system in the era of big data.

The existing Digital Earth systems are currently capable of handling organization, management, modeling, analysis and visualization of the Earth's surface data, but cannot meet the demand posed by the development of Digital Earth in the era of big data. A new Digital Earth 3D grid framework is proposed along with an Earth sphere 3D subdivision model , it enables the seamless organization and management of spatial data across Earth's multiple layers, including the surface, underground, underwater, air and other levels.

Under the 3D sphere subdivision grid framework, a novel system architecture is established. The system architecture integrates a series of popular Digital Earth functions seamlessly, including management and aggregation of spatial data, spatial information services and data sharing, high performance geo-computation, data-intensive analysis, and geoscience modeling and visualization. In this architecture, not only spatial data storage, management and dynamic aggregation are realized by means of distributed heterogeneous spatial database, but also the functions of massive spatial data access, sharing and distribution are provided on the network using cloud service cluster and spatial information grid. Through integration of geoscience big data analysis algorithm library and model library, the architecture also supports universal and intensive remote sensing data fast processing. Furthermore, 3D visualization and spatial-temporal process simulations are achieved based on the 3D subdivision grid model. This architecture has been applied successfully in several fields of Digital Earth such as global climate change simulation and monitoring of natural hazards.

FRENCH SPACE AGENCY PROGRAMMES FOR ACCESS TO PLEIADES AND SPOT 1-5 FOR SCIENCE

Hosford S[1], Fontannaz D[1]
[1]*Centre National D'Etudes Spatiales*

This paper will present the various programmes for obtaining free or low cost access to data from Pleiades and SPOT 1-5 for science use. Operations of the SPOT family of optical imaging satellites has spanned the last three decades with the first, SPOT 1, launched in February 1986 and the last in the family developed by CNES, SPOT 5, completing its operational lifetime at the end of march 2015. This near thirty year operating lifetime has generated over 25 million images that are available free or at low cost for science use through several CNES programmes. The constellation of two Pleiades satellites were launched respectively in December 2011 and 2012 and are currently providing access to single images or stereo and tri-stereo image pairs acquired in a single pass.

This paper will provide a detailed description of the CNES programmes providing access for science use to SPOT and Pleiades imagery. In addition, CNES contributes satellite imagery to several international programmes coordinated via the Committee on Earth Observation Satellites, CEOS (Working Group on Disasters, Space Data Coordination Group for the Global Forest Observing Initiative – GFOI), the Group on Earth Observations (GEOGLAM, GEO supersites and Natural Laboratories – GSNL) and the World Meteorological Organisation (Polar Space Task Group - PSTG) focused on specific thematic areas. Description of these various mechanisms to access commercial satellite imagery for science use will be described including proposal types, access conditions, licensing issues and typical volumes of data that can be provided.

FROM DETECTION OF UNDERGROUND ARCHAEOLOGICAL RELICS TO MONITORING OF WORLD HERITAGE SITES IN DANGER: ONGOING RESEARCH ACTIVITIES IN THE FRAME OF THE ATHENA TWINNING PROJECT

Cerra D[1], Agapiou A[2], Plank S[1], Lysandrou V[2], Tian J[1], Themistocleous K[2], Hadjimitsis D.G. [2], Schreier G[1]
[1]German Aerospace Center (DLR), [2]Cyprus University of Technology

The "ATHENA" twinning project aims at establishing a Center of Excellence in the field of Remote Sensing for Archaeology and Cultural Heritage, through a cooperation between the Remote Sensing Research Laboratory at the Cyprus University of Technology (CUT), the Institute of Archaeological and Architectural Heritage of the National Research Council of Italy (IBAM- CNR), and the German Aerospace Centre (DLR). This paper focuses on the joint research carried out in the first year by DLR and CUT. Different achieved results can be ordered chronologically, according to the stage of exploitation for an archaeological site. First of all, a site of archaeological interest must be discovered and defined: to aid in this process, a quantitative ranking of spectral indices to identify buried archaeological relics in hyperspectral images is proposed. Subsequently, information must be catalogued and stored: in the frame of the Athena project, efforts have been made to organize the information acquired with a spectrometer in laboratory on mosaics tesserae retrieved in Cyprus in a coherent spectral library. Finally, heritage sites must be constantly monitored, and this could be difficult for non-accessible areas such as conflict zones. For this purpose, the first steps have been made towards the automatic detection of damages to cultural heritage sites from space, based on texture descriptors in remotely sensed images.

FROM SATELLITE IMAGES TO AGRICULTURAL SYSTEMS MAPS: A REMOTE SENSING MULTI-LEVEL OBJECT-BASED APPROACH

Bellon B[1], **Begue A[1]**, Lo Seen D[1], Lebourgeois V[1], Leroux L[1], Simoes M[2]
[1]Cirad, [2]Embrapa

In response to the need of large scale spatial information for supporting agricultural monitoring, we present a new remote sensing object-based approach for objective and repeatable agricultural systems mapping at regional level. This approach is in two steps: 1. A segmentation of land units, based only on a remote sensing time series; These land units are assumed to be representative of the human activity and environmental conditions and thus of the in situ agro-ecosystems; 2. A semi-automatic land use classification, performed in each land unit with high-resolution images, to label the land units in terms of agricultural systems.

To produce the land units, a principal component transformation was first applied to an annual dataset of MODIS (MODerate Imaging Spectroradiometer) normalized difference vegetation index (NDVI) images. A series of segmentations were then performed on the principal component images that contain the essential information on the physiognomy and phenology of the cover. An unsupervised evaluation method was used for identifying the optimum segmentation which successfully delineates homogeneous units in terms of agricultural activity, discriminating the different cropland and grassland areas at regional level. Then, for each land unit, cropping systems maps were produced at a field level through object-based analysis of a Landsat8 30m resolution mosaic image, and spectral variables derived from the MODIS NDVI time series, and were validated with in situ data. Finally, a bottom-up spatial analysis of the extracted land use information at field level allowed definitive classification and characterization of the homogeneous regional level land units in terms of agricultural systems.

A map of the main agricultural systems of the Brazilian state of Tocantins, an agricultural expansion region, has been successfully produced for the year 2015 following this approach. This study shows the potential of multi-resolution satellite images to provide valuable baseline spatial information for supporting agricultural monitoring.

FUTURE MULTI-SENSOR EO CONSTELLATIONS

Mostert S[1], Burger H[3], Stanton D[2]
[1]SCS Aerospace Group, [2]Space Advisory Company, [3]SCS Space

Small satellites for Earth Observation have progressed from technology demonstrators to operational constellations. With the advances of satellite technology and the proliferation of commercial constellations, it becomes a question what a future remote sensing constellation should consist of. A number of optical satellite constellations have been proposed, even a combined optical/SAR constellation. But what additional information is being promised, and what are the key questions being addressed by the available additional information.

It is important to address the question of appropriate future investment in space technology for a number of role players. National space programs are impacted, commercial investors are impacted, satellite engineering enterprises are impacted.

This paper will review a number of existing constellation concepts and propose a multi-sensor constellation that will advance the availability of information for a number of key applications, while at the same time address the need for national space programs and sustainable engineering enterprises.

FUTURE NANOSATELLITE-BASED EARTH OBSERVATION - A NOVEL FIRESAT MISSION

Sampson R[1], Anderson P[1], van Zyl R[2], Annamalai L[3], de Villiers D[2], Jooste C[2], Steenkamp L[2]
[1]Clyde Space Ltd., [2]Cape Peninsula University of Technology, [3]CSIR

Clyde Space has developed a low cost satellite system designed to support visual and multi-spectral Earth Observation, enhancing the accessibility of global datasets related to environmental applications. This paper will discuss the "FireSat" nanosatellite Earth observation constellation, which will see the delivery of a space-based Earth observation system to detect bush fires in Southern Africa. Nanosatellite constellations provide cost effective platforms for EO missions that require high temporal resolution. The FireSat payload sensor incorporates a novel means of detecting fires from Low Earth Orbit, which does not depend on more conventional infrared technology. As a model endeavour in skills development and knowledge transfer, the FireSat project involves stakeholders from South Africa, Namibia, Kenya and the United Kingdom, and leverages lessons learned from both the Maritime Domain Awareness "MDASat" mission produced in South Africa in support of Operation Phakisa, and the SeaHawk Ocean Colour Monitoring System in development in the USA and UK. As well as the satellite and payload sensor technologies, which make useful nanosatellite-based observations possible, this paper will explore the data delivery chain, and data interface to the Advanced Fire Information System (AFIS). AFIS has been developed by CSIR in South Africa and currently makes use of Earth Observation satellites from NASA and ESA to detect possible hotspots (fires) on the ground. Polar orbiting satellites provide data 6 times daily, while the Geostationary satellite provides 15 minute updates, detecting fires as small as 50m x 50m, to larger than 300m x 300m.

GEARING UP FOR SENTINEL-3: QUANTITATIVE VALIDATION FOR WATER RESOURCES PROTECTION

Matthews M[1], Kravitz J[3], Griffiths D[2], Faniso Z[4], Hlahane K[3], Bernard S[2], Smith M[2]
[1]*CyanoLakes,* [2]*Council for Scientific and Industrial Research,* [3]*University of Cape Town,* [4]*University of Fort Hare*

The Copernicus mission of the European Space Agency's Sentinel-3 Ocean and Land Colour Instrument is a legacy sensor following on from the Medium Resolution Imaging Spectrometer (MERIS). It's unique spectral, spatial and radiometric characteristics makes it ideally suited to ocean color applications ranging from the open ocean to small inland waters. In this talk, we provide an overview of the activities underway to validate OLCI for applications in South African waters, particularly focussed on small inland water bodies, and high biomass coastal systems. In doing so we demonstrate the considerable value that Sentinel-3 based application can have for a range of applications such as harmful algal bloom monitoring, eutrophication detection and cyanobacteria warning systems. In addition, we also seek to demonstrate the concomitant value of Sentinel-2 high resolution data as a complimentary data source to Sentinel-3, and details its radiometric performance for various targets and situations.

GEO DISASTER RESILIENCE ACTIVITIES

Ivan Petiteville I[2], Aellen V[1]
[1]European Space Agency (ESA), [2]GEO

The increased severity of weather events and rapid urbanization of vulnerable regions around the world have led to growing economic and human losses and environmental damage from disasters. Global hazards (wildfires, earthquakes, floods, etc.) severely and lethally impact people, presenting a major risk to many areas.

Countries and international organisations are required to be more active in risk prevention mitigation and preparedness through improved disaster risk reduction (DRR) policies and programmes. As part of this effort, space agencies have implemented a series of activities aimed at fostering the use of Earth observation (EO) data to support DRR. EO can be processed into accessible and understandable products for civil protection agencies, local authorities, and first responders thus playing a critical role in pre- and post-disaster assessment by providing effective tools to rapidly map natural hazards and assess impacts.

GEO, through key initiatives such as the Data Access for Risk Management (GEO-DARMA), aims at creating synergy across multiple societal benefit areas by bringing together existing information sources and promoting broad international scientific collaboration.

In the area of wildfires there is an increasing amount of data and information being collected at different levels. However, an international initiative to pull resources and information together does not exist. GEO provides a platform for such international cooperation and offers possibilities to explore consolidation of information under GEO's Global Wildfire Information System (GWIS) initiative.

The Geohazard Supersites and Natural Laboratories (GSNL) is promoting advancements in geohazard science over selected sites and will ultimately innovate technologies, processes, and communication models. This will enhance data sharing, scientific collaboration, and capacity building in geohazard science.

The goal of GEO's disaster resilience activities is to provide a comprehensive overview of disasters and to address critical issues related to all phases of a disaster, not only response but mitigation, warning and recovery.

GEO IN SITU FOUNDATIONAL TASK IN 2017-2019

Ochiai O[1]
[1]*Group On Earth Observations Secretariat*

The global, domain-related observing systems are key components of GEOSS. The GEO In Situ Observation Resources Task will analyze the current state, trends, and needs, assess gaps, and develop new scenarios for in situ measurements as they constitute a key element of these global systems that need strengthening. The task will put particular focus on coordination and access to data and will provide various coordination opportunities in order to advocate for new systems, to sustain and strengthen existing and planned ones and to encourage integration and linkages. The main objectives of this activity, with a specific focus on in situ observations, are therefore to support the improvement and coordination of individual in situ observing systems; to strengthen the existing and planned global observation systems characterizing the Earth system domains (Atmospheric, Oceanic, Terrestrial); and to foster and facilitate the development of new global systems, federating existing ones and advocate for observational and informational gaps closure.

GEO´S EFFORTS IN SUPPORT OF SUSTAINABLE URBAN DEVELOPMENT

Pesaresi M[1], **Kemper T[1]**, Obregon A[2]
[1]*European Commission, Joint Research Centre,* [2]*Group on Earth Observations*

Use of Earth observations can promote equity, welfare and shared prosperity for all levels of human settlement, fostering national urban planning and showing land change over time to rethink the Urban Agenda. GEO will advocate the value of Earth observations, engage communities and deliver data and information in support of Sustainable Urban Development by assisting in the development of resilient cities and assessment of urban footprints; in order to make cities and human settlements inclusive, safe, resilient and sustainable through identifying economic externalities, managing environmental, climate and disaster risks, and building capacity to participate, plan and manage based on objective information regarding urban development.

This talk gives an overview on GEO´s activities in support of Sustainable Urban Development focusing on improving coordination of observations and monitoring of human settlements.

GEO'S SOCIETAL BENEFIT AREA "ENERGY AND MINERAL RESOURCES MANAGEMENT": ACHIEVEMENTS AND FUTURE ACTIVITIES

Ranchin T[1], Chevrel S, Obregon A[2]
[1]MINES ParisTech, [2]Group on Earth Observations

Fossil fuel energy use accounts for more than two thirds of greenhouse gas emissions. Earth observations can be used to increase the global share of renewable energy sources such as solar and wind power, in combination with energy efficiency, to help limit a further rise in global temperature, in line with the Paris Agreement. The Global Earth Observation System of Systems (GEOSS) is helping governments and companies to manage energy resources more effectively.

This introductory talk will address activities under the "Energy and Mineral Resources Management" Societal Benefit Area (SBA) of the Group on Earth Observations (GEO) . From the last decade of GEO, success stories will be presented, enhancing the benefits of GEO in this SBA and demonstrating its added value. The activities under development within the new GEO Work Programme (WP) and how they support the GEO Vision will be described. A set of developing projects within the activities of the GEO WP, including a GEO Initiative on renewable energies and a GEO Community Activity on mineral resources, will underline how EO data can support decisions and actions for the benefit of humankind. GEO offers the platform and forum for such essential coordinated, comprehensive and sustained Earth observations and information.

GEOGLAM, THE CANADIAN EXPERIENCE: FROM RESEARCH TO OPERATIONAL IMPLEMENTATION OF AGRICULTURAL MONITORING

Jarvis I[1], Davidson A[1]
[1]*Agriculture & Agri-food Canada*

Globally there is widespread interest in agricultural monitoring revolving around crop type, crop area estimation, near real time crop condition monitoring and yield forecasting. The G20 GEOGLAM initiative is harnessing this interest to develop a global system of systems for crop monitoring. This work is built upon a strong foundation of research, much of which has been coordinated through the development of the Joint Experiments for Crop Assessment and Monitoring (JECAM) research platform. JECAM brings together researchers from a large number of globally distributed, well monitored agricultural test sites that cover a range of crop types, cropping systems and climate regimes. Since 2009 JECAM research has been applied to compare and assess approaches to inform which are most appropriate for operational implementation in any given location. The outcome of these projects will result in a set of best practices that cover the range of remote sensing monitoring and reporting needs, including satellite data acquisition, pre-processing techniques, information retrieval and ground data validation.
This presentation will provide an overview of the JECAM research platform and provide a snapshot of how this R&D has been translated to operational agricultural monitoring in Canada.

GEOGLAM, THE GLOBAL AGRICULTURAL MONITORING INITIATIVE OF GEO, THE GROUP ON EARTH OBSERVATIONS

Deshayes M[1,2], Becker-Reshef I[2,3], Whitcraft A[2,3]
[1]Geo Secretariat, [2]Geoglam Secretariat, [3]University of Maryland

In the context of expanding world population combined with climate change, the challenge of feeding the planet using sustainable agricultural production is more important than ever. This challenge has been recognized by the G20's Action Plan on Food Price Volatility and by the UN through the Sustainable Development Goal #2 of Zero Hunger. GEOGLAM, the Global Agriculture Monitoring initiative (geoglam.org) of the Group on Earth Observations (GEO), contributes to meeting the challenge by promoting the use of Earth observations (EO) for monitoring agriculture toward producing timely, relevant, and actionable information on food supply. This is being implemented globally and regionally with an emphasis on enhancing national capacity to autonomously assess crop conditions. This includes ensuring access to appropriate EO (in coordination with the Committee on Earth Observation Satellites), and empowering analysts to utilize state-of-the-art monitoring methodologies in diverse landscapes, through JECAM, Joint Experiment on Crop Assessment and Monitoring, an international network of agricultural monitoring research sites and teams (jecam.org), or through the Asia-RiCE crop team activity in the ASEAN+3 framework (asia-rice.org).

GEOGLAM has had considerable success through Crop Monitor activities, which rely on EO to generate monthly consensus reports on global crop conditions. At the global scale, GEOGLAM currently operationally produces two monthly reports: the Crop Monitor for AMIS (G20 Agricultural Market Information System), focusing on countries that produce approximately 80% of the world's production of wheat, rice, maize, soybeans, and the Crop Monitor for Early Warning, focusing on countries-at-risk of food insecurity and their crops and drivers (geoglam-crop-monitor.org). At the national level, Crop Monitors have been prototyped for Ukraine, Tanzania and Uganda, with growing interest from other countries. GEOGLAM users are, from global to local levels, UN and other international organisations dealing with food security, national Ministries of Agriculture, and individual farmers through extension services, NGOs and farmer unions.

GEOGRAPHICAL DISTRIBUTION AND PHYSICAL ACCESS TO PUBLIC HEALTHCARE FACILITIES

Mugwena T[1]
[1]*University Of Johannesburg*

According to South Africa's Section 27 of the Constitution every person has the right "to have access to health care services, including reproductive health care". In reality however, this right has not been made available to everyone in the country, largely due to distortions in resource allocation. Many South Africans still do not have proper access to healthcare, reasons ranging from money, distance or travel time to care, long queues at facilities and the service the patients receive at public healthcare facilities. There is a need to improve access to quality and affordable healthcare services in South Africa. Currently the department of health is in the process of launching and implementing the National Health Insurance (NHI) which seeks to pool funds to provide access to quality, affordable personal health services for all South Africans based on their health needs, irrespective of their socioeconomic status. There is limited research that focuses on the physical or spatial accessibility to the Healthcare services. In Some instances ambulances have failed to get to a patient in time due to a spatial barrier, be it lack of proper roads or delay in traffic flow. This paper aims to use GIS to determine the physical accessibility to public healthcare facilities in two municipalities. Using the weighted cost distance tool, spatial barriers to healthcare will be determined as well as the time it takes members of the communities to get to or receive healthcare services. The results of the spatial accessibility will be compared between the rural areas and an urban areas to determine how the variation of accessibility corresponds to the distribution of population with various socioeconomic statuses using the census data.

KEYWORDS: Healthcare access, spatial barriers, cost distance

GEOSHARING ON CONNECTED PLATFORMS – SIMPLIFYING AND ENHANCING THE UTILISATION OF EO DATA FOR EXPERTS AND NON REMOTE SENSING SPECIALISTS

Benz U[1]
[1]*Cloudeo Ag*

With Copernicus and many commercial EO data programmes more and more imagery and geodata are available. Together they have a huge potential to increase the number and quality of GeoServices which retrieve meaningful information from EO data. These GeoServices will be the motor for better utilisation of EO data for government and industry.

The easy access to the data together with processing power are key to enable the development of these GeoServices. Cloud technology and modern distributed data management are an important component to fulfil this requirement. However, a most efficient setup of the workflow may require not only one cloud, but a network of clouds to give the desired fast access to multi-source data, cost efficient storage space for voluminous data and the security for critical business data.

Furthermore, to really increase the utilisation of EO data also non remote sensing specialists need to be able to build meaningful applications on this data. This developer community has to get easy access not only to raw data, but much more to calibrated data and intermediate products. To this end the traditional monolithic value chains need to be broken up. Experts must be enabled to combine their complementary expertise and to cue their analytics without loosing intellectual properties or risking their competitiveness in the market.

Beyond the technical solution, new business, marketing and sales approaches have to be developed to support these new opportunities. Successful strategies of other industries, as standardised products instead of highly customised solutions, pay per use models and sharing concepts have to be evaluated and adapted to the geobusiness.

This presentation will highlight challenges and approaches for solutions based on CloudEO's hands-on experience to create and operate a GeoMarketplace supporting a healthy and sustainable ecosystems for partner and customers, for users and producers of GeoServices.

GEOSPATIAL APPROACH FOR DELINEATING EXTRAPOLATION DOMAINS FOR SUSTAINABLE AGRICULTURAL INTENSIFICATION TECHNOLOGIES

MUTHONI F[1], BAIJUKYA F[1], SSEGUYA H[1], MATEETE B[1], HOESCHLE-ZELEDON I[2], OUKO E[3], MUBEA K[3]
[1]INTERNATIONAL INSTITUTE OF TROPICAL AGRICULTURE, [2]INTERNATIONAL INSTITUTE OF TROPICAL AGRICULTURE, [3]REGIONAL CENTRE FOR MAPPING RESOURCES FOR DEVELOPMENT

Sustainable intensification (SI) is a viable pathway to increase agricultural production and improve ecosystem health. Scaling SI technologies in locations with similar biophysical conditions enhance adoption. This paper employs novel extrapolation detection (ExeDet) algorithm and gridded bioclimatic layers to delineate extrapolation domains for improved maize varieties and inorganic fertilizers in Tanzania. Yields data recorded in on-farm trials was analysed to identify the best performing variety-fertilizer treatment. The ExeDet algorithm generated three maps: (1) the dissimilarity between bioclimatic conditions in the reference trial sites and the target extrapolation domain (Novelty type-1) (2) the magnitude of novel correlations between covariates in extrapolation domain (Novelty type-2) and (3) the most limiting covariate. The novelty type-1 and 2 maps were intersected and reclassified into four suitability classes for the best performing technology. SC719 maize variety grown with YaraMila-CEREAL and YaraBela-Sulfan fertilizers recorded the highest yields in on-farm trials. The resulting suitability map would guide extension agencies in selecting priority sites for scaling out best-bet SI technologies. Precipitation was most limiting factor in largest area of the extrapolation domain. This information is vital when recommending remedial measures to boost crop productivity. The replicability of the method is dependent on availability of long-term trials data.

KEYWORDS: Extrapolation detection tool, Fertilizers, GIS, Improved maize varieties, Spatial targeting

GEOSPATIAL BIG DATA STREAMING: ARCHITECTURAL CONSIDERATIONS

Van Zyl T[1]
[1]*Wits*

Increasingly data of a Geospatial nature presents itself in a form where it is best dealt with in a streaming data context. To this end it is not entirely clear what the best architectural patterns are for dealing with data in these scenarios. In this p[aper we consider specifically the major architectural patterns and how they relate to overcoming the challenges of Geospatial streaming data. To this end in this paper we present our envisioned non-functional requirements and functional requirements as components of a Geospatial streaming platform. We then evaluate how different architectural patterns match these requirements. We argue that given the above evaluation that an event driven architecture with a broker topology is best able to meet our set of non-functional and functional requirements.

GEO-WETLANDS: A NEW FRAMEWORK FOR COLLABORATIVE MAPPING AND MONITORING OF GLOBAL WETLANDS

Strauch A[1], Grobicki A[2], Paganini M[3], Hilarides L[4], Weise K[5], Eberle J[6]
[1]University Of Bonn, [2]Ramsar Convention Secretariat, [3]European Space Agency, ESRIN, [4]Wetlands International, [5]Jena-Optronik GmbH, [6]University of Jena

In November 2016, the GEO-Wetlands initiative has been approved for the 2017-2019 Work Programme of the Group on Earth Observations (GEO). It is a collaborative and distributed effort, building on existing activities, partnerships and projects and trying to funnel currently available funds within the global wetlands community to move towards the establishment of a Global Wetland Community of Practice and a Global Wetlands Observation System (GWOS).

Wetlands are one of the fastest declining ecosystem types worldwide, while at the same time they are hot spots of biodiversity and provide diverse and valuable ecosystem services, such as water supply, hydrological buffering against floods and droughts, and climate regulation through carbon storage. Information on wetlands' extent and their boundaries is often scattered and difficult to find and access, which leads to the fact that wetlands are only partially covered worldwide by policies and management practices. Several Multilateral Environmental Agreements (MEAs) like the Ramsar Convention, the Aichi Targets or the Sustainable Development Goals require improved information and knowledge about the extent, status and trends of global wetland ecosystems. GEO-Wetlands provides a new, global framework for cooperation beyond disciplinary and sectoral boundaries with the goal to fill this knowledge gap in a collaborative effort.

To achieve this, one objective of GEO-Wetlands is to bridge the gap between science and policy by developing an observation system that provides information products and services tailored to the specific needs of users on different levels (e.g. global conventions, or national agencies). Thus, it follows a user and policy driven co-design approach involving partners from science, industry, NGOs, government agencies and global conventions. Once the developed GWOS is in an operational stage, it is expected to improve the capabilities for monitoring the dynamics and changes in wetland ecosystems on a global scale.

KEYWORDS: Wetlands, GEO, mapping, monitoring, science-policy

GEO-WETLANDS: COLLABORATIVE DEVELOPMENT OF A GLOBAL WETLANDS OBSERVATION SYSTEM

Strauch A[1], Grobicki A[2], Paganini M[3], Hilarides L[4], Weise K[5], Eberle J[6], Muro J[1]
[1]University Of Bonn, [2]Ramsar Convention Secretariat, [3]European Space Agency, ESRIN, [4]Wetlands International, [5]Jena-Optronik GmbH, [6]University of Jena

Wetlands provide many essential ecosystem services, but at the same time they are threatened by processes like e.g. urbanization, expansion of farmland and extraction and pollution of freshwater. Due to their heterogeneous characteristics, spatiotemporal dynamics and topographic features, wetlands are often hard to identify, map and monitor, which hinders their efficient conservation and recognition in policy frameworks and management practices. To improve this situation, the Ramsar Convention on Wetlands has been supporting the conceptualization of a Global Wetland Observation System (GWOS) since 2007. Starting from 2011 the Biodiversity Observation Network (GEO BON) of the Group on Earth Observations (GEO) has been coordinating this effort. In November 2016 the new GEO-Wetlands initiative has been approved for the GEO Work Programme 2017-2019. It provides a new collaborative framework for the coordinated implementation of the GWOS.

Several on-going projects contribute different elements to the GWOS development. The "Satellite-based Wetland Observation Service" (SWOS) Horizon 2020 Project is the biggest contributor with a total budget of almost 5 Mio €. It provides a pilot infrastructure for the GWOS by developing a portal and middleware allowing easy access to all kinds of geospatial datasets and ancillary data relevant for the wetlands community. Together with other projects like the ESA GlobWetland-Africa, DLR DeMo-Wetlands & Wetland-Radar, and the JAXA Global Mangrove Watch the SWOS project also contributes mapping and monitoring methods, products, software tools, and documentation.

One of the main objectives of GEO-Wetlands as a global initiative, is to position GWOS as a collaborative framework supporting countries to satisfy their reporting commitments under the Ramsar Convention (to carry out national wetland inventories) and under the Sustainable Development Goals (to monitor wetland extent under SDG Target 6.6). In order to do this, it will be essential to secure additional resources for maintaining and updating the infrastructure after the end of the SWOS project and secure long-term availability of the tools and products developed within the different contributing projects and activities.

KEYWORDS: Wetlands, monitoring, mapping, GEO, cooperation

GIS AND REMOTE SENSING BASED ASSESSMENT OF LAND USE-LAND COVER CHANGES IN THE COASTAL CITY OF LAGOS NIGERIA

Waswa R[1], Idowu T[2]
[1]Regional Centre For Mapping Of Resources For Development, [2]1. Pan African University- Institute for Basic Sciences, Technology & Innovation

Coastal areas have experienced a massive population increase all around the world in recent decades leading to significant changes in land use land cover (lulc) with ecosystem patterns and functions also affected. Lagos Nigeria, a fast rising mega city and de facto commercial and economic capital of the country, characterised by a low-lying coastal strip, swamps, wetlands and lagoon is a typical case. Over the years, there has been a substantial change in the dynamics of land use/ cover of the study area. This study, therefore, assessed these changes with emphasis laid on changes in built-up areas, wetlands, and water bodies due mainly to urbanisation and land reclamation activities over a period of 30years in intervals of 15 years. Landsat images for years 1986 and 2001 were obtained as well as 2016 sentinel 2A images. ENVI software package was used for atmosphere correction, haze detection, layer stacking and classification. A pixel-based supervised classification was used to obtain the LULC maps. Filtering was done to reduce noise in the classification while ArcGIS software was used for assessing the LULC changes and obtaining the statistics across the satellite images representing the epochs under study. It was observed that the coverage of built-up areas increased in leaps (9.52%, 19.0%, and 31.2%) while a loss of 60 to 75% wetland coverage was observed over the 30-year period. Generally, these changes were observed to be mainly due to rapid urbanisation, extensive deforestation and land reclamation activities within the state. This trend poses threats to sustainable development vis-à-vis ecosystem services, health, and livelihoods of the populace as well as disaster management (flooding). It is therefore recommended that a more inclusive and holistic management be adopted for managing the rapid urbanisation, deforestation, wetlands losses and other land use changes in the state.

GLOBBIOMASS - ESTIMATES OF BIOMASS ON GLOBAL AND REGIONAL SCALES

Schmullius C[1]
[1]*University Jena, Department For Earth Observation*

Forest biomass (here understood as above-ground woody biomass [AGB]) is a fundamental biophysical variable describing the amount of woody matter within a forest. It is crucial to human well-being as a source of materials (e.g. for building) and energy (around 90% of the energy consumption in sub-Saharan Africa comes from biomass burning).

The main purpose of the ESA GlobBiomass project is to better characterise and to reduce uncertainties of AGB estimates by developing an innovative synergistic mapping approach in five regions (Sweden, Poland, Borneo, Mexico, South Africa) for 2005-2010-2015 and one global map for the year 2010. All available Earth observation data and validated existing products are being explored for a stratified, operational methodology.

This presentation illustrates 1) provision of improved quantitative biomass maps at regional and global scale, 2) quantified estimates of biomass changes, 3) associated uncertainty maps, 4) validation of the biomass maps and 6) identification of the limitations of current data and methods to estimate biomass. 7) User requirements have been transformed into a complete and consistent set of Product Specifications for the global and regional remote sensing products to be implemented within GlobBiomass . It includes: harmonization of major definitions and classification schemes to be used in the global and regional products and presenting these in a commonly accepted Glossary of Terms.

KEYWORDS: above ground biomass, global forest mapping, forest change, radar-optical synergy, ESA Data User Element.

GLOBELAND30: RECENT DEVELOPMENT AND PROGRESSES

Chen J[1]
[1]*National Geomatics Center Of China/isprs*

GlobeLand30 is a 30-m resolution Global land cover (GLC) data product developed with Landsat-like 30-m remotely sensed imagery and a Pixel-Object-Knowledge (POK) operational mapping approach. It contains ten land cover classes (i.e., water, wetland, artificial cover, cropland, ice/snow, forest, shrub-land, grassland, barren land and tundra) and has two data sets (the year 2000 and 2010) respectively. Since its release on the Sept. 2014, GlobeLand30 has been utilized by about 120 countries and in many Social Benefit Areas.

This presentation will present the latest development of GlobeLand30, including: 1) The data refinement which has led to the adding of 14 second level classes into Globeland30; 1) the preparation of the updating which aims to develop GlobeLand30' 2015's version; 3) the ongoing GEO-led international validation activity which aims to validate 30-m global land cover data set; 4) the concept and vision of Collaborative Global Land community service (CoGland) which aims to deliver "one-stop" community-based service by connecting all available global, regional, and national LCC web services by developing common services.

GP-STAR: GLOBAL PARTNERSHIP USING SPACE-BASED TECHNOLOGY APPLICATIONS FOR DISASTER RISK REDUCTION

Zeil P[1], Samansiri S[4], Post J[5,6,7], Amarnath G[2], Castillo J[3]
[1]International Working Group on Satellite Emergency Mapping, [2]International Water Management Institute, [3]Agencia Espacial Mexicana, [4]Disaster Management Centre, Government of Sri Lanka (DMC), [5]United Nations Office for Outer Space Affairs (UNOOSA), [6]United Nations Platform for Space-based information for Disaster Management and Emergency Response (UN-SPIDER), [7]GP-STAR Secretariat

In response to the calls for voluntary partnerships by UNISDR to support the implementation of the Sendai Framework for Disaster Risk Reduction 2015-2030, the Global Partnership on Space Technology Applications for Disaster Risk Reduction (GP-STAR) was launched during the World Conference on Disaster Risk Reduction in Sendai, Japan, on March 15, 2015. To date the partnership comprises 20 partners and several are in the phase of joining. The partnership committed to fostering the use of Earth observation as well as space-based technologies and applications in the context of the Sendai Framework for Disaster Risk Reduction 2015-2030 (Sendai Framework). In doing so, it will support the implementation of the Sendai Framework through, for example, providing advice to governments, organizations, and projects on the use of space technologies and applications in disaster risk reduction efforts, and the provision of relevant publications and discussion fora. The partnership integrates international, regional and national organizations involved in space science and technologies, Earth observation, disaster risk reduction and civil protection.

The Sendai framework increases the requirements of countries in assessing, evaluating, reducing risks and in reporting. Data, information and services generated from Earth observation (EO) satellites, global navigation satellite systems (GNSS) or telecommunications satellites are now resulting in products that are part of knowledge-based decision-making processes in many countries, sometimes well streamlined in planning processes. But still, the majority of countries, especially developing countries, are not yet sufficiently aware on the beneficial uses of space-based information and its availability for their needs, do have limited access to satellite data, do not possess adequate infrastructures (e.g. Spatial Data Infrastructures) and governance structures and have constraints in human capital and resources to use satellite data. In most cases the use of geospatial information and earth observation data for disaster risk reduction is not institutionalized. Another factor hindering the use of space-based information is the fact that most of the applications are originating from the science domain, most often not fit-for-purpose for operational use. While scientific advancements are essential, it is of at least equal importance to transfer scientific evidence and knowledge meeting operational requirements in disaster risk reduction. Hence, making scientific advancements in this field usable and used where it is needed and feasible.

The presentation will highlight scientific advancements and best practices of the partners while at the same time addressing the view and needs of mandated authorities in disaster risk management at the national level, the entities which have to implement the Sendai framework. The contribution of GP-STAR to facilitate and to support countries in the upcoming years is presented and requirements towards the scientific community are addressed.

KEYWORDS: Disaster risk reduction, Sendai framework, space technology application, Earth observation, development cooperation

GUIDELINES FOR UPSCALING COUNTRY-WIDE WOODY FRACTIONAL COVER - RECOMMENDED AMOUNT OF FIELD PLOTS AND AIRBORNE LIDAR DATA COVERAGE REQUIRED ACROSS SOUTH AFRICA

Naidoo L[1], Mathieu R[1], Main R[1], Wessels K[2], Asner G[3]
[1]Council For Scientific And Industrial Research, Earth Observation Unit, [2]Council For Scientific And Industrial Research, Meraka Institute, [3]Department of Global Ecology, Carnegie Institution for Science

Accurate mapping of woody fractional cover (CC) at the country-wide scale remains challenging due to the large data volumes of sufficiently high resolution. Both field plots and LiDAR datasets serve as representative calibration and validation datasets for country-wide CC mapping using SAR data. This study sought to establish the optimal quantity of field plots and LiDAR coverages required to train a Random Forest model to map CC at a country-wide scale using ALOS PALSAR HH and HV backscatter and DEM ancillary variables. 35% of randomly selected training data, from the five main vegetated biomes (fynbos and thicket, indigenous forest, savannah and grassland) and the savannah biome alone were used to train RF models and validated against a fixed dataset of each of the biomes. This approach assessed sampling representativeness and the optimal number of field plots and quantity (size and number) of LiDAR coverage. Optimal sizes were selected where the modelling results showed the highest accuracy, i.e. the lowest Root Mean Square Error (RMSE). The results have shown that the savannah-only training dataset yielded high accuracies across grasslands, moderate accuracies across thickets but poorer accuracies across indigenous forests and fynbos biomes. From the LiDAR-simulated field plot analysis, it was concluded that a minimum of 500, 1ha field plots, representative of the natural CC range, would be sufficient for effective modelling of CC at the country-wide scale. Additional field plots would improve the overall accuracies only slightly, but incur significant increases in sampling efforts and costs. The analyses suggest that the most frugal LiDAR acquisition strategy would include only four separate 5000ha LiDAR acquisitions distributed across the five biomes. The study found that much less LiDAR data are required to train models than originally expected, provided that the acquisitions are sufficiently diverse in CC and vegetation type.

KEYWORDS: Fractional Cover, LiDAR, SAR

HABITAT USE OF MIGRATORY BIRDS: MOVEMENT ECOLOGY MEETS REMOTE SENSING

Leutner B[1], Kölzsch A[2], Wegmann M[1], Safi K[2], Wikelski M[2], Dech S[3]
[1]Department of Remote Sensing, University of Würzburg, [2]Max-Planck-Institute for Ornithology, [3]German Remote Sensing Data Center, German Aerospace Center (DLR)

Migratory species carry a special ecological role in that they connect biodiversity and ecosystem function across distant regions of our planet. Understanding and monitoring movement patterns and requirements of migratory species is of particular importance to fulfill monitoring and conservation obligations under international conventions and initiatives such as the UN Convention on the Conservation of Migratory Species of Wild Animals.

Within the EO-MOVE project, we model small-scale resource usage of Greater white-fronted geese (Anser albifrons) at stop-over sites during their spring migration from central Europe to northern Russia. Habitat use of the greater white-fronted geese is known to be sensitive to land-use intensity, phenology and and landscape configuration, which calls for the exploitation of remote sensing technologies.

To this end, we collected locational data on several adult geese by means of neckband GPS transmitters and identified stop-over sites based on their movement behavior.

Subsequently, the stop-over sites were characterized regarding their land-use characteristics and landscape context by means of optical and SAR time series data from the Sentinel 1 and 2 satellite missions.
Here we present the first results on modeling animal-environment relationships and discuss the potential for site selection prediction, an important prerequisite for spatially or temporally targeted conservation schemes.

HABITAT VALUES AND PRESSURE INDICATORS FOR PROTECTED AREAS - AN INTEGRATED HABITAT AND LAND COVER CHANGE APPROACH

Brink A[1]

[1]*Joint Research Centre - European Commission*

The integration of land use/cover change (LULCC) data with information on habitats and population density provides the means to assess potential degradation and disturbance resulting from anthropogenic activities such as agriculture and urban area expansion. This study assesses the LULCC over a 20 year (1990–2000–2010) period using freely available Landsat imagery and a dedicated method and toolbox for the Udzungwa Mountains National Park (UMNP) and its surroundings (20 km buffer) in Tanzania. Habitat data gathered from the Digital Observatory for Protected Areas (DOPA)'s eHabitat+ Web service were used to perform ecological stratification of the study area and to develop similarity maps of the potential presence of comparable habitat types outside the protected area. Finally, integration of the habitat similarity maps with the LULCC data was applied in order to evaluate potential pressures on the different habitats within the national park and on the linking corridors between UMNP and other protected areas in the context of wildlife movement and migration. The results show that the UMNP has not suffered from relevant human activities during the study period. The natural vegetation area has remained stable around 1780 km2. In the surrounding 20 km buffer area and the connecting corridors, however, the anthropogenic impact has been strong. Artificially built up areas increased by 14.24% over the last 20 years and the agriculture area increased from 11% in 1990 to 30% in the year 2010. The habitat functional types and the similarity maps confirmed the importance of the buffer zone and the connecting corridors for wildlife movements, while the similarity maps detected other potential corridors for wildlife.

HANDLING OUTLIERS IN REMOTE SENSING PRODUCTS DERIVED BY MODEL INVERSION: A SOUTH AFRICAN CASE STUDY

Liu Z[1], Verstraete M[2], de Jager G[3]
[1]*Cape Peninsula University of Technology,* [2]*The University of the Witwatersrand,* [3]*University of Cape Town*

Outliers, which are essentially unexpected data items, affect the interpretation of model results and measurements alike. They are typically screened out through statistical procedures, though this approach does not provide a real justification for discarding values, other than they appear different from the rest of the data. On the other hand, remote sensing products are frequently derived by inverting a model against measurements, a procedure that naturally leads to the estimation of some 'cost function', a numerical value that quantitatively expresses the ability of the model to 'fit' the data. This paper argues that it may be more meaningful to identify and sift outliers on the basis of this cost function, than solely on the basis of being different from some measure of central tendency. One advantage of this approach is that it will filter out data points with an excessive mismatch between the model and the data, whether or not these appear to be outliers. This approach is demonstrated by analysing specific products derived from NASA's Multi-angle Imaging SpectroRadiometer (MISR) data, though the method is applicable to any result generated through model inversion against observational data, and is therefore of general interest to a wide range of geographical applications.

KEYWORDS: Anisotropy; Cost function; MISR; Model inversion; Outliers;

HERBACEOUS BIOMASS PRODUCTIVITY DYNAMICS MONITORING AND ITS DRIVERS IN SAHELIAN CROPLANDS AND RANGELANDS TO SUPPORT FOOD SECURITY POLICIES

Leroux L[1], Bégué A[1], Lo Seen D[1], Kayitakire F[2]
[1]*CIRAD*, [2]*JRC*

Since the Sahelian population livelihood relies mainly on agropastoral activities, accurate information on biomass productivity dynamics and the underlying drivers are needed to manage a wide range of issues such as food security. This study aims to contribute to a better understanding of these drivers in rangeland and cropland, both at the Sahel and local levels (an agropastoral site in South-West Niger). At the Sahel level, the MODIS Land Cover product was used to extract cropland and rangeland pixels. By analyzing MODIS NDVI trends together with TRMM3B43 annual rainfall (2000-2015), we developed a classification scheme allowing to identify areas of persistent decline/improvement in biomass productivity and to separate rainfall-driven dynamics from other factors. The results showed an overall increase of productivity in the rangeland, and both an improvement and a degradation in the cropland. We found strong evidence that rain is not the only import driver of biomass increase, while the decrease could be attributed chiefly to other factors exclusively or to a combination of both climate- and human-induced factors. At the Niger site level, biomass trends have been put in relation with a set of potential drivers via a RandomForest model, to define which were the explanatory factors of the observed trends. The factor set covered 5 categories: climate, natural constraints, demography, physical accessibility and land cover changes. We highlighted that tiger bushes areas were particularly prone to pressure due to overgrazing and overexploitation of wood, while positive trends were mainly observed near rivers and in fossil valleys where new agricultural practices might have been promoted. The approach developped here could help to delineate areas with decrease in crop and grassland production and thus to assess the vulnerability of the population, but also to target zones with good potential for planning long-term food security policies.

KEYWORDS: Drivers, NDVI trends, Sahel

HETEROGENEOUS SENSOR DATA INTEGRATION: CHALLENGES, APPROACHES AND BENEFITS

Hohls D[1], McFerren G[1], Sibolla B[1]
[1]*CSIR*

Integrating a variety of geospatial observations over wide areas from multiple sources and multiple organisations is a challenging but essential task. Timeous and effective decision making relies on data being readily available and accessible, preferably in visual and numeric formats. OGC Sensor Web Enablement provides a mechanism for standardised HTTP access to, and encoding of, data gathered via differing observation and measurement processes. Sensor data is generated in a variety of formats, is stored in a variety of sources and places, and is updated over a range of time scales. Direct access to raw, or unprocessed, data is a challenge because of the variety of downstream needs. This necessitates processing to map between non-standard and standard sensor data format to enable further processing and display. This mapping, or data translation, tackles the challenge of data reuse by providing automated processes that are built around common- or standards-based formats. It also necessitates transport mechanisms for ad-hoc access to historical data as well as near real-time live data to downstream data consumers. In this paper we deal with our experiences in overcoming some of the challenges of sensor data processing and integration, and describe some of the tools and methodologies needed. We explain the benefits of early adoption of a common data model (CDM) in the data translation process, in order to support the development of tools and techniques that are reusable and/or extensible in different problem domains. We describe the trade-offs that are necessary to enable translation and highlight what steps can be undertaken to ensure the resulting benefits outweigh these. We explore some downstream uses that can be made of sensor data and how the chosen standard or common data format can support rapid geo-visualisation both of real-time and historical data.

KEYWORDS: sensor web enablement, geo-visualisation, data standard, data processing

HIGH RESOLUTION LANDCOVER MODELLING WITH PLEIADES IMAGERY AND DEM DATA IN SUPPORT OF FINE SCALE LANDSCAPE THERMAL MODELLING

Thompson M[1], Moyo L[1], Prinsloo T[2], Bouwer P[2], Brink N[3], Eloff C[4]
[1]Geoterraimage, [2]Pink Matter, [3]DENEL Dynamics, [4]AIRBUS Defense and Space

In the evaluation of air-borne infrared imaging sensors, the use of simulated infrared scenery is a cost-effective way to provide input to the sensor. The benefit of simulated scenes includes control over parameters governing the thermal behaviour of the terrain as well as atmospheric conditions. Such scenes need to have a high degree of radiometric and geometric accuracy, as well as high resolution to account for small objects having different temperatures and/or surface material properties. In support of this, innovative use of tri-stereo, ultra-high resolution Pleiades satellite imagery is being used to generated high detail, small scale quantitative terrain surface data to compliment comparable optical data in order to produce detailed urban and rural landscape datasets representative of different thermal landscape features. These datasets are being generated by integrated modelling of both height and spectral surface characteristics within an object-based modelling environment. This approach provides an operational framework for rapid and highly accurate mapping of building and vegetation structure of wide areas, as is required in support of the evaluation of thermal imaging sensors.

HIGH-RESOLUTION ENHANCED PRODUCT BASED ON SMAP ACTIVE-PASSIVE APPROACH AND SENTINEL 1A RADAR DATA

Das N[1], Entekhabi D[1], Colliander A[1], Dunbar S[1], Yueh S[1]
[1]*Jet Propulsion Laboratory*

The SMAP active-passive soil moisture product (L2SMAP) was discontinued on July 7th, 2015 due to malfunction in the SMAP radar. The active-passive algorithm developed for the SMAP mission is also capable of working with different combinations of coarse resolution passive microwave (L-/C-band radiometer) and high-resolution active microwave (L-/C-band SAR) observations, provided they meet certain criteria. This feature of the microwave active-passive algorithm enables us to work on an Enhanced SMAP-Sentinel high-resolution soil moisture product. We used 1 year (April 2015 to March, 2016) of SMAP L2SMAP product and compared with the SMAP-Sentinel combined product, the initial results show promise. To further investigate and develop the methodology and algorithm parameters, more Sentinel-1A data at a regular time series is desired. Such long time-series of Sentinel data at regular interval is only available on the European continent because unlike other parts of the world the Sentinel C-band SAR consistently observes most of the European countries. For further investigation, we selected a region of Spain to implement the SMAP-Sentinel active-passive algorithm for couple of reasons: 1) The vegetation of the regions is optimum (not very dense) to evaluate the algorithm performance; 2) A core calibration and validation (cal/val) site that contributed to the SMAP mission is also located in this region. From the study over this region in Spain, we will present the methodology to determine snapshot algorithm parameters and compare them with the parameters derived through statistical means, and characterization of errors propagation through the algorithm. Finally, we present the SMAP-Sentinel algorithm outputs and retrieved high-resolution (1 km and 3 km) soil moisture data validated against the core cal/val site in situ observations. This study helps to establish the Enhanced SMAP-Sentinel high-resolution soil moisture product and prepare for the operational implementation that will facilitate product release in the month of March, 2017.

HIGH-RESOLUTION RGB-ONLY IMAGERY WITH OBJECT-BASED IMAGE ANALYSIS IS SUITABLE FOR ESTIMATING CROWN AREA AND ABOVEGROUND BIOMASS IN MIOMBO WOODLANDS

Gwenzi D[1], Tagwireyi P[3], Gara T[4], Mareya H[2], Ndaimani H[2]
[1]Humboldt State University, [2]Dzivarasekwa High School, [3]University of Zimbabwe, [4]University of Twente

Quantification of tree canopy area and aboveground biomass is essential for monitoring an ecosystem's ecological functionalities such as carbon sequestration and habitat provision. Miombo woodlands are vastly existent in third world countries that often lack the resources to acquire lidar data or high spatial resolution multispectral remote sensing data that have been proven to accurately estimate these structural attributes. This study explored the utility of freely available(via Google Maps) high (1-m) resolution Red, Green and Blue (RGB)-only satellite imagery in combination with Object-Based Image Analysis (OBIA) for estimating tree canopy area and aboveground biomass in Miombo woodlands. Using a Geographic Information System (GIS), we randomly established forty-one 225 m2 plots in Mukuvisi Woodlands, Zimbabwe and used RGB-only data with OBIA to estimate tree canopy area in those plots. We also field-measured the height, canopy area and trunk diameter at breast height of all trees that fell in those plots, then used the field data and a published allometric equation to estimate above-ground tree biomass (AGB). OBIA accuracy was high (Jaccard Similarity Index = 0.96). We observed a significant positive linear relationship between AGB and field-measured canopy area ($R2=0.87$, $p<0.003$). We also observed significant polynomial relationships between (1) OBIA-derived canopy area and AGB ($R2=0.56$, $p<0.0001$), and (2) field-measured canopy area and OBIA-derived canopy area ($R2=0.63$, $p<0.0001$), and no significant differences ($t=19.67$, $df=78$, $p=0.28$) between field-measured canopy area ($\bar{x}=187.11m2$, $\sigma=127.03$) and OBIA-derived canopy area ($\bar{x}=163.00m2$, $\sigma=50.08$). Our results provide evidence to that RGB-only data with OBIA is a suitable method for estimating tree canopy area in dry Miombo woodlands for various purposes (e.g. biomass estimation) although its accuracy is relatively low.

HOW COORDINATION OF EARTH OBSERVATION DATA CAN HELP PROTECT THE WORLD'S VANISHING WETLANDS: PROGRESS ON THE SWOS PROJECT

Hüttich C[1], Weise K[1], van Valkengoed E[2], Franke J[3], Thulin S[4], Eberle J[5], Abdul Malak D[6], Strauch A[7], Fitoka E[8], Guelmami A[9], Mino E[10], Flink S[11], Plasmeijer A[12], O'Connor B[13]

[1]Jena-Optronik GmbH, [2]Terrasphere GmbH, [3]Remote Sensing Services GmbH, [4]Brockmann Geomatics Sweden AB, [5]Friedrich-Schiller-University Jena, [6]University of Malaga – European Topic Centre on Spatial Information and Analysis, [7]University of Bonn, [8]Mouseio Goulandri Fysikis Istorias, Greek Biotope Wetland Centre, [9]Foundation Tour du Valat, [10]Unité Technique du SEMIDE, [11]Wetlands International, [12]IUCN European Union Representative Office, [13]UNEP-WCMC

Wetlands are one of the fastest declining ecosystem types worldwide, while at the same time they are hot spots of biodiversity and provide diverse and valuable ecosystem services; such as water supply, hydrological buffering against floods and droughts, and climate regulation through carbon storage. Information on wetland extent, their ecological character and their services is often scattered, underestimated and difficult to find and access, which leads to the fact that wetlands are only partially covered worldwide by policies and management practices.

In this respect, SWOS (a EU Horizon-2020 project) provides monitoring tools and information on wetland ecosystems derived from Earth Observation data.

SWOS is assisting with reporting and monitoring obligations for environmental policies at different scales. SWOS also contributes to the development GEO-Wetlands initiative in close cooperation with the Ramsar and supports the EU-MAES process. Ultimately, SWOS allows for an informed and standardized creation of conservation and restoration measures and indicators which maintain biodiversity and essential ecosystem services.

The most important objective of SWOS is to prepare and install an operational service with users for users. Several user organizations are represented by the SWOS Consortium, and further users are identified at the global (e.g. GEO/GEOSS, Ramsar, EU, CBD), regional, national (e.g. EU member states, WFD, national administrations, Natura2000) and local (protected area managers, scientists, local administrations) level in order to support a multi-level user approach. Service cases will demonstrate opportunities for improved wetland management, planning and decision making.

This paper provides an overview of the SWOS approach and gives insights into how the project is engaging users through the establishment of user groups and the development of service cases. The SWOS portal and mapping software will be presented together with first monitoring and mapping results. More information can be found on www.swos-service.eu.

HOW SATELLITE IS USED FOR ESTIMATING RAINFALL FOR THE SOUTH AFRICAN FLASH FLOOD GUIDANCE SYSTEM (SAFFG) AND THE SOUTHERN AFRICAN FLASH GUIDANCE SYSTEM (SARFFG)

Phakula V[1]

[1]*South African Weather Service*

The World Meteorological Organization defines flash floods as a flood in a small catchment, with the time of concentration less than six hours as a result of intense precipitation. The National Oceanic and Atmospheric Administration defines flash floods as a rapid and extreme flow of high water into a normally dry area, or a rapid water level rise in a stream or creek above a predetermined flood level.

The South African Flash Flood Guidance System (SAFFG) was adapted on the Central American Flash Flood Guidance system which was developed by the US Hydrologic Research Centre (HRC). SAFFG is used today by the South African Weather Service to forecast flash floods operationally. SAFFG uses radar or satellite in areas where there is no radar, to determine the rainfall estimates that falls on a specific area. The hydroestimator (HE) was developed for use in areas where there is no radar coverage. The HE has been improved in such a way that it considers the temperature relative to the surrounding pixels. If a pixel is colder than their surroundings, then it is assumed that, in that area, there will be convective updraft and therefore rainfall will be produced. If it is warm it is assumed to be convectively inactive.

The Southern African Flash Flood Guidance System (SARFFG) was implemented operationally in February 2014 with the National Meteorological Services (NMSs) of the SADC regions. The rainfall input is from satellite and rain gauges. The Global Hydro Estimator (GHE) satellite rainfall estimates is the main precipitation source. GHE uses infrared (IR) sensors of geostationary weather satellites estimating cloud top brightness temperatures and establishing statistical relationships between precipitation intensity and brightness temperatures. The Microwave-adjusted Global Hydro Estimator (MWGHE) Precipitation, uses the CMORPH products available from NOAA/Climate Prediction Center.

KEYWORDS: FFGs, Floods, HE, GHE, MWGHE.

HUMAN SETTLEMENT DATA TO SUPPORT MEASURING AND MONITORING INDICATORS OF SUSTAINABLE DEVELOPMENT GOALS IN SOUTH AFRICA

Mudau N[1]
[1]*Sansa*

This paper demonstrates the contribution of The South African National Human Settlement Layer (SA NHSL) in measuring the ratio of land consumption to population growth. Ratio of land consumption to population growth was identified as one of the indicators to measure progress made on United Nations Sustainable Development Goal (SDG) number 11 which is aimed at making cities and human settlements inclusive, safe, resilient and sustainable. The study utilised human settlement data and census data collected in 2007, 2011 and 2016 to assess the rates of urban and population growth. The human settlement data used was derived using Global Human Settlement Layer-South Africa (GHSL-SA) developed by European Commission Joint Research Centre (EU, JRC) in collaboration with The South African National Space Agency (SANSA). Only two classes; built-up and non-built-up, were mapped for this study. The population data was sourced from census and community surveys. The results show urban and population growth between the years assessment. The ratio of urban expansion and population was derived using the urban and population growth rates. This study shows that human settlement data derived from satellite imagery can contribute towards measuring progress made by countries towards suitable development of the cities.

KEYWORDS: Sustainable Development Goals, human settlement, urban growth, built-up, satellite imagery

HUMAN SETTLEMENT DATA USE ON POWER UTILITY

Mphaphuli T[1], Mudau N[1]
[1]*Eskom Holdings Soc Limited*

The current status of power supply in South Africa is constrained and there is an urgent need of geospatial science technology to assist with better planning of current and future power line infrastructure. Eskom is mandated by government to provide electricity to all South African.

It is a challenging task for Eskom to do field survey to account houses that have been electrified and those that are not yet electrified. Eskom has benefited by using geospatial science technology to assist with electrification planning. South Africa is developing in a rapid speed with metropolitan cities being the biggest electricity consumers followed by local municipality. Using Remote Sensing technology to map all the houses in South Africa is the ideal solution to be used.

Eskom through its contact with SANSA has managed to acquire spot 5 images from 2006 – 2014 and 2015 has spot 6/7. The imagery has assisted with capturing dwellings across the entire country. The project has been running from 2006 to current, and has assisted the organization with electrification planning and load forecast.

Through the use of Eskom spot building count (SBC) and SANSA human settlement data, Eskom has been able to determine load forecasting and electrification planning. It is critical for Eskom to know the areas that require electrical load to be increased due to an increase of settlements in the area, it is also equally important for electrification planning to know the number of houses that require electrification and also to forecast future electricity requirement. SANSA Human Settlement data such as Eskom spot building count and SANSA human settlement data are the only data set that represents the country wide status country housing development.

HYDROLOGY THEMATIC EXPLOITATION PLATFORM (H – TEP): CLOUD - BASED INTEGRATED DATA ANALYSIS AND MODELLING SYSTEM

Menenti M[1], Martinez B[2], Koetz B[3], Fischer P[3], Marin A[3]
[1]*Delft University Of Technology*, [2]*ISARDSAT*, [3]*European Space Agency*

In the coming years, new EO missions will offer a large amount of new data useful for water resource management purposes. These data shall be accessed, processed, analysed and visualized for users and service providers but sometimes these users and providers do not want to invest ICT resources or do not have skills for handling these large volumes of data. The Hydrology Thematic Exploitation Platform (TEP) is an ESA project that aims at providing hydrology services such as flood mapping/forecasting, water quality and level monitoring, hydrology modeling and small water bodies mapping based on EO data. In addition to that, Hydrology TEP will also provide a collaborative framework where scientific users, river basin organisations and service providers could rapidly and easily access to a large number of EO data, integrate their own data and tools (in-situ data, socioeconomic data, analysis tools...) and process their processors (service prototypes, hydrological models, meteorological models) within a user-friendly environment. In summary, Hydrology TEP project wants to build a community and offer to this community an ensemble of services to facilitate and simplify their daily work. A case – study on flooding in the Niger Basin based on a dense time series of Sentinel 1 data and the hydrological model HYPE is used to demonstrate the functionalities of the system.

HYDROMETEOROLOGICAL OBSERVATION SYSTEM AS A BASIS FOR DECISION MAKING IN UZBEKISTAN

Shulgina N[1]
[1]*Hydrometeorological Research Institute Of Uzhydromet*

The Hydrometeorological observation system as a source of integrated objective information is a significant element of realization of sustainable development policy in Uzbekistan.

The most part of obtained data is based on the in-situ observational network in Uzbekistan. One of the main objectives of the network is significant strengthening technology of collection and processing information with aim to get high quality data in support of decision making in various social-benefit areas.

History of regular Hydrometeorological observations accounts more the 100 years in the territory of Uzbekistan. The Centre of Hydrometeorological Service at the Cabinet of Ministers of the Republic of Uzbekistan (Uzhydromet) is responsible to fulfill the systematic hydrometeorological observations and air pollution observations. The observational network consists of about 400 stations and gauging posts. On the base of the observational network data a wide range of activities are carried out on provision of hydrometeorological services for diverse sectors of economy and population including meteorological, agrometeorological, hydrological services.

One of the priority directions is the provision of the hydrometeorological security. In case of emerging dangerous hydrometeorological phenomena a storm warnings are issued and forwarded in a timely manner to the related governmental structures to be a basis for decision making in the natural disasters areas.

Further enhancement of the of Hydrometeorological observation System has to go both ways improving technological level of systematic observations and widening observational network, equipping it with modern devices and instrumentation, achieving more complete coverage of the territory. At that the direct interaction with users of information has a special significance in order to take into account their needs and build the basis for timely and effective decision-making.

IDENTIFYING DRYLAND DEGRADATION NOT CAUSED BY CLIMATE CHANGES

Burrell A[1], Evans J[1], Liu Y[1]
[1]*UNSW, Australia*

Dryland degradation is an issue of global significance as dryland regions play a substantial role in global food production. Remotely sensed data provides the only long term, large scale record of changes within dryland ecosystems. Identifying whether these changes are caused by climate changes or other (usually human related) causes is required to inform remediation efforts. The Residual Trend, or RESTREND, method is commonly applied to satellite observations to detect dryland degradation not caused by climate changes. Whilst effective in most cases, it has been shown that the RESTREND method can fail to identify degraded pixels if severe or rapid degradation has occurred. This paper presents a modified version of the RESTREND methodology that incorporates the Breaks For Additive Seasonal and Trend method to identify step changes in the time series that are related to significant structural changes in the ecosystem. When applied to Australia, this new methodology, termed Time Series Segmentation and Residual Trend analysis (TSS-RESTREND), was able to detect degradation in 5.25% of pixels compared to only 2.0% for RESTREND alone. This modified methodology was then assessed in two regions with known histories of degradation where it was found to accurately capture both the timing and directionality of ecosystem change.

IMAGING SPECTROSCOPY OF FORESTED ENVIRONMENTS: A 20 YEAR PERSPECTIVE

Niemann K[1], Visintini F[2], Quinn G[1], Stephen R[1], McLeod R[1,3]
[1]University of Victoria, [2]CARMS Inc., [3]Geological Survey of Canada

Forest environments have been monitored using remote sensing techniques ever since remote sensing became available to the general public. Recently, we have focused our attention on the use of both optical and microwave data to collect increasingly detailed and sophisticated metrics that describe these environments. The digital optical imagery domain has evolved from initial broad discrete band multispectral data to finely sampled imaging spectroscopy data, so that we now had the ability to move beyond the mere detection of features to performing meaningful, repetitive measurements on those features.

The technological evolution of imaging spectrometers and other sensors has rendered them smaller and less expensive, to the point that they, and the data that they collect, are now more readily available to the research community. Our group has been fortunate to have had access to a range of imaging spectrometer data over the past 20 years. Initially, we were provided with AVIRIS, Hyperion and CASI data acquired over our test forest. More recently we have flown payloads comprised of both a LiDAR and imaging spectrometer installed on fixed-wing and helicopter platforms. This has provided us with the opportunity to collect data over a range of forest conditions through a range of spatial resolutions.

This presentation will provide a summary of our experiences in working with imagine spectrometer data over western Canadian (and U.S.) forests. We have been exposed to issues related to within conifer species classification in socially complex forest canopies, productivity and inputs into ecological modelling, and forest health. Specifically, the presentation will summarize our experiences in working with pre-visual green attack stage of the mountain pine beetle epidemic, species identification, and providing input into ecological demographic modelling. Based on our experience we will provide insight into the potential extrapolation of some of our findings to the space-borne domain.

IMPACT OF LAND USE ON WATER QUALITY IN THE VAAL DAM: SIMULATION OF MANAGEMENT PRACTICES

Abd Elbasit M[1], Yajima H[2], Adam E[3], Weepener H[1], Yoshioka Y[4], Knight J[3], Tswai R[1], Chirima G[1]
[1]Agricultural Research Center-institute For Soil, Climate, And Water, [2]Research Center for Coastal Lagoon Environments, Shimane University, [3]School of Geography, Archaeology and Environmental Studies, , [4]Faculty of Agriculture, Tottori University

Surface water environment is critical from several points of view. Surface water bodies are the major water supply systems for South Africa's domestic, agricultural and industrial water requirements and thus maintaining water quality in surface water bodies is of major concern. These surface water bodies are highly sensitive to various pollutants from both point and non-point sources. The point source water pollution can be traced to a single source where in non-point sources pollutants are diffused across the landscape with the movement of water. The non-point sources are usually associated with climate, landscape processes such as soil erosion, land use management, and human activities within the catchment. Land use changes has an impact on type and volume of pollutants that will be transferred to the aquatic system, and water flow pathways. The projected land use probabilities can be use as indicator for the water quality status. Thus, various management practices can be assessed using simulation methodologies that involve the land use probability matrix and the pollutant load for each class. The land use probability matrix has been generated using National land use data set and MODIS product. On the other hand, the pollutant loads for each class was estimated from previous studies and historical water quality data. Based on the classification of the pollutant sources, the effectiveness of various management practices has been assessed.

IMPACT OF LONG-TERM CHANGES IN THE SPATIAL STRUCTURE OF URBAN GREEN SPACES ON TEMPERATURE: A CASE OF HARARE, ZIMBABWE

Mushore T[1]
[1]*University of Kwa-Zulu Natal*

Urban population of developing countries has been predicted to continue increasing beyond the year 2050. However, rapid urbanization changes the energy balance of an area as natural landscapes are replaced by impervious surfaces which absorb heat. This results in high temperatures whose impact include increases in water and energy demand and thermal discomfort of urban residents. City growth patterns in developing countries especially in Africa places little attention on the importance of green spaces in curbing temperature rises. The objective of the study is thus to investigate the effect of changes in vegetation spatial structure on land surface temperature in Harare between 1980 and 2015 using medium resolution Landsat data. Land use and land cover maps were produced at 5 year time steps and used to derive spatial metrics of vegetation cover for each period. Land surface temperatures were also computed from thermal data of Landsat during the hot seasons of the same periods. Preliminary results showed that number of vegetation patches and size of the largest patches decreased as built-up areas expanded over the years. The findings also showed that mean distance between vegetation patches increased while mean size of patches decreased between 1984 and 2015. Average land surface temperature increased as number and mean size of vegetation patches decreased. The increase in temperature was also in response to increased separation of vegetation patches by impervious surfaces as a result of city growth. The findings are critical in showing that spatial structure of vegetation with respect to impervious surfaces determine the heat mitigation value of urban greenery. The results are also important in informing urban growth plans which ensure spaced out of buildings with large vegetation patches to avoid further increase in temperature.

IMPACTS OF LAND USE ON CARBON DYNAMICS IN IPETUMODU, OSUN STATE, NIGERIA

Atafo O[1], Enaruvbe G[1], Ande O[2]
[1]Africa Regional Institute For Geo-spatial Information Science And Technology (rectas), [2]Institute of Agricultural Research and Training, Moor Plantation,

Rising Carbon in the earth's system has raised concerns due to the role it plays in increasing temperatures and inducing climate change. In this light, it is crucial to investigate the spatio-temporal variation of Carbon stocks stored as vegetation biomass in tropical forests. This study seeks to quantify and estimate changes in the carbon stock of a selected area based on its land uses, over time. Allometric formula was used to determine the biomass of plants so as to determine the total carbon stock in the area. Satellite imageries were acquired for three epochs; 1986 (December), 2000 (January) and 2015(December). Land use classes were extracted from the imageries. Carbon stocks in 1986 and 2000 were estimated on the basis of historical land use change. To estimate the potential carbon stock for the area, projection was made based on the trend in land use. Results showed that present total carbon in the study area is 18,236tonnes/km2 for built-up, 107,247tonnes/km2 for farmland and 349,593tonnes/km2 with a total above ground carbon about 29million tonnes in the area. Historic analysis showed a reduction of about 6 million tonnes between 1986 and 2015. Projection to 2020 estimates that about 1 million tonnes may be lost if current practices are maintained. The study stressed the potential impact of reduced above ground carbon stock to the atmosphere. It recommends halting deforestation rates in the study area to further mitigate global climate change.

KEYWORDS: Climate change, carbon dynamics, land use/cover, biomass, carbon estimates

IMPLEMENTATION OF AN AUTOMATED BURNED AREA MAPPING SYSTEM USING HISTORICAL LANDSAT TM AND ETM+ TIME SERIES DATA IN SOUTH AFRICA

Steenkamp K[1], Goodwin2 N[2], Collett L[2], Flood N[2], Swanepoel D[1], Stegmann R[1], Van den Bergh F[1], Tindall D[2], Wessels K[1], Vhengani L[1], Sibanda P[1], Kleyn L[1], Frost P[1], Hankel M[1]
[1]CSIR Meraka, [2]Department of Science, Information Technology and Innovation (DSITI), Remote Sensing Centre

Fire is a major determinant of tree-grass dynamics and woody vegetation structure in the savanna and grassland biomes of South Africa. Dominant fire regimes are reflected in the historical fire activity providing useful insight into the interactions between vegetation, climate, land use and fire management practices. The CSIR implemented a system developed by the Remote Sensing Centre of the Queensland government, Australia for mapping burned area using historical Landsat TM and ETM+ (1984-2016) time series data. This system utilises spectral (Landsat band 4, and bands 4+5), thermal, temporal and contextual information in a time series approach to perform automated classification of burned areas. Burned pixels, associated with a large decrease in surface reflectance, are identified as negative outliers from a median smoothed time series that serves as a reference. These negative outlier pixels were used to map the spatial extent of change areas using a watershed region growing operation, in which thresholds were optimised to limit under/over growing of regions. A classification tree was developed to attribute burned vs. unburned change objects based on a number of spectral indices. The fire-prone eastern region of SA is covered with 54 Landsat Path/Rows of which six were selected as representative core test sites for optimization of change and region growing thresholds as well as validation of burned area mapping. The number of scenes in the full time series for each Path/Row varied between 465 and 556, from which five scenes were selected for optimization of change detection (n=30). Validation was performed on randomly sampled points generated for scenes selected by using ranked percentiles in the time series of each test Path/Row. Percentiles were based on the fraction of burned area per scene. Results of the validation experiments will be presented.

KEYWORDS: Landsat, time series, burned area, savanna, fynbos, South Africa

ISRSE37
TSHWANE 2017

IMPLEMENTING EO-BASED WETLAND MONITORING IN AFRICA: THE GLOBWETLAND AFRICA PROJECT

Paganini M[1], Tøttrup C[2], Riffler M[3], Wang T[4], Stelzer K[5], Vekerdy Z[4]
[1]European Space Agency, [2]DHI GRAS, [3]Geoville Information Systems GmbH, [4]ITC, Faculty of Geo-Information Science and Earth Observation, University of Twente, [5]Brockmann Consult GmbH

The main objective of GlobWetland Africa (GW-A) is to provide major Wetland Stakeholders involved in the implementation of the Ramsar Convention of Wetlands in Africa with Earth Observation (EO) methods and tools to better assess the conditions of wetlands under their areas of jurisdiction/study, and to better monitor their trends over time. To this end, an open source wetland observing system, referred to as the GW-A Toolbox, is being developed, implemented and validated for a series of geo-information products (Wetland Inventory, Wetland Habitat Mapping, Wetland Inundation Regimes, Water Quality, River Basin Hydrology and Mangroves delineation and characterization) over a number of representative pilot sites across Africa.

The GW-A toolbox unifies proven and stable open source software into a single graphical user interface that will enable the users to perform wetland inventory, assessment and monitoring using freely available Satellite observations primarily from the Sentinel missions of the European Copernicus initiative but also from contributing missions such as US Landsat.

As an ultimate objective GlobWetland Africa will aim to enhance the capacity of the African Stakeholders to develop their own national and regional wetland observatories, and contribute towards the development of the Global Wetlands Observing System (GWOS), a joint initiative of the Ramsar Convention of Wetlands and the Group on Earth Observations (GEO).

IMPLEMENTING THE GROUP ON EARTH OBSERVATIONS DATA MANAGEMENT PRINCIPLES: LESSONS FROM A SCIENTIFIC DATA CENTER

Downs R[1]
[1]*Columbia University*

The Group on Earth Observations (GEO) Data Management Principles (DMP) provide direction for managing geospatial data and related information products and services. Offering opportunities for enabling discovery, accessibility, usability, preservation, and curation, the GEO DMP challenge repositories, such as scientific archives and data centers, to improve practices that foster the use of Earth science data today and in the future. In addition, the Data Management Principles Implementation Guidelines (IG) offer many practical suggestions for implementing the DMP with examples that can inform the consideration of options for improving geospatial data management practices. Implementing such improvements offers value to the users of geospatial data by enabling data providers to support the use of the data products and services that they disseminate. Adopting these improvements also can assist repositories in their efforts to meet the requirements for attaining data repository certification, which offers value for repositories and their stakeholders.The presentation shows how repositories can improve data management practices for geospatial data by adopting the DMP, with examples drawn from experiences of a scientific data center. Current and future opportunities for improving data management practices to attain data repository certification also are described along with practical approaches that repositories can adopt in the short term.

IMPROVING GLOBAL LAND COVER MAPPING BY INTEGRATING EXISTING DATA

Tsendbazar N[1], Herold M[1], See L[2], Frtiz S[2], Mora B[3], Lesiv M[2]
[1]Wageningen University And Research, [2]International Institute for Applied Systems Analysis, [3]GOFC-GOLD Land Cover Office; Wageningen University and Research

Global land cover (GLC) maps and assessments of their accuracy provide important information for different applications such as climate, dynamic vegetation, hydrological and carbon (stock) models. To date, a number of GLC maps have been produced and multiple reference datasets were created for their calibration and validation. These existing maps and reference data have the potential to improve land cover mapping at global scale. For example, Wageningen University tested 5 different methods to integrate recent global land cover maps and reference datasets across Africa. These methods takes the spatial variation in map accuracy and class occurrence into consideration. Our results showed that the regression kriging method performed the best. This method predicts the land cover class occurrence probability based on recent global land cover maps and adjusts the prediction errors using a simple kriging method, which was then further implemented at global scale. Cross validation results further confirmed that the integrated map showed better correspondence (up to 13% regionally and 10% globally) in each of the continents when compared with input global land cover maps. At the International Institute for Applied Systems Analysis, three global land cover maps were integrated with reference data obtained from crowdsourcing activities. The spatial variation in map correspondence was modelled using geographically weighted regression and the land cover class at each pixel was labelled based on a global land cover map with the highest correspondence for that pixel. The results proved that global land cover mapping can be improved by integrating existing maps with a crowdsourced reference dataset. Similar use of existing datasets is now being implemented within the new global land cover activities of the Copernicus Global Land Service. Future efforts should focus on creating global land cover maps that take the requirements and perspectives GLC map users into account.

IMPROVING LAND COVER/LAND USE SUPERVISED CLASSIFICATION WITH THE ADDITION OF TOPOGRAPHIC DATA OVER TSHWANE, PRETORIA

Veramoothea P[1], Breytenbach A[1]
[1]*CSIR*

Organs of state, environmental institutions and academic institutions normally require accurate land cover mapping solutions for informed decision making as well as for monitoring of the environment, both urban and rural. VHR satellite imagery currently allows for detailed feature extraction in a wide range of LCLU applications. Using a multispectral WorldView-2 image, this study implemented two classifiers, the maximum likelihood and ISODATA algorithm, in supervised LCLU classification procedures; one using height information and one without in each case. The image was initially segmented into three primary classes using calculated NDVI values that separated waterbodies and vegetation from non-vegetation. After height differentiation in the built-up and vegetative classes, eight additional primary classes viz. residential, commercial/industrial, transport, trees/woodlot, shrubs/bushes, cultivated grass, wetland, and bare soil were produced using both classifiers. Results showed that although the two selected classifiers performed satisfactorily in each instance, the classification accuracy at a primary level increased significantly when one was able to differentiate between features that are on the Earth's surface from those that are above.

KEYWORDS: Image Classification, Height, Topography, Satellite Imagery, VHR

IMPROVING SATELLITE SKILLS USING BLENDED LEARNING

Prieto J[1], Nietosvaara V[1], Kerkmann J[1], Traeger-Chatterjee C[1], Higgins M[1]
[1]*Eumetsat*

The traditional approach to learning on satellite meteorology involved long and costly face-to-face sessions, which proved to be an efficient way to get familiar with the basic concepts in the field. However, the wealth of topics which have to be dealt with in nowadays training requires a more dynamical approach. The authors name this formula a blended course.

A blended course is a combination of an online phase and a face-to-face phase for training professional satellite meteorologists. The online phase lasts for a few weeks and triggers the communication among course participants from their respective locations. The participants find it convenient to have the flexibility of choosing when to work on their course assignments while continuing the day to day professional duties. They also appreciate the interaction between the instructors and fellow course participants, providing a good foundation for understanding basics of satellite interpretation.

Working on tasks (assignments) awakes interest on learning and results in a very interactive and fruitful classroom discussions after the online phase.

IMPROVING UNDERSTANDING OF AEROSOL PARTICLES AND THEIR IMPACTS IN AFRICA THROUGH REGIONAL MODELLING AND REMOTE SENSING

Garland R[1,2], Horowitz H[3], Engelbrecht C[4], Dedekind Z[1], Naidoo M[1,2], Oosthuizen R[1], Piketh S[2], Sibiya B[1], Engelbrecht F[1,2,5]

[1]Natural Resources and the Environment Unit, Council for Scientific and Industrial Research, [2]Climatology Research Group, North West University, [3]Department of Earth & Planetary Sciences, Harvard University, [4]Agricultural Research Council, [5]School of Geography, Archaeology and Environmental Studies, University of the Witwatersrand

Aerosols can impact the climate directly, through scattering and/or absorbing radiation, indirectly through modifying clouds. Currently, the largest uncertainty in climate models comes from the impacts of aerosols on the radiative balance of the Earth. In addition, aerosol particles are air pollutants that impact on air quality and can have negative health impacts. Africa contains the largest single sources of biomass burning emissions and dust globally, which are large sources of aerosol particles. In addition, Africa contains anthropogenic sources of aerosols, such as vehicles, industries, power generation. An accurate representation of aerosols in air quality and climate models is needed to understand the impacts of aerosols, currently and under future climate change. There are few measurements of aerosol particles in Africa, and thus remote sensing provides an opportunity to provide needed inputs and validation for models.

This presentation will highlight the use and limitations of aerosol products in improving the understanding of aerosols particles and their impact in Africa. Ground-based AERONET products were used to validate the current representation of aerosol particles in the Conformal Cubic Atmospheric Model (CCAM) using the CMIP5 historical emissions inventory over Africa. Over southern Africa, it was found that CCAM simulates the monthly cycle of aerosol optical depth well as compared to AERONET. This study further investigated the sensitivity of the southern African climate and clouds to the presence of aerosol particles and their transport patterns, and investigated using remote sensing to provide comparison to simulated clouds. Finally, the results of a study investigating the use of available datasets on ground-based aerosol concentrations that have been generated through remote sensing and modelling were ground-truthed against measurements in South Africa. It was found that the international datasets underestimated aerosol pollution by, on average, three times. This highlights research gaps on aerosols in Africa.

INDIRECT ESTIMATION OF VEGETATION STRUCTURAL PARAMETERS IN SOUTH AFRICAN FORESTS USING OPTICAL MULTI-VIEW AND LIDAR REMOTE SENSING DATA

Mahlangu P[1,2], Mathieu R[1,2], Wessels K[2,3], Verstraete M, Asner G[4], Naidoo L[1,2], Main R[1,2], Ndyamboti K[1,2]
[1]Ecosystems Earth Observation, Natural Resources and Environment, CSIR, [2]Department of Geography, University of Pretoria, [3]Sensing Unit, Meraka Institute, CSIR, [4]4Department of Global Ecology, Carnegie Institution for Science, Stanford

Forest structure can be characterized by various quantifiable parameters, such as canopy cover, tree height, and canopy volume, which are required for assessing biophysical processes, forest management and conservation, particularly within the framework of international treaties and national programs. Light Detection and Ranging (LiDAR) technology has proven useful to estimate the structural parameters; however it remains limiting due to its high cost and is difficult to apply at national or regional scale. The availability of quasi-simultaneous multispectral and multi-angle measurements provided by the Multi-angle Imaging SpectroRadiometer (MISR) offers unique opportunities compared to traditional nadir-viewing remote sensing instruments. In addition, the recent production of MISR High Resolution (HR) data products at 275m resolution may improve the estimation of forest structural parameters. Our objective was to determine whether the two MISR HR data products (Bidirectional Reflectance Factor (BRF) 36 band and Rahman-Pinty-Verstraete (RPV) 12 band) could improve estimates of mean tree height (Hmean), canopy cover (CC) and canopy volume (CV) across two structurally diverse sites (savannah in the Kruger National Park and indigenous forest and plantations in the Kwa-Zulu Natal) in South Africa. We used airborne LiDAR to train and validate Random Forest models testing and comparing a number of MISR HR BRF and RPV band scenarios. The preliminary results showed that the full 36 band BRF data consistently showed a higher accuracy in modelling the structural parameters, with coefficient of determination (R2) for Hmean of 0.80, for CC of 0.51, and for CV of 0.49. Similar patterns emerged for all the RPV bands, which yielded the best results for two parameters: Hmean (R2 = 0.69) and CC (R2 = 0.60). These findings demonstrated the promise of multi-angle data for estimating forest structure across our forest types.

KEYWORDS: structural parameters, LiDAR, MISR HR, Random Forest

INFLUENCE OF DRY SNOW COVER ON TANDEM-X DATA OVER A HIGH LATITUDE BOREAL FOREST ECOSYSTEM

Panagiotopoulou D[1], Brown I[1,2]
[1]*Department of Physical Geography, Stockholm University*, [2]*Bolin Centre for Climate Change, Stockholm University*

Snow is a crucial component to the hydrology of northern forest ecosystems. Snow cover variability is affected by climate change at both spatial and temporal scale. In this paper, we investigate the effect of dry snow cover on TanDEM-X interferometric synthetic aperture radar (InSAR) data over a boreal forest ecosystem. Eleven bistatic HH-polarised TanDEM-X pairs were acquired over a sub-arctic forest site in Norway. In bistatic configurations of TanDEM-X mission temporal decorrelation is negligible. Hence, the primary hypothesis was that volume scattering within the snowpack is the dominant decorrelation effect on coherence in the absence of attenuation in forest canopies. The forest consists of Norway spruce, Scots pine, and birch species. Open terrain areas (bogs, fens, clear-cuts) were also analysed and the importance of ground contributions was evaluated. This study demonstrates the sensitivity of TanDEM-X products to varying weather conditions (air temperature, cloud cover, snow fall, wind drift), which in turn govern snowpack dynamics. It is observed that backscatter is lowest for temperatures around 0 ℃ and it increases with decreasing temperatures. The results indicate significant X-band SAR signal penetration into the crown layer for temperatures below -10℃ and into the ground surface in the absence of near-surface snow melt events. Daytime warming temperatures and overnight cloudy conditions prior to the acquisitions are associated with snow metamorphosis. Near-surface snowmelt events are evidenced by substantial reduction in both coherence and backscatter. It is concluded that for very cold and dry conditions X-band SAR penetrates the canopy layer of low biomass sub-arctic boreal forests and the underlying snow-ground composite dominates the signal. Moreover, coherence and backscatter are highest for forests and lowest for open terrain, emphasizing the significance of volume decorrelation effects from snowpack. Altogether the results emphasise the promising potential of TanDEM-X to provide important insights into snow-covered boreal forest environments.

INFLUENCE OF ENVIRONMENT ON THE SPATIAL VARIABILITY OF DENGUE OUTBREAKS AT DIFFERENT SCALES IN SOUTH PACIFIC AND NEW-CALEDONIA

Mangeas M[1], Menkes C[1], Teurlai M[1], Descloux E[2], Mercier A[3], Zellweger R[4]
[1]Ird, [2]Department of Internal Medicine and Infectious Diseases,TerritorialHospital,, [3]SPC, [4]London school of hygiene and tropical medicine

Context and Objectives: Dengue is a complex disease resulting from the interaction of human, biological, environmental, geographical and socio-economic factors. In this paper, we assess the potential influence of environment (including climate) on the spatial variability of dengue outbreaks at different scales in New-Caledonia. This work attempted to make a link between published studies (Descloux et al, 2012 ; Teurlai et al, 2015) and ongoing projects, each of them assessing the risk of outbreak at a specific scale (South Pacific level, municipal level and city of Nouméa). The purpose was also to understand how the environmental risks interact and if there exists forcing from large scale to smaller scale.

Methods: Environment (e.g. human habitat, vegetation, climate) were mainly characterized by low resolution (NCEP climate data, 1°) to high resolution (spot 5, 5m) remote sensing data, according to the spatial scale. Then a statistical analysis is performed in order to select the most relevant variables capable to explain the epidemic dynamics. Finally, at each scale, a statistical model is set for estimating the spatial variability of dengue outbreak.

Results: At South Pacific scale, focus was made on the climate and quality of health systems. We discover that usually outbreaks started in French Polynesia before spreading to the other countries. Endemic or epidemic situations related to the countries would appear to depend on temperatures. At municipal scale, climate variables and socio-economic variables seemed to be the most relevant variables. At the city scale, the 2008/09 epidemic was spatially structured, with clusters of high and low incidence neighborhoods. Dengue incidence rate was positively associated with unemployment, vegetation coverage and proportion of old houses. Hence, climate seemed to be the main driver at large scale when human behavior became more and more relevant at smaller scale.

INSAR FOR OIL FIELD DEFORMATION MONITORING DURING CYCLIC STEAM STIMULATION

Wessels J[1], **Staples G[2]**, Jones K[1]
[1]PCI, [2]MDA

The focus of this study was to use InSAR to monitor surface deformation that occurs when bitumen is extracted using a process termed Cyclic Steam Stimulation (CSS). CSS results in significant volumetric strain of the reservoir, so deformation is of interest.

The approach entailed the installation of corner reflectors on pipeline-pilings which respond to the surface deformation induced by the CSS process. RADARSAT-2 UltraFine data were acquired every 24 days. PCI-developed InSAR algorithms were applied to the data to estimate millimeter-scale surface deformation. A reservoir dilation model was used to calculate the amount of surface heave and the model results were compared with the InSAR measurements. The model had a number of variable parameters of which the initial pore pressure, the failure pore pressure, and the dilation between the initial and the final pore pressures are the most significant.

There was good agreement between the InSAR and model with-respect-to surface heave or subsidence, but not with the magnitude of the deformation. To better understand why the magnitude differed, two wells were analyzed. For study well F07, using the standard parameters for the dilation calculation, the heave model only predicted ~30 mm of heave versus ~120 mm of InSAR-measured heave. Good agreement between the model and InSAR was obtained if the dilation prior to fill-up was increased to account for the larger depleted zone of a late-cycle CSS well. For study well U, the heave model predicted over 110 mm of heave versus ~80 mm from InSAR. Good agreement was obtained if the dilation at fill-up was eliminated since the steam was injected into a new reservoir.

IN-SITU LONG-TERM AND MULTIDISCIPLINARY OBSERVATION NETWORKS AND SATELLITE ECOLOGY CONCEPT FOR TERERSTRIAL BIODIVESITY AND ECOSYSTEM

Muraoka H[1]
[1]*Gifu University*

Recent challenges of ecosystem science involves detection of changes in ecosystem functions and hence services under climate change. Among the numbers of ecosystem research (ecological, biogeochemical and micrometeorological means) taken so far in the Asia-Pacific and Oceania regions, up-scaling in space and time by satellite imagery, GIS and simulation models has been crucial for our broader understandings on the environmental changes and their possible influence on sustainability. Carbon cycle and budget, tree water use and hydrological cycle, and primary production have been the major forest ecosystem functions, and it is no doubt that they play key roles in earth system and biodiversity. Recent global agenda on these theme involves to link such observations with regional and global climate observations in order to find and predict their interactive changes which influence societal security. Promoting our collaborative long-term observations and in-situ experiments would be useful to identify the major observational parameters, methodology and data analysis for our understandings and prediction of environmental changes in regional scale. In this paper, in order to discuss our further collaborative observation from in-situ to space, we introduce our long-term and multi-disciplinary observations (leaf ecophysiology, canopy spectrum, eddy covariance) in forest ecosystems in Japan, by applying "super-site" concept. Then we discuss these observation needs by referring to the recent activities of the Group on Earth Observations.

KEYWORDS: Forest ecosystem, Long-Term Ecological Research, In-situ and satellite observations, Observation networks

IN-SITU OBSERVATIONS: COORDINATION NEEDS AND BENEFITS

Plag H[1], Maso J[2], Team of the Foundational Task on GEOSS In-Situ Earth Observation Resources[3]
[1]Old Dominion University, [2]CREAF, [3]GEO

Information derived from in-situ observations (i.e., all ground, water- and airborne observations excluding space-borne observations) is crucial for a better understanding of processes in the Earth system and built environment, the development and effectiveness of policies, the planning of actions, and the monitoring of progress towards goals, as well as for the support of space-based observations. Examples of the application of coordinated Earth observations are the policy development for the implementation and monitoring of the SDGs, the documenting of global change and its impact on humanity, the sustainable use of resources, governance of risks and the reduction of disasters, and the prediction of environmental conditions (such as air quality, weather, droughts, floods, seasonal climate in support of agriculture, sea level, and changes in environmental health). Providing the full range of information that could be derived from in-situ measurements often requires cross-discipline coordination and data integration and coordination across national boundaries. Systems for in-situ observations are diverse and there is no single global group responsible for their overall coordination. A report prepared in 2016 by the GEO Foundational Task "GEOSS In-Situ Based Earth Observation Resources" found that despite considerable progress in recent years there is still a need to improve coordination on regional and global basis. Considering the wide range of technological and organizational approaches, the diversity in conditions impacting the development and operation of infrastructure, funding, and data sharing, and the multitude of players, coordination of in-situ Earth observations is not an easy task. The report concludes that GEO and its regional organizations could play an important role in convening a cross-discipline coordination process and makes a number of recommendations how to proceed. Importantly, the Foundational Task on GEOSS In-Situ Earth Observation Resources is asked to thoroughly assess different approaches to cross-discipline coordination and their potential benefits.

INTEGRATING FIELD-BASED AND REMOTE-SENSING TECHNIQUES TO MONITOR THE PHENOLOGY OF SOUTHERN AFRICAN SAVANNAS

Whitecross M[1], Archibald S[1], Witkowski E[1]
[1]*University Of The Witwatersrand*

Phenology – the study of the timing of biological events – enables scientists to monitor the 'pulse of our planet'. We are faced with an uncertain climatic future and monitoring the changes in the phenology of ecosystems has been highlighted as an important tool for understanding how these systems may be affected in the future. Savannas are a key biome in southern Africa; however, few long-term phenological datasets currently exist for these ecosystems. The green-up phenology of trees and grasses in a broad-leaved savanna in the Nylsvley Nature Reserve (NNR), South Africa, was monitored in relation to rainfall at weekly intervals over three seasons (2012-2014). Each observed season experienced different environmental conditions from early-onset average rainfall (2012), to late-onset high (2013) or low rainfall (2014). Early-greening by trees was only observed in the late-onset, low rainfall season. The field data collected were then used to inform an analysis of remotely-sensed vegetation greenness and identify the threshold NDVI value where greening occurred in the field. This enabled a regional analysis of early-greening across seven broad-leaved savanna sites with a similar species composition to the NNR. By comparing the dates of greening to the onset dates of rainfall events >15mm using the Tropical Rainfall Measuring Mission (TRMM) precipitation datasets, we found that early-greening shows a latitudinal trend across southern Africa with more frequent and longer periods of early-greening at sites closer to the equator. The next challenge is to develop and implement the ground-based Southern African Phenological Network (SAPheN) to expand the collection of phenology data in savannas and improve the links between field-based and remotely-sensed phenological datasets. Monitoring and understanding phenological cycles in these highly seasonal savannas will assist in our understanding of which changes fall within natural cycles and which are being driven by changing climates in the future.

INTEGRATION OF MULTIPLE REMOTE SENSING OBSERVATIONS TOWARDS ASSESSING FOREST CARBON STOCKS OF 2010

Santoro M[1], Cartus O[1], Mermoz S[2], Bouvet A[2], Le Toan T[2], Ericsson A[3], Carvalhais N[3], Avitabile V[4], Herold M[5], Schmullius C[6]

[1]Gamma Remote Sensing, [2]CESBIO, [3]Max Planck Institute for Biogeochemistry, [4]Joint Research Centre, [5]Wageningen University, [6]Friedrich Schiller University

Estimates of forest carbon stocks are highly debated because of the large difference between existing estimates and the large uncertainties associated with the estimates. With the availability of observations from multiple platforms acquired around the year 2010, we have developed an approach to estimate forest aboveground biomass and thereof carbon stocks combining SAR, LiDAR and optical observations together with auxiliary datasets from forest inventories, climatological variables and ecosystems classifications (see Cartus et al., this conference). Here, we present current results from our investigations on the reliability of the biomass estimates and compare with existing datasets and estimates. The spatially explicit dataset of forest aboveground biomass and carbon stocks is the first of its kind that was obtained with a single retrieval approach, tuned locally to account for the spatial variability of forest structure.

The estimates and related uncertainties have been obtained with a pixel size of 25 m. Nevertheless, this is not the target resolution given that each of the remote sensing datasets used was only indirectly related to biomass and often the observations were not free from processing artifacts. To decrease such external contributions and provide a tighter range of uncertainties, spatially averaged estimates are also considered, between 50 m and 1,000 m are generated. The spatial distribution of biomass appears to be well captured, with the largest values in tropical forests and in temperate forests of the Pacific Northwest, Chile and South Australia. Carbon stocks of the northern hemisphere are in line with existing observations from on ground surveys. In the tropics, the estimates appear to be in agreement with previous mapping activities except for Southeast Asia where we are currently estimating less biomass. Validation will help to understand the reliability of the results. At the conference, the current status of the investigations will be presented.

INTEGRATION OF OPTICAL AND SYNTHETIC APERTURE RADAR (SAR) IMPROVES HERBACEOUS BIOMASS ESTIMATION USING SENTINEL-1 AND 2 IN THE SAVANNA ECOSYSTEM, SOUTH AFRICA

Ramoelo A[1], Cho M[1], Mathieu R[1], Main R[1], Naidoo L[1], Dziba L[1]
[1]*Council for Scientific and Industrial Research , Pretoria, South Africa*

Herbaceous biomass is a key variable to understand productivity of rangeland or grazing areas. Drought has negatively affected rangeland ecosystem reducing grazing and browsing potential – hence impacting on people's livelihood through reduced livestock production. The development of remote sensing techniques to estimate biomass has progressed and mainly based on vegetation indices, which often saturates during peak productivity. Combining optical and synthetic aperture radar (SAR) at high spatial resolution would be crucial for validation of the near real-time monitoring of herbaceous biomass with low spatial and high temporal resolution. The objective is to test the whether the integration of optical and SAR based indices could improve the estimation of herbaceous biomass in the Lowveld area, North-Eastern part of South Africa. Field data was collected during growth to peak productivity period (December to March). Sentinel-1 and -2 data for the closest field data collection dates were collected. To estimate herbaceous biomass, simple and machine learning techniques were tested. The results indicate that integrating optical indices (e.g. red edge position, simple ratio, normalized difference vegetation indices) and SAR backscatter improves the estimation of herbaceous biomass using machine learning techniques. The use of optical and SAR provides a potential of monitoring grass or herbaceous during wet and dry season. The information about herbaceous biomass is not only important for assessing food availability and carrying capacity for grazing animals, but also important for determining fire risk for potential hazard preparedness in the arid savanna.

KEYWORDS: Herbaceous or grass biomass, optical, SAR, rangelands, savanna

ISRSE37
TSHWANE 2017

INTEGRATION OF SATELLITE RAINFALL DATA AND CURVE NUMBER METHOD FOR RUNOFF ESTIMATION UNDER SEMIARID WADI SYSTEM

Osman Adam E[1], Ahmed M[2], Tesfamichael S[1], Ahmed F[4]

[1]Department of Geography, Environmental Management & Energy Studies, Faculty of Science, University of Johannesburg, [2]Senior Researcher, ARC, South Africa, [4]School of Geography, Archaeology and Environmental Studies, University of Witswatersrand

Arid and semiarid catchments in general require effective management strategies for better planning and land management. Surface runoff quantification of ungauged semiarid catchments is an important challenge in hydrology. A 7586 Km^2 Wadi El Hawad catchment is located in a semiarid region in central Sudan. Annual 250 mm rainfall is the ultimate source of water supply for the catchment. The catchment lacks the main inputs of hydrological modelling, climate and environment data. The objective is to parameterize hydrological characteristics and estimate runoff using model that suit the scarce geospatial information. The USDA Natural Resources Conservation Service (SCS) runoff curve number (CN) model is used to estimate runoff. Landsat data is used to generate landcover and soil information. ASTER 30 meter DEM data is used for catchment parametrization. Five days rainfall event (2 inches) is used as an input for the model. Data were compiled and processed in GIS. The model has resulted in estimating three runoff depths based on Antecedent Moisture Conditions (AMC). The dry AMC-I, estimated 0.14 inches, medium AMC-II resulted in 0.62 inches and wet soil AMC-III yielded 1.17 inches. Runoff depths of AMC-II and AMC-III indicate the possibility of having small artificial surface reservoirs that could provide water for domestic and small household agricultural use. An attempt has been made to overcome the problem of ungauged catchment's scarce data. Tropical Rainfall Measuring Mission (TRMM) satellite rainfall data is fed into SCS CN model to estimate surface runoff. Five days satellite rainfall (mean 1.03 inches) is used. The spatial variability of satellite rainfall has resulted in runoff variations. Only 1.7% area recorded runoff ranging between 0.8 to 0.4 inches within AMC-II condition, which is considered potential runoff. The advantage of using satellite rainfall data over the lumped rainfall data is the spatial variation.

KEYWORDS: Runoff, Semiarid, Wadi, TRMM, SCS-CN

INTERACTIVE TOOLS FOR ASSESSING GRAZING IMPACTS ACROSS AUSTRALIA'S ARID RANGELANDS USING VARIABILITY IN NON-PHOTOSYNTHETIC GROUND COVER

Barnetson J[1,2], Phinn S[1], Scarth P[1,3], Denham R[3]
[1]Joint Remote Sensing Research Research Program - University of Queensland, [2]Department of Environment and Natural Resources, [3]Department of Science, Information Technology and Innovation

Tools that assist land managers and policy makers to make informed decisions about ground cover management across the arid rangelands of the Northern Territory of Australia are needed. The open-source Python computing language was used to develop a dashboard of tools that explore the dynamics of ground cover change across the region. Dense time series analysis of the Landsat based fractional ground cover data set, in particular the non-photosynthetic component, involved the development of an automated time series change point detection method. Interactive use of the dashboard allows ground cover growth cycles between change points to be automatically identified, segmented and illustrated. Periods of greatest grazing impact produced by climate variability are identified and a measure of climate accounted grazing impact can be visualised. The method was successful in detecting and illustrating both long term (> 3 years) and short term (< 1 year) growth cycles and their identified impacted periods. Users can explore and select areas including those that have been assessed against field measurements of ground cover indicating a relationship between both field and remotely sensed ground cover. The exploration of big-data sets such as the Landsat archive is invaluable to gaining a better understanding of the temporal and spatial dynamics of ground cover and its relationships to climate and management. Practitioners across the study area will know be able to do so utilising tools developed specifically for arid and semi-arid environments.

INTRODUCTION TO THE NEXTGEOSS CONCEPT

Catarino N

Running from the end of 2016 until mid-2020, NextGEOSS will provide a streamlined, European hub for Earth Observation data discovery and retrieval, catering for European institutions as well as private sectors needs in terms of Earth Observation data, ICT and services. It will engage global communities of data providers and GEOSS users, promoting pilot activities aiming for innovative GEOSS powered applications for multiple societal areas. In this presentation a general overview of the NextGEOSS concept, approach, and methodologies will be given.

INVESTIGATING FOOD ENVIRONMENTS IN SELECTED AREAS OF TSHWANE

Schmitz P[1], Cooper A[1], Tshabalala N[1], Whittle T[1]
[1]*CSIR Built Environment*

A food environment is a combination of physical, biological, social and psychological factors that affect the eating habits and patterns of people. The food environment is determined by the availability, affordability and access to the food required for a nutritional diet. The characteristics influencing the food environment include the nature of the food retailers; transport network; physical barriers such as rivers, mountains, highways and industrial areas; cultural factors such as food taboos and the socio-economic profile such as poverty. The food environment influences what people eat, by constraining what consumers can purchase.

We report here on two preliminary investigations of food environments in Tshwane, South Africa. The one study was conducted in an informal settlement called 'Marry Me' in Soshanguve in the northern part of Tshwane. Informal settlements and food insecurity are a global concern and both are part of the South African community. Informal settlements do not have any form of planning and the structures are made of plastics, zinc material and hardboards.

The second study was conducted in Pretoria Gardens, an old, formal, low to medium-income suburb situated in north-central Tshwane. The physical barriers impeding access to Pretoria Gardens include a ridge, a railway line, a major road and commercial business district, an unused railway line and open spaces.

The results show that there is only one fruit and vegetable stall that provides healthy food to the residents of 'Marry Me' that is within a 400m walking distance, which is insufficient to provide nutritious healthy food to the community. Future research may include other variables such as food pricing, food quality, and be done in a larger scale to better understand the food environment in a township.

INVESTIGATING SAMPLING METHODS FOR VALIDATING LAND COVER PRODUCTS

Dlamini S, Mayekiso S, Ngcofe L

The results of accuracy assessment of remotely sensed data depend on the designed sampling methods used. Therefore, choosing proper sampling method is absolutely critical to generating an error matrix that is representative of the entire classified image. Sampling is the process of obtaining information from selected parts of the product (in this land cover product), with the aim of making general statements that apply to the quality of the product. Using the wrong sampling design can be costly and yield poor results.

Technological developments over the years have allowed for sampling techniques to be automated through GIS and RS software packages. This paper looks at two automated sampling methods which are Random points and Fishnet sampling. Random point sampling creates a specified number of random points within a given area while fishnet sampling creates regular grids and then places a point at the center of each grid.

This investigation of automated sampling method was to assess and analyse the impact of number of sample point distribution towards accuracy assessment of land cover mapping and also to determine the minimum required number of sample points for accuracy assessment. The study was conducted in the Eastern Cape Province of South Africa covering 3225D and 3127C square degree block.

The number of sample point investigated for land cover mapping validation was 25, 50, 100 and 900, points. The results revealed that 100 sampling points in both automated methods could sufficiently cater for all land cover classes available in the study area. While the 900 sample points could not bring improvement.The lesser sample points (25 and 50) reduced the number of land cover classes that were able to be assessed thus reducing the representation of classes.

LACO-WIKI: AN ONLINE TOOL FOR LAND COVER VALIDATION AND AREA ESTIMATION

See L[1], Perger C[1], Dresel C[1], Laso Bayas J[1], **Fritz S[1]**, Weichselbaum J[2]
[1]*International Institute For Applied Systems Analysis,* [2]*GeoVille Information Systems GmbH*

The validation of land cover and land use products derived from remote sensing is part of a quality assurance process in which the overall accuracy and the accuracy of individual classes is determined using ground reference data. Increasingly, reference data are derived from expert interpretation of very high resolution satellite imagery and aerial photographs due to the high costs associated with field data collection as well as open access to imagery through providers such as Google Earth and Bing. There is currently a lack of open tools available for land cover validation since validation processes are usually undertaken in-house with tools and procedures specific to a particular institute or organization. For this reason, we have developed the LACO-Wiki tool for online land cover and land use validation. A workflow of four simple steps has been encapsulated in the LACO-Wiki system as follows: (i) upload a raster or vector map for validation where the data can be categorical or continuous; (ii) create a validation sample, which can be random, stratified random or systematic; (iii) carry out the validation session by interpreting the sample using Google Earth, Bing imagery, OpenStreetMap, Sentinel-2 or images supplied by the user via a Web Map Service; and (iv) calculate the confusion matrix and accuracy statistics, presented as a report and/or raw data for downloading. In addition to validation, the tool is currently being modified to add area estimation functionality. The tool is aimed at a variety of users from researchers and students to map producers and commercial enterprises. In addition to land cover validation, the vision behind LACO-Wiki is to become an open access repository for calibration and validation data that can be used by the land monitoring community to improve future land cover products.

LAND COVER CHANGE DETECTION WITHIN GRANITE QUARRIES USING REMOTE SENSING

Moeletsi R[1], Tesfamichael S[1]
[1]*Department of Geography, Environmental Management and Energy Studies, University Of Johannesburg, Auckland Park, 2006*

Environmental impacts caused by mining activities on the ecosystem is a major issue on sustainable development and resource management. Quarrying, like any other mining activity has the potential to disrupt the environment and its related ecosystem, and thus needs mitigation that is informed by continuous monitoring. Remote sensing technique is a science of obtaining information about an object without being in contact with it. The techniques has been widely used in monitoring land cover changes while there is a need to intensify its utility in monitoring mining related environmental degradation. This paper aims at utilizing remote sensing techniques to assess and quantify environmental impacts caused by granite quarrying activities in Brits area of the North West province, South Africa. Quarries in this area were sampled based on their spatial coverage (> 1 ha) and by maintaining sufficient distance between them. Landsat images for the year 1998 and 2015 were acquired from the United State Geological Survey. Each image was classified using supervised classification and subsequently compared with each other to determine land cover change within the quarries. Results indicated a significant change in land cover around the quarries the study time period. Granite quarry area increased by approximately 1175 ha. This increase was accompanied by an increase in water bodies that are necessary for quarry activities. The increase in quarrying activity led to decreases in vegetation (a loss of 1308 ha) and bare land (a loss of 18 ha) around the quarry areas. These findings show the effect that granite quarrying has on environmental variables. In addition, the study demonstrates the potential of freely available Landsat data for continuous monitoring of granite quarrying. This potential should further be demonstrated by similar studies after satisfactory findings of which can be recommended for adoption routine monitoring purposes.

KEYWORDS: change detection, remote sensing

LANDSAT TIME SERIES ANALYSIS – THE IMPACT OF FOREST ECOSYSTEM HISTORY ON BIODIVERSITY

Graf W, Magdon P
[1]*University of Goettingen*

Biodiversity of forest ecosystems highly depends on heterogeneous forest structure and composition. Vegetation dynamics and management/conservation actions over time determine the current state of forests. Thus, we investigated the relationship of ecosystem history and biodiversity in temperate forests in Germany and addressed the following questions. Can trends, changes in trend and disturbances be detected in Landsat time series of temperate forests from 1985 to 2015, and do they affect herbal layer plant species diversity?

This study is embedded in the "Biodiversity Exploratories" – one of the leading long-term, large-scale projects on functional diversity. The three study sites in temperate mixed forests comprise 150 plots (100x100m) with low to high land use intensity, including biodiversity monitoring.

We used yearly time series of Landsat 4, 5, 7 and 8 from 1985-2015 to investigate ecosystem history. Images from the USGS (3365 images) and ESA (747 images) archives were combined to gain continuous time series of vegetation indices. We analyzed changes in trend with BFAST (Verbesselt 2010), monotonic trends with Mann-Kendall test (Mann 1945) and the relationships between diversity indices and trend parameters.

We found highly significant upward trends in NDVI time series for all plots and significant changes in trend for 5 times series. The changes can be explained by management actions and weather events. We found first indications on the relationship between trend parameters and biodiversity, demanding for further research.

Our analyses showed that Landsat time series of USGS and ESA archives can be used to analyze ecosystem history in temperate forests and contribute to a better understanding of forest dynamics and their impact on biodiversity.

LARGE-SCALE INDICATIVE MAPPING OF SOIL RUNOFF

Panidi E[1], Trofimets L[2], Sokolova J[1], Kunaeva E[3]
[1]Saint Petersburg State University, [2]Orel State University, [3]Pushkin Leningrad State University

We estimate in our study the relationships between the quantitative parameters of relief, the soil runoff regime, and the spatial distribution of radioactive pollutants in the soil. The study is conducted on the test arable area located in basin of the upper Oka River (Orel region, Russia). Previously we collected rich amount of the soil samples, which make it possible to investigate the redistribution of the Chernobyl-origin cesium-137 in the soil material and as a consequence the soil runoff magnitude at the sampling points.

Currently we are describing and discussing the technique of large-scale mapping of the soil runoff based on the cesium-137 radioactivity measurement in the different relief structures. Key stages are the allocation of the places for soil sampling points (we used very high resolution space imagery as a supporting data); soil samples collection and analysis; calibration of the mathematical model (using the estimated background value of the cesium-137 radioactivity); and automated compilation of the map (forecasting map) of the studied territory (the digital elevation model is used for this purpose, and the cesium-137 radioactivity is forecasted using quantitative parameters of the relief). Produced maps can be used as a support data for precision agriculture and for recultivation or melioration purposes.

LASER SCANNING IN SAVANNAS AND OTHER RANGELANDS FOR ESTIMATING BIOMASS AND OTHER VEGETATION STRUCTURAL PARAMETERS

Schaefer M[1], Sims N[2], Zhou Z[3], Abbott B[4], Vanderduys E[4]

[1]University Of New England, [2]Commonwealth Scientific and Industrial Research Organisation , [3]Data61, Commonwealth Scientific and Industrial Research Organisation, [4]Commonwealth Scientific and Industrial Research Organisation

Measurement of forageable biomass in grassland ecosystems is an important task for providing spatial information on forage mass as well as nutrition quality of vegetation available for livestock. The Compact Biomass Lidar (CBL), a terrestrial laser scanning (TLS) system has been deployed at the Spyglass Beef Research Facility near Charters Towers in Queensland, Australia, to develop non-destructive methods for grass biomass estimation that reduce the need for traditional biomass harvesting using destructive methods.

At each field site, thirteen laser scans from CBL were collected alongside traditional field measurements such as; destructive biomass harvesting of grass, grass height, handheld normalized difference vegetation index (NDVI) measurements, tree and ground cover intercept methods as well as tree stem diameter.

This research presents a case study at a single cattle station in Queensland Australia, where methods and analysis of terrestrial laser scanning data have been developed to estimate grassland biomass. The initial findings are at a plot by plot level are promising, however more research in other complex grassland regions around Australia during different seasons are needed to refine this work.

ISRSE37
TSHWANE 2017

LESOTHO LAND COVER DATABASE (2015)

Latham J[1]
[1]*Fao - Food and Agriculture Organization of UN*

The Land Cover Database of the Lesotho for the year 2014-2015 was prepared in the framework of the FAO Lesotho Resilience Strategy. Owing to the increasing frequency of climate-induced agricultural emergencies, FAO has designed an emergency and resilience programme to promote climate smart agricultural technologies throughout Lesotho.

The availability of detailed and reliable Land Cover Database is identified as one of the main requirements to support DRR and agriculture monitoring in the country. FAO implemented the Lesotho land cover database based on integral use of innovative geospatial technology to respond to the needs of the country related to: land cover change assessment, land degradation and erosion analysis, agriculture monitoring, area frame and statistics, etc. It provided technical assistance as the executing agency in close cooperation with all national parties.

The development of very high resolution Land Cover database has multiple uses and benefits including strengthening the national capacity to undertake consistent land mapping and assessments using standards, cutting edge technology and tools.

The immediate benefits are the provision of improved, timely and reliable information for decision making in the Ministry of Agriculture and Food Security (MAFS); Land Administration Authority (LAA); Ministry of Forestry and Land Reclamation (MFLR); Bureau of Statistics (BOS) and other national and multilateral organizations engaged in agricultural rehabilitation, poverty alleviation and food security programmes.

The ultimate beneficiaries are the Basotho's people in particular the rural people which suffer from significant food insecurity. They will benefit from better targeted programmes and policies to rehabilitate and further develop the agricultural sector, and hence reduce poverty and vulnerability to food insecurity. Through the provision of better information, short-term emergency assistance needs, and also longer-term rehabilitation and development needs, will be fulfilled more efficiently and effectively, ultimately reducing food insecurity, vulnerability and poverty.

KEYWORDS: Land Cover, Land Degradation, DRR, Agricultural Rehabilitation

LEVERAGE THE GOVERNMENTAL OPEN DATA INITIATIVES: STATUS IN AFRICA

Mogire O[1], Wafula M[2]
[1]*South Eastern Kenya Univesity*, [2]*JKUAT*

A sustained data revolution is needed to drive social, economic and structural transformation in African. Successful open data initiatives involve more than just putting datasets online but make data more accessible and re-useable. Open data initiatives can be led by governments proactively releasing data they hold, by private actors or civil society. To guarantee availability of open data over long-term and impact oriented, it must be rooted in clear policies, strategies and consistent global data management approach. In developing these policies, governments aim to stimulate and guide the publication of government data and to leapfrog sustainable developments. However, Open data is still in its very early stages in Africa without the support of clear policies and strategies. Again no country has truly demonstrated clear political leadership with defined policies and strategies on open data. Africa's many initiatives are presently resting on shallow foundations, at risk of stalling or falling if political headship or community pressure diminishes because they are not based on clear policies. The policies can be improved by collaborating with other organizations, focusing on the impact of the policy, stimulating the use of open data and creating a culture in which publicizing data is incorporated in daily working processes. Currently there is a multiplicity of open data policies at various levels of government, whereas very little systematic and structured research has been done on the issues that are covered by open data policies, their intent and actual impact in Africa. This paper looks the extent to which policies and strategies have been adopted, implemented,and evaluated in various governments and institutions handling data in Africa in order to leverage open government data initiatives.The paper outlines benefits accrued from open data policies.It also maps out specific challenges associated with legal, political, social,economic,institutional, operational and technical dimensions in open government data.

LEVERAGING TRADITIONAL FIELD SAMPLING, SPECTRAL UNMIXING, AND ALLOMETRIC MODELING IN MULTI-SCALE QUANTIFICATION OF WOODY BIOMASS IN KALAHARI SAVANNA ECOSYSTEMS

Meyer T[1]
[1]*University Of Texas*

Savannas are complex arid or semi-arid biomes, covering approximately 20 % of the terrestrial surface of the earth and providing substantial ecosystem services, such as carbon storage, firewood, rangeland and protection against soil erosion. This knowledge gap and its importance are only amplified by increasing anthropogenic influences, increasing human population, and increasingly varying climate change and the ongoing scientific debate regarding the role of savannas as a carbon sink or source. A key factor in understanding carbon cycling is the role of woody vegetation. To overcome the challenges associated with the quantification of biomass across large tracts of land, this work took on the challenge to extract biomass using a two-dimensional passive remote sensing system. Extensive field-derived measurements, including fractional cover of green vegetation (GV), non-photosynthetic vegetation (NPV), soil, vegetation structure and species for all woody plants, were collected in both wet and dry seasons over 15 sites along a 950km transect. These in situ data were used to develop, test, and validate a new method, the Spectral Line Point Intercept Transect – SLPT, to obtain spectral information efficiently across long distances. Multiple validation methods were leveraged to test the performance of various spectral unmixing techniques in heterogeneous savanna environments to derive fractional cover of GV,NPV, soil and shade at structurally different sites and in different phenological conditions (wet vs. dry season). The fractional cover of shade derived using Multiple Endmember Spectral Mixture Analysis (MESMA) and the Moderate Imaging Spectroradiometer (MODIS) Enhanced Vegetation Index (EVI) products were used to derive a relationship (ϕ) that identifies fractional cover of woody vegetation for a given MODIS pixel. By combining the MESMA, ϕ, and the allometric relationships described above, this research was able to develop a modelling approach that successfully quantifies biomass and woody vegetation cover across the Kalahari region.

LIDAR- BASED PREDICTION OF ARTHROPOD ABUNDANCE AT THE SOUTHERN SLOPES OF MT. KILIMANJARO

Ziegler A[1], Otte I[1], Röder J[1], Detsch F[1], Appelhans T[1], Brandl R[1], Nauss T[1]
[1]*Philipps-University Marburg*

Biodiversity research is often limited by the inverse relationship between grain (i.e. spatial resolution) and extend. Statistical upscaling mostly fails in heterogeneous landscapes if no spatial explicit variables are considered. One solution rests in the effective integration of in-situ and remote sensing observations. In the framework of the DFG RU KiLi (Kilimanjaro ecosystems under global change: Linking biodiversity, biotic interactions and biogeochemical ecosystem processes) aerial LiDAR (Light Detection And Ranging) observations have been conducted at the southern slopes of Mt. Kilimanjaro. The LiDAR dataset was utilized to derive structural parameters of the vegetation to predict the abundance of eight arthropod assemblages with several models. For the model training of each arthropod assemblage, different versions (extent, filters) of the LiDAR dataset were provided and evaluated. Furthermore the importance of each of the LiDAR-derived structural parameters for each model was calculated.

The best input dataset and structural parameters were used as predictors of the abundance of arthropod assemblages. Results across seven different landcover classes and eight arthropod assemblages exposed, that LiDAR-based predictions were best feasible for "Orthoptera" (average R^2 over all landuses: 0.14), while considering landcover classes, "disturbed forest" showed best results (average R^2 over all assemblages: 0.20), whereas "homegarden" was the least predictable (average R^2 over all assemblages: 0.04). Differentiated by arthropod-landcover pairs, the results revealed distinct differences and the R^2 values diverged clearly. If model settings were optimized for only one arthropod taxa, R^2 reach values up to 0.55 ("Orthoptera" in "disturbed forest"). The analysis of the importance of each structural parameter for the prediction revealed that about one third of the 18 used parameters were among the most important ones for the prediction of all assemblages. This strong ranking of parameters implies the need for a conscious selection of predictor variables in future research.

LIMITS TO GROWTH IN THE ALBANY THICKET BIOME: VEGETATION TRENDS AND PATTERNS FROM EARTH OBSERVATION PRODUCTS OVER THE ADDO ELEPHANT NATIONAL PARK AND EZULU GAME RESERVE

Palmer A[1], Gwate O[1]
[1]*Agricultural Research Council*

Developing and testing new models of evapotranspiration (ET) and net primary production (NPP) is essential if we are to provide evidence-based values of annual NPP and ET for rangelands in protected areas. These values will support interventions that ensure sustainability of these protected areas in relation to alternative land-use options currently challenging the value of rangelands over water demanding crops and land cover types. The potential role of Albany Thicket in carbon sequestration and climate regulation has resulted in substantial state expenditure on restoration. Defending the sustainability of a fully functional Albany Thicket biome over alternative land-use options (e.g. biofuel production) requires improved evidence-based methods for estimating ET, backed up by ground measurements of ET and NPP. Estimates of ET are also important in defining the health of an ecosystem and the quantity of water used by the vegetation when preparing a catchment-scale water balance. Models using both earth observation and weather station data were run for the Albany Thicket in the Addo Elephant National Park (AENP) and the Ezulu Game Reserve (EGR). We derived ET and NPP values for the Albany Thicket under a range of different land cover conditions. Modelled ET compared favourably with the measured ET data from eddy covariance on EGR and enabled us to model total annual ET in the absence of a complete year of eddy covariance data for AENP. Using this calibration, we show how the annual ET for three Albany Thicket condition classes (intact, sparse and degraded) in the AENP has changed under increasing elephant population in AENP.

LINKING GLOBAL SCALE REMOTE SENSING AND IN SITU MONITORING PROGRAMMES – GEO AND ILTER

Haubold H[1]
[1]*Environment Agency Austria*

Remote sensing applications rely on sound in situ data for calibration and validation/verification, as a reference background, and as a complementary source of information enabling holistic environmental monitoring approaches. This study investigates the specific opportunities arising from new operational links between two global programmes focussing primarily on remote sensing and on in situ observations, respectively, namely GEO (Group on Earth Observation) and ILTER (International Long Term Ecological Research), a network encompassing 42 national LTER networks across the globe. ILTER recently became a participating organisation of GEO.

Satellite based services deliver a whole range of biophysical parameters. ILTER sites collect both biodiversity related data and abiotic parameters, and ILTSER (the S stands for Socioecological) platforms also consider the human dimension regarding the changes of ecosystems, and possible degradation of their ability to provide ecosystem services to society. Therefore, the network can help to interpret biophysical parameters measured via remote sensing in terms of their ecological and societal relevance. Another specific example are phenology services benefitting from time series of soil temperature measurements to verify satellite derived SBD (Season begin day) across biogeographical gradients. Especially in Northern latitudes, the vegetation season starts before the snow cover completely melts. Only soil temperature measured by sensors under the snow cover can reliably determine the exact dates. Also land surface temperature is measured at almost all ILTER sites. Currently, long time series of ILTER data are used as an independent data set to validate a new global satellite based land surface temperature product. A range of further opportunities was worked out.

ILTER involvement in GEO can be three-fold: (1) Central calibration, validation/verification facility, (2) In-situ data provider for the creation of integrated data products, and (3) User of remote sensing derived products to better fulfil its own mission.

KEYWORDS: In-situ monitoring, environment, society, ecosystem.

LONG –TERM MONITORING OF THE GLACIERS IN WORDIE BAY, ANTARCTIC PENINSULA, USING MULTI-MISSION SAR TIME SERIES

Friedl P[1], Seehaus T[2], Wendt A[3], Braun M[2]
[1]German Aerospace Center (DLR), [2]University of Erlangen-Nuremberg, [3]Bavarian Academy of Sciences and Humanities

The Antarctic Peninsula is one of the world`s most affected regions by Climate Change.

We present results of a study at the former Wordie Ice Shelf, located at the south-western side of the Antarctic Peninsula. As the ice shelf disintegrated in a series of events since the 1970s, only disconnected tidewater glaciers have remained today. Due to the loss of the buttressing force of the ice shelf, the former tributary glaciers reacted with an acceleration of their flow speeds. While the loss of the ice shelf itself does not affect sea level, the increased outflow of the glaciers and the associated mass loss contribute to global sea level rise. All former studies conducted at Wordie Bay so far covered only relatively short investigation periods. Hence it was not well known how long the process of adaption to the changing boundary conditions exactly lasts and how it is characterized in detail.

We provide measurements of glaciological parameters (e.g. flow speeds, elevation change, grounding line positions) for the glaciers in Wordie Bay. For this purpose large datasets of previously active (e.g. ERS, Envisat, ALOS PALSAR, Radarsat-1) as well as currently recording SAR sensors (e.g. Sentinel-1, TerraSAR-X, TanDEM-X) were processed with different techniques and combined with data from other sources (e.g. optical images, airborne laser altimeter and ground penetrating radar data).

The high temporal resolution of our data reveals new insights into the glacier`s behavior after the disintegration of the former ice shelf. Therefore we are able to show that changes and adaption processes at Wordie Bay still go on. This is in contrast to the assumption of earlier publications which proposed that some of the glaciers in Wordie Bay may be approaching a new equilibrium.

KEYWORDS: Cryosphere, Glacier, Ice Shelf, SAR

LONG-TERM ATMOSPHERIC AEROSOL CHARACTERISTICS OVER NAMIBIA

Formenti P[1], **Piketh S[2]**, Namwoonde A[3], Klopper D[2], Adesina J[2], Holben B[4]
[1]LISA/CNRS/UPEC/UPD/IPSL, [2]North-West University, [3]SANUMARC/UNAM, [4]NASA/GSFC

The South East Atlantic off Namibia is the crossroad of aerosols of distant and local origins (biogenic, anthropogenic, biomass burning, sea salt and mineral dust) from continental and marine sources, with significant differences in terms of physico-chemical and optical properties, water affinity, scale and height of transport. The region is also characterized by an almost permanent, wide-spread stratocumulus deck topping the marine boundary layer as a result of low sea-surface temperature. It is therefore an ideal natural laboratory to investigate the interaction of clouds, aerosols and radiation which is highlighted as key climate uncertainties in the recent Intergovernmental Panel on Climate Change (IPCC) assessment report (Boucher et al., 2013).

In this general context, in this presentation we report on long-term measurements of aerosol concentrations and optical properties that have been performed since 2011 at two ground-based sites, Henties Bay (22.09°S, 14.26°E, elevation: 20m) and Gobabeb (23.56° S, 15.04° E, elevation: 405 m), representative of coastal marine and desert remote environments, respectively.

LONG-TERM LAND COVER AND LAND USE CHANGES IN THE KILOMBERO FLOODPLAIN, TANZANIA

Amler E[1], Kirimi F[1], Steinbach S[1], Thonfeld F[1], Hentze K[1]
[1]*University of Bonn, Remote Sensing Research Group*

Population growth and economic development drive the rapidly accelerating food demand and conversion from natural to cultivated land taking place in East Africa over the past decades. Wetlands are commonly viewed as future bread baskets. However, land cover information of the region is often inconsistent, outdated, incomplete or simply not available.

The objective of this study is to map land cover and land use (LULC) of the Kilombero catchment in Tanzania with the aim of understanding the seasonal wetland dynamics and long-term changes. Analysis was done at catchment and wetland scale. For the catchment scale, temporal compositing from Landsat scenes from 1993-1995, 2003-2005, and 2013-2015 was carried out to generate cloud free composites from this tropical region for the dry and rainy season respectively, and bi-temporal metrics (e.g., maximum, minimum, percentiles) were calculated. A two level classification approach was adopted. Level 1 distinguished water, bare soil, forests and other natural and agricultural vegetation. In level 2, the dynamic nature of the classes between the wet and dry seasons was introduced. Additionally, a time series classification with the higher resolution RapidEye images was performed at the wetland scale. The results show static and dynamic LULC changes over the last 2 years.

LULC for the wet and dry season were well identified and the changes between the two seasons clearly distinguished. At the wetland scale, the flooding extent in the rainy season was delineated whereas the dry season showed vast amounts of bare land. Such kind of information forms a basis upon which the agricultural managers can make decisions on increased and sustainable agricultural production to enhance food security.

KEYWORDS: wetlands, LULC, long-term changes, seasonality, Landsat

LONG-TERM LAND MANAGEMENT EFFECTS ON RANGELAND VEGETATION COVER IN THE STRZELECKI DESERT

Fisher A[1,2], Letnic M[2], Horn G[3]
[1]*University of Queensland,* [2]*Univeristy of New South Wales,* [3]*Office of Environment and Heritage*

The Strzelecki Desert is a large sandy desert in arid Australia's south east, with shrub covered dunes and grassy inter-dunal swales. It is divided into areas with different long-term land management practices. The building of a dingo-proof fence along the state border of New South Wales (NSW) in 1914-1917, combined with intensive trapping, shooting and poisoning has made the dingo rare on the NSW side of the fence where sheep grazing is common. Cattle grazing is common on both sides of the dingo fence, while large areas not subjected to commercial livestock also exist, such as Sturt National Park.

The removal of the apex predator from western NSW has had a dramatic ecological impact, with extensive field evidence of a trophic cascade. Decreased dingo abundance is associated with increased kangaroo abundance, increased abundance of feral foxes, and decreased small mammal abundance. Interpretation of aerial photography (1948-1999) has revealed an increase in dune shrub cover above that experienced outside the dingo fence. Understanding the spatio-temporal patterns of the ecological change is required to improve land management practices, and to assist a new project that plans to re-introduce locally extinct mammals.

We have examined the pattern of vegetation cover in the Strzelecki Desert using Landsat seasonal fraction cover time series (1987-2016). This data, provided by the United States Geological Survey, processed by the Joint Remote Sensing Research Program and distributed through the Terrestrial Ecosystem Research Network Auscover facility, reveals the continuing encroachment of shrubs on the dunes of western NSW. The data also show the contrasting vegetation cover dynamics of dunes and swales, and their response to larger sporadic rainfall events, across the different land management areas. Although the vegetation dynamics are complex and influenced by several factors, the satellite data provide a valuable context for ongoing ecological research.

LOW-COST, USER-DRIVEN SOIL MAPPING WITH AN INTELLIGENT 3U HYPERSPECTRAL EARTH OBSERVATION CUBESAT

Martinez P[1], Estebanez Camarena M[1]
[1]*University of Cape Town*

A sustainable and fruitful agriculture and land management framework must take into consideration the specific conditions and properties of soils used for agricultural production. Different soils require different treatments and bad practices (such an inappropriate use of fertilizers) can lead to environmental contamination and inefficient agricultural productivity. Yet, consistent soil information is not always available to farmers.

Hyperspectral Earth observation offers an alternative to the costly, labour intensive and time consuming conventional ground-based grid sampling of soils. However, the usual size of hyperspectral sensors and the vast amount of information to be downloaded normally restrict this capability to large satellites and high-performance ground stations. Highly skilled image processing experts are also needed to process the images in order to extract useful information products for end users. All these requirements limit the reach of the technology to a reduced number of users, mostly well-resourced, large-scale farmers.

This project aims to give some African nations and other communities with limited economic resources and ground infrastructure and expertise the means to perform low-cost user-driven soil analysis independently.

For this purpose, a 3U CubeSat is proposed and given the capacity to perform on-board the image processing tasks traditionally performed on the ground. The CubeSat can be built for under $100 000 and the associated onboard software tools are open source. The final product will be a compressed GIS map layer representing only the relevant information for a specific need. Alternatively, the user can download an image in a selected wavelength or the whole hyperspectral data cube.

Ultimately, this will improve agricultural productivity and reduce negative environmental impacts of ill-informed farming practices, thus contributing to the future of food security and environmentally sustainable agricultural practices.

KEYWORDS: Sustainable agriculture, Soil mapping, CubeSat, Low-cost, Image processing

LTHE DYNAMIC GLOBAL LAND COVER LAYER AT 100M RESOLUTION FROM COPERNICUS GLOBAL LAND

Buchhorn M[1], Smets B[1], Van De Kerchove R[1], Herold M[2], Tsendbazar N[2], Verbesselt J[2], Fritz S[3], Lesiv M[3]
[1]VITO - Flemish Institute For Technological Research, [2]University of Wageningen, [3]International Institute for Applied Systems Analysis

The Copernicus Global Land Service is the component of the European Copernicus service which ensures a global systematic monitoring of the Earth's land surface. It provides bio-geophysical variables in near real time describing the daily state, and changes in state, of vegetation, land surface processes and is currently preparing the release of a Moderate resolution Dynamic Global Land Cover layer.

In this presentation the methodology and rationale behind the Moderate resolution Dynamic Global Land cover layer is explained. This layer complements several global land cover 'epoch' datasets which have been created at medium (and high) spatial resolution during the last decade by providing a yearly dynamic land cover layer at 100m resolution. We will present the validated sub-product covering continental Africa for the year 2015. To build this global land cover layer, 100m spatial resolution PROBA-V data is used as primary EO data. Data fusion techniques are applied for areas with insufficient 5-daily PROBA-V100 m data and daily 300 m datasets are fused in. Next, time series metrics together with ancillary data sets (e.g. other Copernicus global land service biophysical products) are used in a supervised classification approach. Finally, at a third level, we build upon the success of previous global mapping efforts and focus on the improvement in areas where the thematic accuracy of the respective maps was insufficient to perform the final classification of each pixel. The map uses a hierarchical legend based on the United Nations Land Cover Classification System (LCCS). Compatibility with existing global land cover products is hereby taken into account, and extended by providing several cover layers. Training and validation data has been collected from multiple sources, among others by using existing reference datasets (e.g. GOFC-GOLD) and by running a crowdsourcing campaign through Geo-wiki.

KEYWORDS: global land cover, classification, Copernicus

MALARIA RISK ASSESSMENT THROUGH REMOTE SENSING AND MULTI-CRITERIA EVALUATION IN CENTRAL HIGHLANDS OF MADAGASCAR

Rakotoarison H[1], Piola P[2], **Rakotomanana F[1]**
[1]CELSIGS, Epidemiology Unit, Institut Pasteur de Madagascar, [2]Epidemiology Unit, Institut Pasteur de Madagascar

Indoor residual spraying is the adopted strategy for malaria control in the Central Highlands and Fringe regions of Madagascar. Remotely sensed data analysis combined with Multi-Criteria Evaluation become crucial to target priority areas for intervention. We aim to provide a decision making tool for the National Malaria Control Program.

Satellite images were used to update land cover information using object based image analysis method, NOAA and MODIS for temperature and rainfall data. Multi-Criteria Evaluation was performed by weighted linear combination to obtain the gradient of malaria transmission risk. Factor weights were determined by pair-wise comparison based on literature review and expert knowledge. Fuzzy set theory was used to perform the factors weighting. All the process was compiled in a semi-automatic plugin working in an open source software. Comparison of risk magnitude between two consecutive years was performed to assess the environmental change.

Two models of malaria risk gradient are available for 2014 and 2015. The updated land cover map showed suitable breeding sites for mosquito responsible of malaria transmission in Central Highlands with an accuracy of 84%. Risk change detection between 2014 and 2015 showed an increase of risk magnitude, with 1.3% and 7% respectively for low and high risk groups. We observed a decrease of 1.1% for the very low risk, 2% for moderate risk groups and 1.3% for class with very high groups. However, 87.4% of the area of interest remains unchanged in the risk status.

It is crucial to focus the indoor residual spraying efforts according to the risk gradient. This allows to increase the effectiveness of the intervention targeting areas with the most need, as well as to optimize financial and logistical resource management.

MANAGEMENT EFFECTS ON GROUND COVER "CLUMPINESS": SCALING FROM FIELD TO SENTINEL-2 COVER ESTIMATES

Scarth P[1]
[1]*Joint Remote Sensing Research Program*

Significant progress has been made in the development of cover data and derived products based on remotely sensed fractional cover information and field data across Australia, and these cover data sets are now used for quantifying and monitoring grazing land condition. With the advent of higher spatial resolution data, such as that provided by the Copernicus Sentinel 2 series of satellites we have the ability to look beyond reporting purely on cover amount and more closely at the operational monitoring and reporting on spatial arrangement of cover and its links with land condition.

We collected high spatial resolution cover transects at 10 and 20 cm intervals over a number of grazing trials and across critical reef catchments in Queensland, Australia. Spatial variance analysis were used to determine the cover autocorrelation at various support intervals. Very high resolution orthophoto mosaics were collected using a Go Pro camera and structure from motion software along the transects to link with satellite imagery. Coincident Worldview 3 and Planet Labs imagery were acquired over several of these sites, and Sentinel-2 and Landsat imagery was collected and processed over all the sites providing imagery with multiple ground resolutions.

We show that the effects of grazing intensity are apparent in variograms produced using PlanetLabs 3m imagery but the improved spectral characteristics of Sentinel 2 data at the lower 10m resolution allow us to retrieve similar information on spatial arrangement effects in response to management. We show that the spatial arrangement and temporal dynamics of cover are important indicators of grazing land condition for both productivity and water quality outcomes. The metrics and products derived from this research will assist land managers to prioritize investment and practice change strategies for long term sustainability and improved water quality, particularly in the Great Barrier Reef catchments.

MAPPING AND MONITORING OPEN WATER BODIES IN AN AGRICULTURAL AREA USING SAR INTERFEROMETRY AND POLARIMETRY

Kemp J[1]
[1]*Department of Geography & Environmental Studies, Stellenbosch University*

South Africa is a water-scarce country in which more than 60% of all available water is used for the irrigation of crops. This places enormous strain on available water resources, and nessecitates careful monitoring and accounting of available water and its use in the country.

Small farm dams provide water for irrigation and other agricultural activities. There is, however, growing concern and evidence that small farm dams serve as sediment traps, especially in dryland agricultural areas. This means that not only are these dams losing their water storage capacity, but that they run the risk of serving as multiple point sources of sediment to other reservoirs downstream. There is consequently a need for a reliable methodology to identifiy and map small farm dams so that their water levels and quality can be assessed and monitored.

Traditionally, optical satellite imagery is used for mapping water resources – a task which is enabled through the spectral range and resolution of multispectral sensors, but challenged by the spectral complexity of standing water. The use of SAR backscatter has been used in the past, but also suffers from some ambiguities in open water body detection related to issues such as the SAR incidence angle and local wind conditions.

In this study, high-resolution TanDEM-X SAR data collected in 2014 and 2015 is used to explore the value of interferometric and polarimetric information for mapping small farm dams in the Berg River Catchment in the Western Cape, South Africa. Object-based image analysis is combined with machine learning classification algorithms and feature selection to test the capacity of these SAR techniques for detecting and monitoring small farm dams. The results provide a basis for how such SAR data can be combined with optical data for further improvements in accuracy and efficiency.

KEYWORDS: water bodies, SAR, interferometry, polarimetry

MAPPING AND MONITORING WOODY VEGETATION COVER FOR NAMIBIA USING LIDAR TRAINING DATA, MACHINE LEARNING AND ALOS PALSAR DATA

Wessels K[1], Mathieu R[2], van den Bergh F[1], Main R[2], Naidoo L[2]
[1]CSIR-Meraka, Earth Observation Science and IT, [2]CSIR-NRE, Ecosystems, Earth Observation

Namibia has been experiencing undesirably high rates of bush encroachment which has negative economic impacts on livestock production and potential positive opportunities for biomass-based energy production. Current de-bushing efforts are carried out without reliable national maps of woody vegetation to aid planning and policy. The objective of our research was to develop a system which can map and monitor woody vegetation at national scale. The system generated training and validation data from a variety of airborne LiDAR data sets that were acquired for the purpose of utilities and transportation infrastructure planning. Estimates of fractional woody vegetation cover (higher than 1m) were derived from the LiDAR canopy height model for each 25m, L-band ALOS PALSAR pixels in the training and validation sets. ALOS PALSAR backscatter (HH, HV, 2010 and 2015 global mosaic), associated textural features (explanatory variables) and the fractional woody cover were used in a Random Forest model (i.e. bag of regression trees) to predict and map woody cover for northern Namibia (2010, 2015). The accuracy of the cover maps were assessed using a 30% subset of the LiDAR data. The initial woody canopy cover outputs of the system achieved an R^2 of 0.67, a mean absolute error of 13% and RSME of 18%. Although the backscatter of the ALOS PALSAR global mosaics were pre-processed to correct for terrain variation, the woody cover on steep slopes of hills were overestimated and required additional processing. Changes due to de-bushing were clearly captured between 2010 and 2015, while subtle increases in woody cover were harder to validate. The initial products have proven very useful in estimating the distribution of woody cover across northern Namibia. The system has undergone continuous refinement with experiments on alternative SAR features and LiDAR pre-processing being conducted at operational scales.

MAPPING COASTAL AQUACULTURE PONDS IN ASIAN HOTSPOTS WITH HIGH SPATIAL RESOLUTION SENTINEL-1A/B SAR DATA

Ottinger M[1], Clauss K[1], Wagner W[3], Kuenzer C[2]

[1]Department of Remote Sensing, University of Wuerzburg, [2]German Remote Sensing Data Center (DFD), Earth Observation Center (EOC), German Aerospace Center (DLR), [3]Department of Geodesy and Geoinformation, Vienna University of Technology

Aquaculture is one of the fastest-growing animal food production sectors worldwide, the main protein source in many countries and offers great potential for global food security. While capture fisheries production stagnated over the past years, aquaculture is already contributing nearly half of the total volume of aquatic foods in human consumption today. Rising demand and international trade has driven the rapid expansion of aquaculture with a remarkably increase from 20 Mio tones in 1994 to 74 Mio tones produced in 2014. Asia alone generates 90 percent of the total global aquaculture volume which is mainly produced by pond systems in fertile coastal environments. On the other hand, the development of aquaculture has often led to large-scale land use change, destruction and loss of coastal wetlands, and pollution of waters and soils.

We present an earth observation based approach to detect aquaculture ponds in study sites along the coast of China and Vietnam. The great advantage of active microwave instruments for aquaculture mapping is their all-weather, day and night capabilities which applies particularly for cloud-prone coastal areas. Topographic information derived from DEM data and coastline proximity were used to identify potential aquaculture areas. High spatial resolution time series of the free and open data acquired by the European Sentinel-1A and Sentinel-1B radar satellites were collected and processed to obtain temporally filtered images. Segmentation was applied to the filtered data to extract image objects and its features such as shape and size. The different backscatter response of the pond components (dikes and enclosed water surface) and its distinct rectangular structure allow for separation of aquaculture areas from other natural water bodies. Mapping of aquaculture ponds can effectively support sustainable development in the coastal zone and help to protect its valuable ecosystems.

MAPPING DISTRIBUTION OF WATER HYACINTH IN RWANDA USING MULTISPECTRAL REMOTE SENSING IMAGERY

Mukarugwiro J[1], Newete S[1], Adam E[1], Byrne M[1]
[1]*University Of Witwatersrand*

Water hyacinth (Eichhornia crassipes (Mart.) Solms) is an invasive aquatic macrophyte associated with major negative economic and ecological impacts in Rwanda and other east African countries since the plant's establishment in the country in the 1960s. Reliable estimates of water hyacinth distribution and its extent are required to determine the severity of the problem and identify the most infested water bodies requiring management action. To provide such estimates we processed and analyzed remotely sensed imagery based on Landsat 8 and Sentinel-2 images of Rwanda. A total of X points of geographical coordinates were also collected from each site of water hyacinth infestation to ground truth the remote sensing data. Water hyacinth coverage was quantified using a supervised and unsupervised image classification approach. Water indices such as NDWI (Normalized Difference Water Index), continuum removal of water absorptions and average reflectance in the SWIR (shortwave infrared) inputs were used to differentiate water hyacinth from other aquatic species using the high water content of its leaves. Results confirmed different levels of water hyacinth infestations in three major Rwandan rivers the Mukungwa, Nyabarongo and Akagera and most of the lakes in the Eastern province of the country. Among the lakes of the Akagera National Park, Lake Hago is the most infested, followed by Lake Rwanyakizinga. While among the lakes from Bugesera, Ngoma and Kirehe districts, Lake Rweru is the most infested compared to other lakes. These findings will assist the government of Rwanda, policy makers and their partners to put in place sustainable methods that can be used to manage and control the water hyacinth invasion.

KEYWORDS: Extent of water hyacinth, Multispectral, classification, NDWI, Ground-truthing

MAPPING FRACTIONAL WOODY COVER IN SEMI-ARID SAVANNAHS: DATA MINING BULK-PROCESSED LANDSAT AND ALOS PALSAR DATA

Higginbottom T[1], Symeonakis E[1], Mayer H[2]
[1]*Manchester Metropolitan University,* [2]*Philipp University of Marburg*

Effective monitoring of the Earth's ecosystems requires the availability of methods for quantifying the structural composition and cover of vegetation. This is especially important in heterogeneous environments, such as semi-arid savannahs, which are naturally comprised of a dynamic mix of tree, shrub, and grass components. The fractional coverage of woody vegetation is a key ecosystem attribute in savannahs, particularly given current concerns over the invasion of grasslands by shrub species (i.e. shrub encroachment), or the over-exploitation of woody biomass for fuelwood. Remote sensing has a clear role to play in monitoring semi-arid environments, and in recent years, the number of both spacebourne sensors and imagery acquired has increased dramatically allowing for data mining-based investigations. In this study, we investigated the potential of optical and radar-based remote sensing data for mapping woody canopy cover in sub-Saharan Africa savannahs, using the Limpopo Province of South Africa as a case study. A total of 92 variables were compiled, consisting of 90 Landsat spectral variability metrics and two PALSAR backscatter layers. These variables were used as input to a Random Forest-based work-flow that tested the impact of sensor combinations, seasonality, and scale on resulting predictions. Results showed that models at a 120 m scale produced considerably more accurate results than finer resolutions. PALSAR variables were consistently the most important predictors, but alone produced poor modelling accuracies. Using Landsat metrics, dry season data was the best predictor followed by annual, with wet season the worst performer. The bulk processing of the Landsat archive to generate spectral variability metrics provides a rapid method for mapping savannah woody cover. The potential for multi-sensor applications, incorporating ALOS PALSAR data, also offers improvements to monitoring efforts.

MAPPING GROWING STOCK VOLUME GLOBALLY USING ALOS PALSAR L-BAND SAR MOSAICS, OPTICAL CANOPY DENSITY MAPS AND ICESAT GLAS

Cartus O[1], Santoro M[1]
[1]*Gamma Remote Sensing*

While many studies have documented the sensitivity of, in particular, long wavelength radar backscatter observations to forest variables such as growing stock volume (GSV) or aboveground biomass (AGB), the large-scale application of spaceborne SAR for mapping forest resources and changes thereof over time faces a number of challenges, such as the sensitivity of the measurements to changing environmental conditions (e.g., freeze/thaw, moisture variations), forest structural differences, and the limited availability of insitu data for locally adaptive calibration of models, relating backscatter to GSV/AGB.

We present a new retrieval method, which bases on an algorithm first presented in Santoro et al. (2011), and which exploits the information provided by global optical canopy density maps (Hansen et al., 2013) and ICESAT GLAS lidar for spatially adaptive calibration of semi-empirical models, relating L-band backscatter to GSV, while minimizing the need for insitu data. We discuss the performance and limitations of a GSV retrieval using the global mosaics of ALOS PALSAR L-band backscatter released by JAXA for forest areas in the boreal, temperate, sub-tropical and tropical biomes. When comparing against regional GSV maps and insitu data, we find that the algorithm allows for capturing the major differences in the relationship of L-band backscatter to GSV across the different biomes and imaging conditions. Significant differences in the sensitivity of L-band radar at high ranges of GSV are observed. The modeling and data analysis indicate that with a single observation of L-and backscatter, a retrieval of GSV beyond ~200 m3/ha is not feasible in the tropics, whereas in temperate forests GSVs well beyond the saturation levels assumed so far seem possible (i.e., up to 500 m3/ha), albeit with high levels of uncertainty at the full resolution (~25 m) of the PALSAR mosaics.

KEYWORDS: growing stock volume, aboveground biomass, L-band SAR, global mapping, ESA DUE GlobBiomass

MAPPING PADDY RICE IN ASIA - A MULTI-SENSOR, TIME-SERIES APPROACH

Clauss K[1], Ottinger M[1], Wagner W[3], Kuenzer C[2]

[1] *Department of Remote Sensing, Institute of Geography and Geology, University of Wuerzburg,* [2]*German Remote Sensing Data Center (DFD), Earth Observation Center (EOC), German Aerospace Center (DLR),* [3]*Department of Geodesy and Geoinformation, Vienna University of Technology*

Rice is the most important food crop in Asia and the mapping and monitoring of paddy rice fields is an important task in the context of food security, food trade policy, water management and greenhouse gas emissions modelling. Asia's biggest rice producers are facing increasing pressure in terms of food security due to population and economic growth while agricultural areas are confronted with urban encroachment and the limits of yield increase. At the same time demand for rice imports is increasing, spurred by global population growth.

Despite the importance of knowledge about rice production the countries official land cover products and rice production statistics are of varying quality and sometimes even contradict each other. Available remote sensing studies focused either on time-series analysis from optical sensors or from Synthetic Aperture Radar (SAR) sensors. We try to address the sensor specific limitations by proposing a paddy rice mapping approach that combines medium spatial resolution, temporally dense time-series from the optical MODIS sensors and high spatial resolution time-series from the Sentinel-1 A/B SAR sensors.

We developed a method to use MODIS time-series and a one-class classifier to create medium resolution rice maps (doi:10.3390/rs8050434). In a next step we used these medium resolution rice maps to mask Sentinel-1 Interferometric Wide Swath images, which limits the amount of data to process and allows efficient rice mapping over larger areas. The high resolution rice masks are then created by segmentation of multi-temporal SAR images into objects, from which backscatter time-series are derived and classified. We created 10m resolution rice-maps that also allow seasonality extraction, given enough Sentinel-1 acquisitions. This method allows concurrent, accurate and high resolution mapping of paddy rice areas from freely available data. Results of our paddy rice classification will be presented for selected study sites in Asia.

MAPPING PRIMATE HABITAT AND OIL PALM PLANTATIONS USING LANDSAT OLI IMAGERY AUGMENTED WITH UNMANNED AUTONOMOUS SYSTEM PHOTOGRAPHY

Szantoi Z[1]
[1]*European Commission*

Conservation of primates' habitat is threatened by change in land use/land cover, due to the logging of their primary forest habitat, and such cover conversion to oil palm plantations. Such relatively rapid changes require frequent monitoring and it is very important to fast conservation interventions. Medium resolution satellite imagery (e.g., Landsat), however, provide data at a 30 m spatial resolution, permitting identification only general land cover classes, and limits the detection of small-scale deforestation. Here we combine Landsat imagery with very high resolution Unmanned Aircraft System (UAS). The UAS imagery was used as 'UAS truthing' data to train image classification algorithms. Our results show that UAS data can be used to help discriminate similar land cover/use classes (oil palm plantation vs. reforestation vs. logged forest) with consistently high identification of over 75% on the generated thematic map, where the oil palm detection rate was as high as 89%.

MAPPING PROSOPIS GLANDULOSA (MESQUITE) INVASION IN ABA ISLAND, SUDAN USING HIGH-RESOLUTION WORLDVIEW-2 IMAGERY AND MACHINE LEARNING CLASSIFIERS

Adam E[1]
[1]*Wits University*

Prosopis glandulosa (mesquite) is one of 44 Prosopis species of which 40 are native to the Americas. Both accidental and deliberate introductions of Prosopis outside of their ecological range have occurred, making them a significant alien weed. Over the last decade, new-generation multispectral imagery such as the RapidEye, Sentinel and WorldView (WV) series, with high spatial and spectral resolutions, have been used to monitor ecosystem changes as a result of climate change, agriculture, deforestation and spread of invasive species. This study investigates the ability of WorldView-2 imagery in mapping the invasion of P. glandulosa and coexistent indigenous species in Aba Island, Sudan, using the random forest and support vector machines as classifiers. Our results show that the eight-band multispectral WV-2 imagery is able to detect and distinguish P. glandulosa effectively from the three coexisting indigenous species of acacia, with an overall accuracy of 86% at 2 m spatial resolution. This result shows that high-accuracy can be achieved with the easily-available multispectral WV-2 sensor. This high-accuracy provides the possibility for economically-feasible mapping of the distribution and spread of invasive alien plants with similar accuracy as that of hyperspectral imagery.

KEYWORDS: Prosopis glandulosa; Invasive species; WorldView 2; Image classification; Random forest; Support Vector Machine.

MAPPING SMALLHOLDER AGRICULTURE USING SIMULATED SENTINEL-2 DATA; OPTIMIZATION OF A RANDOM FOREST-BASED APPROACH AND EVALUATION ON MADAGASCAR SITE

Lebourgeois V[1], Dupuy S[1], Vintrou E[1], Ameline M[1], Butler S[1], Begue A[1]
[1]*Cirad*

Thanks to the recent launch of Sentinel-2 satellite, access to a fine monitoring of the crops is now possible for smallholder agricultural systems. However, many constraints still need to be addressed, such as time-series incompleteness in cloudy environments, high intra- and inter-fields spatial variability, and large volume of data.

In this paper, we propose 1. to test a combined Random Forest (RF) classifier applied to multisource satellite data with an object-based approach, in order to produce land use maps at different levels (from cropland to crop type) ; 2. to optimize the performance of the RF classifier by reducing the number of input variables. The test site is a small agricultural zone in Madagascar. RF is applied by using two approaches: classical (by classifying the different levels from the whole learning database) and hierarchical (by classifying first the cropland and non-cropland and classifying crop types inside the cropland only). The importance (mean decrease accuracy measure) of the different variables is analyzed according to their source (high spatial resolution (HRS) time series simulating Sentinel-2 data, PLEIADES images, DEM), and category (reflectance, texture, spectral indices, auxiliary data). The contribution to the classification accuracy of each source andcategory of variables is also tested.

Results showed that optimizing the RF classifier allowed dividing the number of variables to be extracted by 1.5 to 6, depending on the classification level. Using a hierarchical approach improved the classification results whatever the level of classification and allowed to obtain 64.4% overall accuracy for crop type classification. The HRS time series derived spectral variables revealed to be the more discriminant, with a slight advantage of spectral indices over reflectances. Next step is the application of the optimized approach to a Sentinel-2 time series acquired over a large area in the context of the Sen2-Agri demonstration phase.

MAPPING SOIL MOISTURE CONTENT USING SENSTINEL-1 AND SENTINEL-2: CASE STUDIES FROM KENYA

Greifeneder F[1], Khamala E[2], Sendabo D[3], Zebisch M[1], Farah H[3], Notarnicola C[1]
[1]Eurac Research, [2]LocateIT Ltd, [3]Regional Centre for Mapping of Resources for Development (RCMRD)

Soil moisture content (SMC) is an important element in the global cycles of water, energy, and carbon and an indicator for natural hazards such as drought, floods and land-slides. In Africa, up to date reliable SMC data that adequately responds to national and sub-national level applications is often lacking. This study, therefore, leverages the synergy between in-situ and Earth Observation techniques to derive SMC at sub-national level in five counties in Kenya. These counties, namely Elgeyo Marakwet, Kajiado, Machakos, Narok, and Uasin Gishu have crop and livestock production as their dominant sources of livelihoods, with drought increasingly becoming a major threat.

Satellite remote sensing is one possibility for production of quick, direct, spatially continuous measurements of SMC. The new Copernicus family of Sentinel satellites affords us the aforementioned in levels of detail that adequately serve local to national purposes. The active Synthetic Aperture Radar (SAR) instrument aboard Sentinel-1 (S1) provides backscattering intensities that are directly related to the water content within the top soil layer. However, these measurements are influenced strongly by other surface parameter like roughness, vegetation, and topography. One of the important advantages of an active radar system is that these measurements are available independent of the Sun's illumination, clouds, and other atmospheric conditions. Sentinel-2 (S2), with its optical sensor, delivers detailed information on current surface conditions like land-use/land-cover and vegetation status.

To exploit the information gathered from these two satellites in a complementary way, we introduced a data driven machine learning approach based on Support Vector Regression (SVR). During the training phase, known combinations of S1 backscatter, S2 reflectance, and in-situ measured SMC (using time-domain reflectometry or TDR probes) were used to construct an empirical model that subsequently estimated SMC. Quantitatively, the estimated SMC showed a high correlation with the collected reference data (R=0.7, RMSE=4%).

MAPPING URBANIZATION BY MEANS OF ESA RADAR IMAGERY - THE SAR4URBAN PROJECT

Marconcini M[1], Üreyen S[1], Metz A[1], Esch T[1], Paganini M[2]
[1]German Aerospace Center (DLR), [2]European Space Agency - ESRIN

Starting from the beginning of the years 2000, more than half of the global human population is living in urban environments and the dynamic trend of urbanization is growing at an unprecedented speed. In this framework, an effective monitoring of urban sprawl represents a key issue to analyze and understand the complexity of urban environments and ensure a sustainable development of urban and peri-urban areas. To this purpose, the ESA SAR4Urban project aims at implementing a novel service that allows to automatically and reliably derive maps of past and current extent of urban areas by means of archived ERS/ASAR and novel Sentinel-1 data, respectively.

The rationale of the intended approach is that given a series of multi-temporal images for a given study area, the temporal dynamics of urban settlements are sensibly different than those of all other non-urban classes. Accordingly, after pre-processing all the images available over a region of interest in the selected time interval, for each pixel we derive key temporal statistics like the backscattering temporal mean, minimum, maximum, standard deviation, etc. Heterogeneity features are also extracted to ease the detection of lower-density settlements and, finally, specific unsupervised classification schemes are applied to ERS/ASAR and Sentinel-1 data, respectively.

Output of SAR4Urban include the 2002-2003 urban extent map of entire Africa derived from ASAR WSM data, as well as the urban extent maps of Athens, Beijing, Los Angeles, Mexico City, Atlanta and the Pearl River Delta derived from ERS-1/2 PRI and ASAR IMP scenes. Moreover, the current built-up extent of both these and several African cities will be delineated by means of Sentinel-1A imagery. Experimental results are extremely promising and confirm the great potential of ESA SAR data for mapping urbanization over time.

KEYWORDS: Urban Remote Sensing, Urban Extent Mapping, Radar, ERS, ASAR, Sentinel-1

MOBILE ATMOSPHERIC SENSING

Wang L[1], Huang Y[1]
[1]*School of Remote Sensing and Information Engineering*

Atmospheric quality dramatically deteriorates over the past decades around the metropolitan areas of China. Due to the coal combustion, industrial air pollution, vehicle waste emission, etc., the public health suffers from exposure to such air pollution as fine particles of particulates, sulfur and carbon dioxide, etc. Many meteorological stations have been built to monitor the condition of air quality over the city. However, they are installed at fixed sites and cover quite a small region. The monitoring results of these stations usually do NOT coincide with the public perception of the air quality. This paper is motivated to mimic the human breathing along the city's transportation network by the mobile sensing vehicle of atmospheric quality. To obtain the quantitative perception of air quality, the Environmental Monitoring Vehicle of Wuhan University (EMV-WHU) has been developed to automatically collect the data of air pollutants. The EMV-WHU is equipped with GPS/IMU, sensors of PM2.5, carbon dioxide, sulfur dioxide, anemometer, temperature, humidity, noise, and illumination, as well as the visual and infrared camera. All the devices and sensors are well collaborated with the customized synchronization mechanism. Each sort of atmospheric data is accompanied with the uniform spatial and temporal label of high precision. With the EMV-WHU, constant collection of the atmospheric data along the Luoyu Road of Wuhan city has been conducted at the daily peak and non-peak time for half a year. Experimental results demonstrated that the EMV is very efficient and accurate for the perception of air quality. It is promising for the aerial and emergent air quality monitoring over the sky of big cities, if EMV-WHU be miniaturized for the unmanned aerial vehicles(UAV) in the future.

KEYWORDS: mobile atmospheric sensing, Environmental Monitoring Vehicle (EMV), synchronization, spatial analysis

MODELLED HABITAT SUITABILITY OF A MALARIA CAUSING VECTOR (ANOPHELES ARABIENSIS) RELATES WELL WITH HUMAN MALARIA INCIDENCES IN ZIMBABWE

Gwitira I[1], Murwira A[1], Masocha M[1], Zengeya F[1], Mutambu S[1]
[1]*Unviersity of Zimbabwe*

Accurate modelling of the geographic distribution of disease vectors is an important step towards developing strategies for effective control of vector borne diseases. In this study, we used maximum entropy (Maxent) to develop a spatially explicit model to predict the habitat of a malaria causing vector, Anopheles arabiensis, based on key environmental factors. Our results show that altitude combined with isothermality, temperature seasonality, annual precipitation and precipitation of the wettest month can be used to successfully model habitat suitability of Anopheles arabiensis. Based on these five key factors, our results show that areas that are highly suitable for Anopheles arabiensis are generally in the north, northeast, south and south eastern parts of Zimbabwe. In fact, our results show that all the five factors had AUC values >= 70% which is classified as good for predictive purposes. The results of our Maxent model overally show AUC values of 0.84 for training and 0.88 for test data. In addition, our results also show that the habitat suitability model positively correlated ($p<0.05$) with malaria incidences recorded at health facilities for the period 1997, 1998 and 1999 although the correlations are weak. Our results suggest that Anopheles arabiensis habitat suitability can be used as an indicator of malaria incidences.

KEYWORDS: Bioclimatic, Anopheles, Habitat, Maxent, AUC, Malaria

MODELLING CARRYING CAPACITY IN THANDA GAME RESERVE, SOUTH AFRICA USING LANDSAT 8 MULTISPECTRAL DATA

Oumar Z[1], Oosthuizen Botha J[1], Adam E[2], Adjorlolo C[3]
[1]KZN Department of Agriculture & Rural Development, [2]University of Witwatersrand, [3]South African National Space Agency

Rangelands which consist of grasslands, shrublands and savannahs are used by wildlife for habitat and are the main source of forage for livestock. The assessment and monitoring of rangeland condition is one of the most important factors for rangeland scientists in order to calculate the carrying capacity of livestock with consideration for coexisting wildlife. This study assessed the potential of Landsat 8 multispectral bands and broadband vegetation indices to model tree equivalents (TEs) and total leaf mass (LMASS) at Thanda Game Reserve using partial least squares regression (PLSR). The PLSR model predicted TEs with an R^2 value of 0.76 and a root mean square error (RMSE) of 1411 TE/ha using an independent test dataset. LMASS was predicted with an R^2 value of 0.67 and a RMSE of 853 kg/ha on an independent test dataset. The predictive models were then inverted to map TEs and LMASS over the study area. The modelled TEs and LMASS layers were integrated with conventional grazing and browse capacity models to map carrying capacity over the Game Reserve. The study indicates the potential of Landsat 8 multispectral data in carrying capacity modelling. The result is significant for rangeland monitoring in Southern Africa using remote sensing technologies.

MODELLING COFFEE (COFFEEA ARABICA) CROP CONDITION IN HETEROGENEOUS AGRICULTURAL LANDSCAPES USING AN INTEGRATED HIERARCHICAL MULTISENSORY APPROACH

Chemura A[1], Mutanga O[1]
[1]*University Of Kwazulu Natal*

Coffee (Coffea arabica) is an important commodity crop that can be produced over large areas in tropical areas. Long term monitoring of growth vigour and for pest and disease incidences is therefore an important aspect of crop management. Current field-based crop assessment methods are subjective, expensive and arduous, and thus remote sensing can be used to alleviate these challenges. However, since the crop is in the field for many years and throughout the year, long-term crop monitoring is difficult using a single sensor as its availability or quality is not guaranteed due to revisit times and cloud cover. In this study we evaluated the utility of combining MODIS 250m, Landsat 8 OLI and Sentinel 2 data into an integrated monitoring model for coffee plantation conditions focusing on growth patterns at landscape scale for two distinct sites. The different sensors produce spatial resolution and spectral integrity mediated profiles that can be reconciled for sequential monitoring of coffee conditions between seasons. Although optimum results are produced by the vegetation indices derived from Sentinel 2 sensor, curves produced by both Landsat 8 OLI and MODIS 250m NDVI can be adjusted to provide data for times where each sensor may not be available or available in insufficient quality. In addition, local factors play a significant role in determining differences between sensor performances and have to be accounted for in integrating multiple sensors in modelling coffee conditions with remote sensing.

KEYWORDS: multi-sensory data, spatial integrity, coffea arabica, crop condition monitoring

MONITORING ABOVEGROUND WOODY BIOMASS STOCKS IN DIFFERENT FOREST TYPES WITH L- BAND ALOS 2 PALSAR 2 DATA

Ndyamboti K[1,2], Mathieu R[1,2], Wessels K[2,3], Naidoo L[1,2], Main R[1,2], Mahlangu N[1,2]

[1]Ecosystems Earth Observation, Natural Resources and the Environment, CSIR, Pretoria 0001, South Africa, [2]Department of Geography, Geomatics and Meteorology, University of Pretoria, Pretoria 0028, South Africa, [3]Remote Sensing Research Unit, Meraka Institute, CSIR, Pretoria 0001, South Africa

The mapping of forest aboveground biomass (AGB) stocks is critical for understanding the terrestrial carbon budget. In this study, we investigated the potential of multi-seasonal, dual-polarized (HH and HV) L-band ALOS 2 PALSAR 2 data to model woody biomass across different forest types (i.e. savanna, plantations, and indigenous forests) in South Africa. We used the Random Forest (RF) regression algorithm to determine the extent to which (i) backscatter intensity, (ii) texture metrics, and (iii) polarization ratios could be used to model AGB in each forest type. All three image variables (backscatter intensity, image textures, and polarization ratios) were first used independently then jointly in the modelling process in order to determine the predictive strength of each variable and their combinations. We further investigated the extent to which topography influences AGB retrieval. In situ data consisted of 56 × 1 ha plots distributed across the three forest types. Our preliminary results show that backscatter intensity (HH and HV) exhibits good relationships with woody biomass in savannas (R2 = 0.64, RMSE = 13 t/ha) and plantations (R2 = 0.59, RMSE = 51.1 t/ha), with moderate performance in indigenous forests (R2 = 0.49, RMSE = 17.6 t/ha). A comparative analysis of the two PALSAR 2 polarisations shows that HV backscatter performs slightly better than HH. Polarization ratios also exhibit promising results whereas texture parameters show the least correlation to AGB. The use of forest plots positioned at slopes greater than 15 degrees reduces model performance with the R2 value dropping by as much as 10% in plantation forests. Overall, the study demonstrates the potential of PALSAR 2 data to monitor AGB stocks in different forest ecosystems in Southern Africa.

KEYWORDS: Aboveground biomass, ALOS 2 PALSAR 2, forest, multi-season, Random Forest

MONITORING AND MODELLING VEGETATION PHENOLOGY FROM SPACE

Verstraete M[1]
[1]*University Of The Witwatersrand*

Phenology is the study of the timing of significant ecological events marking the development of living organisms in response to seasonal forcing from climatic variables such as light, temperature or humidity. Many plant and animal species exhibit yearly cycles of growth, development, reproduction, etc., and the timing of these events is crucial to the functioning of the environment as a system. Seasonal changes in the terrestrial biomass are important enough to modify the concentration of CO2 in the atmosphere. Climate changes are expected to disturb these cycles and perturb the life cycles of inter-dependent species. Monitoring phenology is critical to understand the nature, extent and impact of these changes, and hence to inform evidence-based policies to address this issue. While phenological records have been collected in situ for centuries, satellite-based remote sensing offers a unique opportunity to monitor key phenological events on a broad range of spatial scales (from local to global), over periods of multiple decades, and using standardized approaches to facilitate comparisons. This presentation reviews recent scientific and technological developments in this field, focusing on the interplay between modeling and observing, and underscores the advantages and challenges associated with this approach.

MONITORING COASTAL STORM EROSION AND RECOVERY USING SATELLITE IMAGES

Liu Q[1], Trinder J[1], Turner I[1]
[1]*The University of New South Wales*

We use three sets of high-resolution satellite images to monitor a major East Coast Low storm event that occurred in early June 2016 and caused widespread beach erosion along the south east coastline of Australia. Focusing on the 3.6 km long Narrabeen-Collaroy embayment where detailed and immediate pre- and post-storm surveys obtained by airborne Lidar and drone photogrammery were available for ground-truthing, experimental images consist of a WorldView-3 and a PlanetScope image before the storm, as well as a Pleiades image after the storm. Firstly, a Region Of Interest (ROI) is automatically generated using the Normalised Difference Water Index and morphological operations. Superpixel segmentation is then applied to divide similar pixels into small segments called superpixels. Subsequently, a supervised classifier is applied to distinguish water, sandy beach and land areas, using position, intensity and shape features of each superpixel. The instantaneous waterlines are then extracted based on the classified images.

Since the instantaneous waterline location is affected by the astronomical tidal elevation and wave conditions at each image acquisition time, a waterline elevation model is employed to determine the corresponding elevation of the waterline. Empirical parameterizations are used to estimate wave setup and swash oscillations. Consequently, the horizontal locations of the instantaneous waterlines are corrected to shorelines at Mean Sea Level. The results show that the extracted shorelines are able to reveal and quantify the significant beach erosion that was observed. The waterline elevation model is shown to be a useful approach for the wave and tidal correction of the shoreline. In addition, spatial differences along the beach due to the differing degree of wave exposure during the storm are also identified and quantified. This case study indicates the suitability of high-resolution satellite images to monitor short-term shoreline changes efficiently and accurately.

KEYWORDS: shoreline monitoring, satellite images, waterline elevation model

MONITORING DIMENSIONS OF BIODIVERSITY IN A MEGA-DIVERSE REGION OF SOUTHERN AFRICA: FROM TRAITS TO COMMUNITIES TO ECOSYSTEMS

Wilson A[1], Merow C[4], Frye H[2], Silander J[2], Slingsby J[3]
[1]University at Buffalo, [2]University of Connecticut, [3]South African Environmental Observation Network (SAEON), [4]Yale University

The United States National Aeronautics and Space Administration (US NASA) regularly conducts 'field campaigns,' to collect airborne and field observations of Earth's physical, chemical, and biological systems. We are developing plans for a notional field campaign to measure and monitor the distribution and abundance of biodiversity across the Greater Cape Floristic Region (GCFR) of southwestern South Africa. The GCFR includes several formally defined biomes, including the fire-prone Fynbos shrubland and arid Succulent Karoo biomes, with species diversity rivaling that of mega-diverse tropical rainforests in an extremely compact area (300x700km). Off the coast, South Africa's oceans include both temperate and tropical marine biomes with high net productivity and endemism. At the interface of land and sea are extensive freshwater ecosystems that rival the GCFR in levels of species endemism and are a critical source of water in a water scarce nation. The GCFR also contains one of the highest proportions of species of global conservation concern; extinction risk studies suggest that GCFR species are among the most vulnerable to climate change over the next 50 years. The field campaign would include collection of new high resolution (~20m) hyperspectral imagery from AVIRIS-NG, PRISM, and HyTES spectrometers, combined with the LVIS laser altimeter. In combination with the rich historical data and well-developed ecological understanding in this region, these new observations would enable detailed interdisciplinary exploration into the drivers and mechanisms of change including feedbacks from changing biodiversity to regional climate, disturbance, post-fire recovery, freshwater provisioning, and other ecosystem services. As the field campaign is developed, we are looking for feedback and suggestions to maximize the scientific utility of the data collection and identify opportunities for collaboration in the ISRSE community.

MONITORING EXTREME HYDROLOGICAL EVENTS USING L-BAND MICROWAVE: THE CASE OF DROUGHTS AND FLOODS

Al Bitar A[1], Wigneron J[2], Kerr Y[1], Al-Yaari A[2], Fan L[2], Mialon A[1]
[1]CESBIO, [2]ISPA

In this presentation we show the capabilities of L-Band observations from microwave radiometers provided by Soil Moisture and Ocean Salinity (SMOS) mission in the developments of regional scale early warning systems for the predictions of agricultural droughts and heavy rainfall floods. The intensification of extreme events frequency and intensity at global scale was one of the early predictions of Intergovernmental Panel on Climate Change (IPCC) that is being confirmed currently. This emphasis the need of improved disasters early warning systems based on remote sensing. We consider here the case of flood prediction from heavy rainfall and agricultural drought predictions. For the first case we show how the flood early warning systems based on the curve rank approaches are leveraged using current soil moisture conditions from L-Band radiometers. The proposed algorithm modifies the forecasted probabilities of flood events with the use of soil moisture percentiles. We show how the number of hits is improved in flood prone areas by comparing the outputs to existing flood events databases from Dartmouth Flood Observatory. For the second case we show how a root zone soil moisture can be derived from L-Band soil moisture namely from SMOS and SMAP missions. The product is then used to derive a drought index at global scale. We show how the use of the drought index can provide an early warning information compared to vegetation or optical sensors based index. Also an extension of drought monitoring to forest fire monitoring is demonstrated. A focus is made on the 2015 and 2016 events. Last we show the teleconnections between the depicted flood and drought events in the light of the current ENSO. As a conclusion this work shows how L-band observations is improving our knowledge on extreme events and providing early prediction algorithm which has a high societal interest.

MONITORING FOREST DEGRADATION IN THE TROPICAL DRY FORESTS OF SOUTH-CENTRAL AFRICA

De Cauwer V[1], Stellmes M[2], Knox N[1], Roeder A[3], Revermann R[4]
[1]*Namibia University Of Science And Technology (nust),* [2]*University of Berlin,* [3]*University of Trier,* [4]*University of Hamburg*

Forest degradation is generally considered as a process whereby the ecosystem value of a forest area decreases. In Africa, this process is driven by fire, grazing and harvesting of forest resources. Detection and monitoring of forest degradation is complicated; for practical reasons, most studies focus on indicators that can easily be measured, especially forest cover through remote sensing. The use of forest cover as a measure of degradation is appropriate in areas where no silvicultural practices affecting forest cover take place, as in most of south-central Africa. Detection of a decreasing forest cover is however not evident in the open forests in the border area of Angola and Namibia. Tree canopy cover ranges between 10% and 40% which makes the detection of forest versus non-forest areas already difficult. This study aimed to identify areas with a relatively lower tree cover and basal area compared to their surroundings as those areas are mainly characterised by secondary vegetation lacking the important fruit and timber trees. Phenological variables derived from time series of MODIS images for the period 2000 to 2012 were related to tree cover and basal area derived from ground-based studies to identify degraded areas. A model was tested that integrated climatic, topographic or anthropogenic related variables with the MODIS data. Anthropogenic related variables include indirect measures of human impact, such as the distance to settlements and roads. Results confirm a decreasing trend in the basal area and forest cover related to the mean annual rainfall. Degraded areas were identified along this gradient and degradation trends within the period 2000-2012 were analysed for some of the degraded areas. Mapping forest degradation with moderate resolution images provides a benchmark for future or more detailed assessments in this mainly remote and inaccessible area.

KEYWORDS: forest cover, basal area, ecosystem model, MODIS

MONITORING INUNDATION DYNAMICS WITH THE NASA-ISRO SYNTHETIC APERTURE RADAR (NISAR)

Chapman B[1], Kellndorfer J[2], Saatchi S[1], Simard M[1], Siqueira P[3]
[1]Jet Propulsion Laboratory, [2]Earth Big Data, LLC, [3]University of Massachusetts

The NASA ISRO Synthetic Aperture Radar (NISAR) mission is scheduled for launch in 2021 by the ISRO GSLV Mk. 2 launch vehicle, and is now in its implementation phase. NISAR will carry both an L-band and S-band Synthetic Aperture Radar (SAR), and will utilize the "SweepSAR" technique to achieve both high resolution (3-10m) and wide swath coverage (240 km image swath). While NISAR will be fully polarimetric, the standard land observing mode will be dual polarization (HH and HV). Operations are planned to consist of acquiring L-band SAR data during every orbit over most land surfaces, as well coastal and polar sea regions. Since NISAR will have a 12-day exact repeat orbit, most land areas on Earth will be imaged twice every 12 days. S-band data can be acquired simultaneously with L-band; however, S-band data collection will be more limited.

This data set will be immensely valuable for monitoring inundation dynamics. For decades, it has been recognized that L-band SAR data is sensitive to not only open water but to inundated, vertically emergent vegetation as well. Because cloud cover, rain, and solar illumination are not relevant to observational success, L-band SAR can be used to observe inundated conditions at any time, limited only by the observational capabilities of the instrument. NISAR's twice-in-12-days observational scenario will produce an unprecedented data set for understanding inundation dynamics for globally distributed wetlands. Since NISAR will have a free and open data access policy, this imagery will be available to the entire wetlands community.

In this presentation, we will discuss the NISAR accuracy requirement for detecting inundated conditions, the preliminary algorithm, and possible methodologies for validation.

This work was partially performed at the Jet Propulsion Laboratory, California Institute of Technology, under contract with the National Aeronautics and Space Administration.

KEYWORDS: NISAR, SAR, inundation, wetlands,L-band

MONITORING KEY LANDSCAPES FOR CONSERVATION WITHIN THE GLOBAL LAND COMPONENT OF THE COPERNICUS LAND SERVICE: THE AFRICAN CONTINENT

Szantoi Z[1]
[1]*European Commission*

Protecting African biodiversity continue to be an enormous challenge. During the past decades the continent lost its flora and fauna at a shocking rate. Pressure on natural resources has increased significantly and it is not showing any signs of slowing down due to the ongoing demographic boom and economic activities. Thus, Key Landscapes for Conservation such as protected areas have to be constantly monitored to assess their effectiveness and adjust their protection level. However, not only such areas, but also larger ecological (and economical) zones and the wildlife corridors between them should be monitored as these are having an even higher level of pressure and thus, habitat loss. The main objective of this monitoring activity is to provide detailed land cover and use information on selected areas globally. The monitoring will focus particularly within the domain of the sustainable management of natural resources. For such areas of interest, including protected areas or hot spots for biodiversity and land degradation, land cover and land cover change products will be developed and distributed through the Copernicus Global Land Monitoring Services program.

MONITORING MINING INDUCED LAND USE AND LAND COVER CHANGES IN EMALAHLENI MUNICIPALITY, SOUTH AFRICA USING REMOTE SENSING

Mhangara P[1], Adjorlolo C[1]
[1]*South African National Space Agency (SANSA)*

Mining is the mainstay of the South Africa economy and its contribution to the Gross Domestic Product (GDP) and job creation is well established. Satellite based remote sensing can play a significant role in mapping and monitoring the land use and land cover changes in mining districts. High resolution satellite remote sensing has been proven to be an invaluable tool in monitoring mining developments and the associated human settlements dynamics. Whereas the economic benefits of mining are irrefutable, the long term negative externalities that accompany mining are often unqualified. Land use competition between mining, agriculture, urban development and ecosystem services is evident in coal mining areas of Mpumulanga Province in South Africa. Quantifying the land use and land cover changes provides critical insights in qualifying the role of mining in the shifting land use patterns and its undesirable impacts on prime agricultural land and detrimental impacts of pristine ecosystems. Using Emalahleni Municipality, located in Mpumalanga Province in South Africa as a case study, the objective of this study was to quantify the land use and land cover changes in the area using satellite imagery and to examine the impacts of mining on agriculture, urban development and ecological sustainability. The preliminary results of this study indicate that the growth mining was positively interlinked with urban development pointing to the economic benefits of mining. The results also indicate that mining growth was increasingly encroaching on agricultural land posing a threat to food security.

KEYWORDS: Mining, Satellite-based remote sensing, agriculture, land use and land cover change, urban development

MONITORING OF HUMAN PRESENCE AND EXPANSION IN THE SAHELO-SAHARAN ZONE – A CASE STUDY FOR NIGER BASED ON MULTI-SENSOR EARTH OBSERVATION

Esch T[1], Bernd A[2], Pinchon S[2], Rabeil T[3], Uereyen S[1], Wegmann M[4]
[1]German Aerospace Center (DLR), [2]Noé Conservation, [3]Sahara Conservation Fund, [4]University of Würzburg

The global human population is continuously growing, with the highest dynamics occurring in Africa, Asia, and Central and South America. The resulting expansion of built-up area and man-made infrastructure into the natural and semi-natural hinterland is a key challenge to nature conservation and environmental management. Hence, a precise knowledge on the location of human settlements and infrastructure inside or in the vicinity of high-value and/or vulnerable biomes is vital for the development of effective strategies to protect wildlife against encroachment from anthropogenic activity.

A promising approach to detect and monitor man-made structures is the analysis of Earth observation (EO) imagery. However, considering human settlements layers (HSL) for developing countries, the capability of existing data sets to accurately delineate small and scattered towns and villages or industrial complexes is quite limited. Therefore, recent EO-based initiatives have started with the aim to provide more accurate HSL based on high resolution EO imagery. One of these new data sets is the Global Urban Footprint (GUF) layer that has been derived from very high resolution radar imagery collected by the German TanDEM-X mission in the years 2011-2013. The GUF data represents a binary raster layer indicating built-up and non-built-up areas in a spatial resolution of 75 m at global scale.

In this study we analyse the potential of TerraSAR-X/TanDEM-X data and the GUF layer in combination with additional satellite imagery from sensors such as Envisat-ASAR and Landsat to monitor the development of man-made infrastructures in Niger which are threatening the ecosystem intactness. Various time series collections of EO data sets from multispectral as well as radar-based satellite systems for Niger are used together with available in-situ data to detect the man-made built-up structures and their spatiotemporal development from 2007 to 2016. The results obtained demonstrate the importance of Earth Observation data for conservation.

MONITORING OF OIL SPILL FIRE EVENTS USING MODIS SATELLITES

Martinez P[1], Ogungbuyi M[1], Eckardt F[1]
[1]*University of Cape Town*

Ten thousand ground based oil spill events were recorded for the Niger Delta of Nigeria in the period from 2007 – 2015. These oil spill events are recorded by the National Oil Spill Detection and Response Agency (see www.oilspillmonitor.ng). Spills from oil facilities (i.e. flowlines, oil wells, flow stations, and pipelines) are caused by operational error, equipment failure and by acts of deliberate sabotage or illicit tapping of oil pipelines. The latter two causes of oil spills are often accompanied by explosions and fires. This study aims to identify whether accidental or deliberate oil fires may be sufficiently large to be detected by 250 m MODIS fire products. Examining the MODIS record vis-à-vis the Nigerian spill data base in a Geographical Information System (GIS) suggests that ground spill events from the data base and MODIS active fires are spatiotemporally correlated. We demonstrate that some events are detected by MODIS. Additional remote sensing evidence includes burnt vegetation, as well as oil spill leaks. The ultimate goal of this work is to develop a near-real-time warning system which identifies oil spill fires from MODIS fire products and provides warning signal for oil seepage and fire events.

MONITORING POST-MINING REHABILITATION USING MULTI-SENSOR SAR DATA: FIRST RESULTS

Haupt S[1], Engelbrecht J[1], Kemp J[2]
[1]*CSIR Meraka Institute*, [2]*Department of Geogrpahy and Environmental Studies, Stellenbosch University*

Mining has severe environmental and social consequences and post-mining rehabilitation needs to take place before land re-zonation can occur. Post-mining rehabilitation is the process in which previously mined areas are returned to some degree of its natural state or to a condition that can sustain an intended post-mining land use. Common rehabilitation practices, specifically in open-pit mines include backfilling; and the subsequent re-vegetation. The process of backfilling by nature leads to varying degrees of ground settling over both space and time. This can lead to differential settlement rates which can be harmful to surrounding settlements as well as the environment. To comply with legislation governing mine closure, mining companies have developed deformation monitoring systems for areas where surface deformation is a high probability. Conventional field-based approaches have been employed but are inefficient over large areas. To overcome these limitations satellite based monitoring techniques such as Differential Synthetic Aperture Radar Interferometry (dInSAR) has been employed. DInSAR is known to be an effective operational tool to detect measure and monitor surface deformation up to a millimeter level of accuracy. However, its application for monitoring of backfill settlement dynamics is largely untested. This paper presents the first results of a study to determine to what extent dInSAR techniques can detect and monitor backfill settlement in rehabilitated open pit mines in South Africa. The results of this study can be used to guide the design of monitoring system for the long-term operational monitoring of areas subject to post-mining rehabilitation.

KEYWORDS: Differential interferometry, SAR, surface deformation, backfill settlement

MONITORING STRUCTURE AND ABOVE-GROUND CARBON STORAGE IN AFROTEMPERATE FOREST PATCHES, KWAZULU-NATAL, SOUTH AFRICA

Fisher J[1]

[1]*University Of Witwatersrand*

Although Afrotemperate forest patches within KZN grassland matrixes are small (<10 km^2), they contribute substantially to carbon storage. Old growth trees (DBH >20 cm) are responsible for > 95% of above ground carbon (AGC) in these forests, containing as much as 250 MgC/ha in pristine forests. These forest patches are vulnerable to/ under threat from alien invasive plants and unsustainable utilisation by neighbouring impoverished communities. However, forest terrain and dense vegetation makes carbon stock measurement and monitoring difficult. We use small footprint, discrete return LiDAR (Light Detection and Ranging) and high resolution (7 cm) aerial imagery, in conjunction with field based structural measurements and biomass estimates, to measure and map 250 ha of eastern mistbelt forest. This is a first step towards identifying the minimum remote sensing requirements to monitor these small extent forests as an operational natural resource management tool. This research is the first instance of AGC being mapped for an entire forest patch in heterogeneous Afrotemperate forests. Results will be discussed in the context of global threats to small forest patches, as well as competing land-uses and conservation requirements. Mapping and monitoring the structure and AGC density in these forest patches can support management decisions, and potential inclusion in a payments for ecosystem services programme such as REDD+.

MONITORING THE IMPACTS OF EL NIÑO ON THE EXTENT OF CULTIVATED FIELDS USING SAR DATA AROUND THE AGRICULTURAL REGIONS OF THE FREE STATE, SOUTH AFRICA

Ngie A[2], Tesfamichael S[2], Ahmed F[1]
[1]University Of The Witwatersrand, [2]University of Johannesburg

There is continuous challenge in combating food security from erratic climatic phenomena such as lack of regular rainfall episodes during required farming seasons. Remote sensing has offered vital support in the monitoring of such scenarios and informs relevant authorities for better decision making. While optical sensors measure the greenness of vegetation to enable monitoring of its status, their usage is constrained by the continuous cloud cover during crop growth seasons in sub Saharan Africa. Synthetic aperture radar data (SAR) are on the other hand capable of penetrating clouds and are sensitive to the structure and moisture content of target features, thereby providing complementary information for monitoring crop cultivated fields. This study sought to evaluate the sensitivity of Sentinel -1 SAR data to the status of cultivated crop fields that experienced varying rainfall amounts between 2014 and 2016 growing seasons as a result of El Niño induced drought in 2014/2015. Known sample farms were delineated from Google Earth image. Backscatter values of HH and HV polarizations as well as ratio of the two polarizations were extracted per farm. Various statistical tools were used to compare the extracted backscatter values of all dates. Preliminary results showed an overall difference in SAR backscatter sensitivity to cultivated crop fields before and after the El Niño phenomenon. While these results are encouraging for areas that experience clouds during growing seasons, further improvements can be expected by factoring in other variables such as topographic and moisture conditions of farms.

KEYWORDS: Synthetic aperture radar, El Niño, Monitoring crop cultivated fields

MONITORING URBAN CHANGE IN THREE DIMENSIONS

Nghiem S[1], **Small C[2]**
[1]*NASA Jet Propulsion Laboratory,* [2]*Lamont Doherty Earth Observatory, Columbia University*

Rapid urbanization has occurred across the world in recent decades. We present innovative methods using multi-sourced satellite observations to monitor urban change in three dimensions (3D), including lateral expansion and vertical build-up. A new index called the Continuous Infrastructure Index (CII) has been developed and implemented with a combination of radar data (e.g., SRTM SAR and Sentinel-1A and 1B SAR data) and spectral data (e.g., Landsat and Sentinel-2 MSI data). CII allows a versatile way to identify urban sprawl with a transition from urban core to suburban extending to peri-urban and rural areas rather than an artificial binary abrupt demarcation of urban versus non-urban status. With the use multi-sourced data, the CII is more rigorous and yet more flexible than using a single type of satellite data for urban extent monitoring. To monitor vertical build-up, we have developed and implemented advanced methods based on physical characteristics of multi-sourced remote sensing signatures, rather than simple numerical correlation analyses. Radar signatures over urban areas are determined by the number, size and height of building structures; i.e., the total building volume. Using radar bakscatter measured by satellite scatterometer (e.g., QuikSCAT 1999-2009), the patented Dense Sampling Method (DSM) with the Rosette Transform (RT) have been demonstrated and applied to observing 3D changes of megacities and less-populated cities across the world. Moreover, spectral data allow the identification of building shadows in urban areas, and are thus useful to delineate vertical patterns. We present results for a number of selected cities in different continents with both the lateral and vertical features, not only in one given year but also the rate of change in a decadal time scale. These results reveal astounding real estate developments that may lead to economic booms or busts, and formations of mega urban agglomerations that may become geopolitical hot spots.

MULTI-SATELLITE OBSERVATION OF AFRICAN FIRES AND CHARACTERIZATION OF BURNING PATTERNS

Ichoku C[1], Pereira G[1,2], Ellison L[1,3], Schroeder W[1,4], da Silva A[1]
[1]Nasa Goddard Space Flight Center, [2]Federal University of São João del-Rei (UFSJ), [3]Science Systems & Applications, Inc., [4]University of Maryland

The African continent is estimated to contribute well over 50% of the total global carbon emissions from open vegetation fires, as seasonal burning is widespread in many African landscapes. These are mostly anthropogenic fires used for agricultural and related purposes. Over the past couple of decades, satellite remote sensing has increasingly facilitated regular observation of these fires. In addition to detecting the fires under suitable (non-cloudy) conditions, some sophisticated satellite sensors can also provide quantitative estimates of the fire radiative power (FRP), which is a measure of their relative size or intensity. The sensors that have been able to do this on a daily or sub-daily basis over Africa are: MODIS on Terra and Aqua, VIIRS on Suomi-NPP, SEVIRI on Meteosat-8, -9 and -10. These observations show that fire distribution is related to landcover types, with savanna being by far the most burned across Africa. However, a detailed analysis of fire observations from MODIS shows that fires observed in regions generally classified as one type may actually be burning a different type embedded in the dominant landcover type. For instance, a fire located in an area generally classified as forest may actually be burning a small farmland within a forest. This has enabled us to generate a new biomass-burned landcover product, which is suitable for various applications. We have also conducted a detailed comparative analysis of FRP data from MODIS and VIIRS, gridded to various spatial resolutions (between 1x1 km and 1°x1°), and found that, although they have a general agreement, there are a number of discrepancies related to differences in native footprint sizes, observation view angles, grid resolutions, and fire regimes. Results of these analyses provide valuable insights into the appropriate handling of satellite-retrieved fire datasets for various environmental applications.

KEYWORDS: Africa, burn patterns, Fire Radiative Power, MODIS, VIIRS

ISRSE37 TSHWANE 2017

MULTI-SCALE REMOTE SENSING OF MANGROVE STRUCTURE, BIOMASS AND TOTAL CARBON STOCKS FROM INSAR, HIGH RESOLUTION STEREO AND AIRBORNE LIDAR DATA

Fatoyinbo L[1], Lagomasino D[1], Feliciano E[1], Lee S[1], Simard M[2], Trettin C[3]
[1]NASA Goddard Space Flight Center, [2]Caltech/NASA JPL, [3]USDA Forest Service

Measuring and monitoring forest aboveground biomass (AGB) and carbon stocks (C) has become increasingly important because of its relevance to international climate negotiations, national climate change adaptation and mitigation programs in developing nations as well as the importance of forest Carbon stocks in global C cycle studies. Vegetated coastal ecosystems, also called Blue Carbon ecosystems are highly efficient carbon sinks and have been shown to play a significant role in ameliorating the effect of increasing global climate change by capturing significant amounts of carbon into sediments and plant biomass. In fact, current studies suggest that mangroves and coastal wetlands annually sequester carbon at a rate two to four times greater than mature tropical forests and store three to five times more carbon per equivalent area than tropical forests. Mangrove-lined estuaries and coastal ecosystems are significant to global biogeochemical processes and disproportionately to their land cover regulate the structure, productivity and function of adjacent coastal ecosystems.

In this talk, we will give an overview of recent efforts to quantify mangrove forest 3-D structure, composition and change at high resolution globally in the context of estimating forest biomass and blue carbon stocks. Our presentation covers field and remotely sensed investigations and describes unique remotely sensed datasets produced and collected at NASA, with an emphasis on recently collected airborne Lidar, Interferometric Radar and High resolution Stereo data from the Americas, Africa and South Asia. In addition, we will highlight new remote sensing methods, which have permitted high-resolution 3-dimensional mapping of forest structure and aboveground biomass stocks in blue carbon ecosystems.

MULTI-SENSOR INTEGRATED AGRICULTURAL DROUGHT MONITORING AND ASSESSMENT

Zhang X[1], Chen Z[1,2], Chen N[1,2]
[1]State Key Laboratory of Information Engineering in Surveying, Mapping, and Remote Sensing (LIESMARS), Wuhan University, [2]Collaborative Innovation Center of Geospatial Technology

Droughts severely affect the plant growth, crop yields and global food production. The enhanced risk of droughts is even more relevant in the context of the globally changing climate. This study aims to analyze the typical droughts occurring globally and quantify the drought severity, which is directly correlated with the loss in crop yield.

For this purpose, multi-sensor integrated approach was adopted. In particular, long term remote sensed/assimilated datasets of precipitation, root zone soil moisture and vegetative states are combined to calculate the Process-based Accumulated Drought Index (PADI). PADI is a newly proposed drought assessment method which integrates drought evolution with crop phenology at the same time.

In this study, the evolution process of an agricultural drought was divided into four different phases, including Latency, Onset, Development, and Recovery. These four phases describe how a meteorological drought evolve into an agricultural drought in a quantitative manner. Then, crop growth stages and soil moisture deficit coefficients are integrated with the prior drought evolution process. We believe that the impact of an agricultural drought not only depends on its own severity, but also relates to the ground crop conditions. However, existing drought models like SPI or VegDRI have not considered this point. Moreover, the prior drought condition should also be taken into account when evaluating current drought condition, as agricultural drought impact is an accumulative impact. Building off this feature, PADI also provides the accumulated drought impact on crops.

We will report the methodology of PADI, applications over China and U.S., and its merits when comparing with current drought models, including SPI, SPEI, VCI, PDSI, VegDRI, and USDM. The advantage of PADI in indicating drought impacts on crop yield loss will be demonstrated in detail. Findings from this study will help agricultural drought mitigation.

KEYWORDS: Agricultural drought, Drought index, Impact, Multi-sensor.

MULTITEMPORAL ANALYSIS OF TROPICAL AREAS USING MODIS DATA

Bonifaz R[1], Velasco V[1]
[1]*Instituto De Geofísica, UNAM*

Tropical forests occupy less than 7% of the terrestrial surface, yet they contain more than half of all plant and animal species. Although tropical forests have limited spatial extent, they play critical roles in global exchanges of energy, in biogeochemical cycling and as reservoirs of biodiversity. During the past, large areas of tropical forest have been modified by human activities such as agriculture, managed forestry and urbanization Such transformations of land use and land cover are of concern because, among other things, they often contribute to increased levels of greenhouse gases and carbon dioxide and result in reduced biodiversity. Moderate Resolution Imaging Spectroradiometer (MODIS), instruments having high temporal resolution, but typically moderate or coarse spatial resolution have been collecting data for more than 15 years. Here, MODIS 250m vegetation indices data is used to analyze the temporal behavior with both the NDVI (Normalized Difference Vegetation Index) and the EVI (Enhanced Vegetation Index) sixteen-day cloud free composites profiles. By means of wavelets and Fourier series approximation approach, multitemporal analysis is performed in order to quantify, qualify and understand indicators of temporal changes and where they occur. The study area is known as the Mayan forest of southern Mexico and Northwest Guatemala. The Mayan forest, one of the most important northern tropical areas in the continent, is an area with a large expanse of undisturbed tropical vegetation including several biosphere reserves but also the area has been subjected to drastic changes mainly deforestation for pasture conversion.

KEYWORDS: Multitemporal remote sensing, land cover monitoring, wavelets analysis, Fourier Series analysys, modis vegetation indices

MULTITEMPORAL SPACEBORNE SAR DATA FOR URBANIZATION MONITORING

Ban Y[1]
[1] *KTH Royal Institute Of Technology*

More than half of the people on the planet now live in cities. By 2050, the world is expected to add an additional 2.5 billion urban dwellers, with nearly 90 percent of the increase concentrated in Asia and Africa. Rapid urbanization changes the surface of our planet raising questions of sustainability, ecological functionality, environmental pollution, loss of ecosystem services and decreased quality of life in urban regions. Thus, accurate and timely information on the spatiotemporal patterns of urbanization and the resulting environmental impact is of critical importance to support sustainable urban development.

The objective of this research is to evaluate multitemporal spaceborne SAR data for global urban mapping and urbanization monitoring using a robust Urban Extractor. The automatic urban extraction is based on spatial indices and Grey Level Co-occurrence Matrix (GLCM) textures, an existing method and several improvements i.e., SAR and optical data preprocessing, enhancement, and post-processing. Ten cities around the world were selected to represent developed and developing cities in various environmental conditions. Multitemporal Sentinel-1A IW SAR and Sentinel-2A MSI data were acquired during the vegetation season to maximize the difference between urban and rural areas. Historical ENVISAT ASAR and ERS-1/2 SAR data were also collected for urbanization monitoring. This research demonstrates that multitemporal Sentinel-1A SAR and Sentinel-2A MSI data is very promising for urban extent mapping and urbanization monitoring at global scale. Urban areas and small towns could be well extracted using multitemporal Sentinel-1 SAR, Sentinel-2A MSI data and their fusion. Compared to the urban extraction results from historical SAR data in 1995 and 2005, the spatiotemporal patterns of urbanization could be analyzed.

KEYWORDS: Multitemporal, Sentinel-1A SAR, Urban Extractor, Urbanization, Monitoring

NASA FIRE SCIENCE AND APPLICATIONS: TECHNOLOGY, SATELLITES, AIRBORNE DATA, AND MODELS

Soja A[1], Ambrosia V[2], Friedl L[3]
[1]NIA / Nasa Langley Research Center, [2]NASA AMES; California State University - Monterey Bay (CSUMB) , [3]NASA Head Quarters

NASA supports fire research and the application of fire data, models and technology in many cross-cutting Earth Science programs to include Terrestrial Ecology, Carbon Cycle and Ecosystems, Climate Variability and Change, Atmospheric Composition, Interdisciplinary Science and Applied Science. In this presentation, we will discuss NASA Missions that have data that could support fire research, land management, fire recovery and active firefighting. We will also provide several examples of the successful use of NASA satellite and model data in fire science research and the application of those data.

NASA has an Applied Science program element, Wildland Fires, that specifically targets the use of NASA data in 'customer organizations' or communities that manage related fire science, which includes regional, national or global active firefighting, rules and regulatory communities, air quality, ecosystem protection, disaster and recovery organizations, fire weather, fuels and other modeling. Currently there are nine active projects that support pre-, active-, and post-fire applications in 'stakeholder' organizations. For instance, the US Forest Service actively uses MODIS and VIIRS active-fire detection data to locate and track fire during the fire season. Another application of NASA data is preparing post-fire recovery tools to aid ecosystem rehabilitation and manage flood/erosion control for the Burned Area Emergency Response (BAER) teams. Additionally, an international team in making active fire detection and fire risk data available to local communities in remote regions in Peru, Brazil, Madagascar, and Indonesia in an effort to protect critical environments.

NATIONAL FOREST MONITORING IN AFRICAN COUNTRIES FOR REDD+: CAMEROON AND MALAWI

Gomez S[1], Haeusler T[1], Enssle F[1]
[1]*GAF AG*

African countries committed to the UNFCCC REDD+ policy process require national forest mapping/monitoring systems (NFMS) adjusted to their national circumstances. This paper examines the challenges of transferring the policy and user requirements to Earth Observation (EO) based technical specifications and addressing the variability in forest types and land use change assessment will be noted. Specific country examples from Cameroon and Malawi will be presented. The NFMS in these countries started with the baseline year 2010 and aims to provide a continuous monitoring of forest cover and forest cover changes from 2015 onwards using multi-sensoral and -temporal satellite data (Sentinel-1/2, Landsat 8), which are validated with VHR optical data. The advent of the Sentinel-2 data series enhanced dramatically the utilisation of dense time series of multi-temporal satellite imagery to resolve problems caused by phenology changes of forest canopies between the seasons. Furthermore, this data is also needed to monitor forest degradation which can be better detected by assessing forest canopy disturbances with high frequency time series. Based on Sentinel 2 data and integration of Landsat 8 imagery the automatic processing chain for the NFMS is comprised of geometric, radiometric and topographic pre-processing steps and an iterative classification procedure that includes a rule based correction system which yields to thematic accuracies above 85%. Due to the data volume generated with the application of Sentinel for the near real time forest monitoring it is necessary to utilise cloud processing in the operational systems and this further enables the user community to be directly involved in different aspects of the processing chain.

KEYWORDS: REDD+ MRV, African Countries, GFOI

NATIONAL SATELLITE MONITORING SYSTEMS FOR REDD+: THE FAO APPROACH

Jonckheere I[1], Giaccio S[1]
[1]*FAO of the United Nations*

The UN-REDD Programme, a collaborative partnership between FAO, UNDP and UNEP launched in September 2008, supports developing countries to develop capacity to REDD+ and to implement it in-country.
The programme works at both the national and global scale, through support mechanisms for country-driven REDD+ strategies and international consensus-building on REDD+ processes.

The UN-REDD Programme gathers technical teams from around the world to develop common approaches, analyses and guidelines on issues such as measurement, reporting and verification (MRV) of carbon emissions and flows, remote sensing, and greenhouse gas inventories. Within the partnership, FAO supports countries on technical issues related to forestry and the development of cost effective and credible MRV processes for emission reductions. While at the international level, it fosters improved guidance on MRV approaches, including consensus on principles and guidelines for MRV and training programmes. It provides guidance on how best to design and implement REDD+, to ensure that forests continue to provide multiple benefits for livelihoods and biodiversity to societies while storing carbon at the same time. Other areas of work include national forest assessments and monitoring of in-country policy and institutional change.

The outcomes about the role of satellite remote sensing technologies as a tool for monitoring, assessment, reporting and verification of carbon credits and co-benefits under the REDD+ mechanism are presented in this paper. The different open-source tools and software options used by FAO Forestry Department are described, which allow the developing countries to generate their own national data which can be used then for reporting (e.g. to UNFCCC).

KEYWORDS: MRV, Satellite Land Monitoring System, Degradation, Deforestation, open source, webportal

NATURAL RESERVE “PALUDE LA VELA” (TARANTO, ITALY): REQUALIFICATION STUDY BASED ON SATELLITE DATA ANALYSIS

De Giglio M[1], Ribezzo D[1]
[1]*University Of Bologna*

Wetlands are transition areas between land and water and are recognized as important ecosystems with high vegetal productivity. Among them, the Regional Reserve "Palude La Vela" (Taranto, Italy) is a WWF area characterized by a rich biodiversity that hosts important species of flora. It is also considered a significant ecological stopover and wintering area for many rare aquatic species, included in Directive 927437CEE "Habitat", and numerous bird species, included in Dir. 7974097CEE "Birds". The reserve requires an urgent redevelopment and preservation plan due to its being subject to numerous stress sources from the proximity of a huge and debated steel plant, a refinery, a cement industry and its presence inside a former aquaculture facility.

This work is based on the analysis of Aster satellite images, acquired in 2000, 2007 and 2014. Working from previous changes, the aim is therefore to predict a future scenario, that will be useful for future management and area safeguards.

An accurate classification through the algorithm of maximum likelihood, the most commonly used for wetland mapping, allowed to discriminate the vegetation present in each image. Considering the confusion between the spectral responses of the different cover classes, processing required numerous ancillary sources and detailed consultations with the WWF technicians. After the class identification, the application of change detection provided an identification and quantification of the changes that affected the analyzed area during the whole period considered. In order to hypothesize the Reserve status evolution, a scenario was built to a defined date with the integration of the models of the most significant disturbances (roads, buildings, canals and waterways, air and water pollution measurement points).

Finally, a proposal for redevelopment was made, looking for the most appropriate activities for the reserve, respecting its history and biodiversity.

KEYWORDS: Wetlands, Classification, Change Detection, Change Prediction

NDWI-BASED TECHNIQUE FOR DETECTION OF CHANGE DATES OF THE GROWING SEASONS IN RUSSIAN SUBARCTIC

Panidi E[1], Tsepelev V[2]
[1]Saint Petersburg State University, [2]Russian State Hydrometeorological University

Previously we conducted number of studies devoted to the investigation of regional climate change in Northern Russia (namely the taiga and tundra zones of the European part of Russia). We postulated the sparseness of the ground meteorological observation network in northern regions as a main study issue, which makes it impossible to monitor and evaluate the regional features and differences in the climate change (i.e., differences at the regional scale). We identified statistically significant relationship between the trends of the surface air temperature and of the shrub vegetation productivity. In addition, in the context of study goal we proposed the principle of the vegetation productivity estimation during the growing seasons, which are the periods when the surface air temperature exceeds +5°C and +10°C and phytomass growth is affected by different weather conditions. Finally, we proposed to use the Normalized Difference Water Index (NDWI) as a marker of growing seasons change.

Currently, we are describing and discussing the technique for automated analysis of the NDWI annual graphs. We designed the enhanced approach of detection of the growing seasons change dates. This approach allows to detect the beginning and ending dates for summer season (when the temperature is above +10°C) in addition to the detection of beginning date for spring season and ending date for autumn (when the temperature is above +5°C), as it described by some authors. We are describing the general algorithm for the analysis of NDWI annual graph and some inconsistences and collisions in the graph, which are possible when the graph is derived from the time series of satellite imagery.

NEAR REAL-TIME FOREST DISTURBANCE DETECTION USING SENTINEL-1 TIME SERIES

Mermoz S[1], Bouvet A[1], Le Toan T[1], Koleck T[1], Villard L[1]
[1]*Cesbio*

Forests act as a carbon source through deforestation and degradation and as a carbon sink through regrowth. Efforts to monitor and map changes in the forest using Earth Observation have been increasing in the past decade, driven for example by the reporting needs outlined in the REDD+ program.

Up to now, monitoring forest dynamics was hampered by availability of satellite time series with concurrent high temporal and spatial resolutions. Hansen et al. (2013) implemented an annual forest monitoring system based on 30m Landsat optical imagery, whereas Hargrove et al. (2013) developed a near real-time forest monitoring system based on 250m MODIS time series. With Sentinel-1A and -1B, launched in April 2014 and 2016, multi-temporal series of SAR imagery (5.3 GHz, C-band) are expected at an outstanding time interval of 12 days or 6 days, providing globally an unprecedented amount of open data. In this study, we specifically assess the potential of Sentinel-1 time series to monitor forest disturbances in near real-time at 10m resolution.

Using C-band Sentinel-1 SAR data, detecting forest disturbances is hampered by several causes including the variable contrast between forest and disturbed areas, which can be obscured by speckle effects, and the effects of surface conditions after disturbances. In the presence of those different noise sources, detecting forest disturbances from Sentinel-1 data is a challenging task that cannot be tackled without a comprehensive methodology based on time series data. The method we developed combines appropriate preprocessing of SAR data well adapted to change detection, masking areas that may induce misdetections or false alarms, followed by time series analyses and change detection methods.

Our results over Africa (South Africa, Malawi), South America (Brazil) and Asia (Vietnam) indicate that the near real-time detection of deforestation events, in particular illegal deforestation hotspots, is possible with dense Sentinel-1 time series.

NEXT-GENERATION SAR MISSIONS - RELEVANCE TO GLOBAL ENVIRONMENTAL MONITORING AND INTERNATIONAL ENVIRONMENTAL CONVENTIONS

Rosenqvist A[1]
[1]*solo Earth Observation (soloEO)*

Within the next 5-7 years, a suite of new Synthetic Aperture Radar missions are scheduled for launch, ranging from X-band (TerraSAR-X 2nd generation, COSMO-SkyMed 2nd generation), C-band (Radarsat Constellation Mission, Sentinel-1C/D), S-band (NISAR-S, NovaSAR-S), L-band (ALOS Next-generation, SAOCOM-1A/B, TangoSat, Tandem-L, NISAR-L) and P-band (BIOMASS). Apart from enhanced performance and novel technical solutions, several of the missions also feature systematic acquisition strategies which are designed to assure the collection of spatially and temporally consistent data over the Earth's land areas. The objective of this paper (and invited session) is to assess how these missions can contribute to the information needs (implicitly or explicitly) posed by international environmental conventions (such as e.g. the Ramsar wetlands convention, UNFCCC/REDD+ and CBD), the international science community (climate change, carbon cycle) and organisations involved in environmental conservation. And whereas these missions almost excusively are developed independently from each other, we also aim at investigating the potential for coordination of observations, science and applications.

KEYWORDS: Systematic acquisition strategies, Next-Generation Synthetic Aperture Radar (SAR), international environmental conventions

NOVEL ENVIRONMENTAL MONITORING AND IMAGERY EXPLOITATION CONCEPTS OF THE URTHECAST CONSTELLATIONS

Beckett K[1]
[1]*Urthecast*

UrtheCast is currently developing two revolutionary Earth Observation constellations. UrtheDaily will monitor the entire global landmass every day with 5-metre resolution Optical imagery. OptiSAR will provide high-frequency revisit to targets with high-resolution Optical and SAR imagery. The imagery, as well as imagery from other platforms, will be made available to users and applications through the UrthePlatform for geoanalytics exploitation.

The UrtheDaily constellation is ideally suited to global-scale environmental monitoring, providing high-quality imagery at an unmatchable price-for-performance ratio, eliminating the traditional acquisition requesting, planning and dissemination cycle, allowing users to rely on receiving data that their applications depend on. Our research is exploring how high frequency imaging leads to new classes of products that tracks vegetative growth and stress, detects subtle and sudden environmental changes, and monitors human activity.

The OptiSAR Optical platform is comprised of a high-resolution dual-mode (0.5m pushbroom and video) optical camera, and a lower-resolution MetCam (acquires atmospheric parameters needed for accurate reflectance products). The OptiSAR SAR platform is comprised of a dual-band (X and L) SAR instrument providing 1-metre class X-band imagery and 5-meter class fully polarimetric L-band imagery, combined with a low-resolution CloudCam (to map out cloud cover). A total of 16 satellites, with each SAR satellite leading an Optical satellite by about two minutes, will be placed in two orbit planes to maximize revisit. Our research is exploring fusing SAR and Optical imagery using classification and machine learning techniques.

These innovative remote sensing systems will enable the reliable generation and timely delivery of truly unique and unparalleled environmental monitoring information, data fusion exploitation and geoanalytics products. The UrtheCast constellations are expected to lead to significant advances and benefits in a wide range of environmental monitoring applications such as precision agriculture, forestry management, water conservation, natural resources monitoring, change detection and humanitarian response.

NSIGHT-1 : A 2U EARTH OBSERVATION NANOSAT IN THE QB50 CONSTELLATION

Malan F[1], Wiid H[1], Burger H[2], Visagie L[1]
[1]SCS Aerospace Group, [2]SCS Space

nSight-1 (QB50 AZ02) is one of two 2U size satellites developed in South Africa as part of the international QB50 project. In addition to its FIPEX (Flux-Φ-Probe Experiment) science payload, nSight-1 carries the South-African developed "Gecko" RGB matrix camera for Earth Observation. We present the development process and initial results from the nSight-1 mission, paying specific attention to its role as Earth Observation technology demonstrator.

Our modular imaging hardware was adapted to fit the severely space-and-power constrained 2U form factor enforced by the QB50 mission. Image compression and imaging modes were optimized to work with the available UHF/VHF data downlink that not only has to serve the imaging payload, but also the FIPEX science payload. Attitude determination and control is handled by an ADCS stack developed by CubeSpace in South Africa. A dedicated ground station was set up in the Western Cape province to service nSight-1's daily data downlink needs.

nSight's Gecko camera features a unique modular design that includes integrated high-speed flash memory and an FPGA processor. These features provide adaptability and pave the way for being used in future larger multispectral satellites with more powerful optics.

When ISRSE-37 is held in May 2017 nSight-1 will be operational in space, assuming launch and deployment according to the current schedule. This should allow us to report on commissioning and our first images taken of Earth.

OAK (QUERCUS PETREA) TREE PHENOLOGY VARIABILITY MONITORED BY DIGITAL CAMERAS ONBOARD A UAV: A COMPARISON OF INDICES

Berra E[1], Gaulton R[1], Barr S[1]
[1]*Newcastle University*

The aim of this study is to assess the potential of imagery acquired from an Unmanned Aerial Vehicle (UAV) to track spring phenology of oak at individual tree level. Eighteen UAV flights were carried out over a deciduous and evergreen woodland from February to August 2015, from which a temporal series of 5 cm spatial resolution radiometrically calibrated orthomosaics was generated. The digital numbers (DNs) were used to calculate Green Chromatic Coordinate (GCC) and were also corrected to surface reflectance and to Normalized Difference Vegetation Index (NDVI); allowing therefore a comparison of the ability of these two metrics to monitor the spatio-temporal variability in leaf phenology at tree level. Time series of GCC and NDVI were calculated per tree following a manual delineation of the tree crowns. The time series of NDVI and GCC were fit by logistic functions in order to extract the start of spring season (SOS) date, which were compared against visual observations of leaf-out. The estimates derived from the visual assessment of leaf-out revealed differences of up to three weeks between individual oak trees growing very close together (<15 m), pointing towards a large intraspecific variability. The GCC estimates for SOS (RMSE = 4 days, R^2 = 0.75, Bias = 1 day, N = 20) were found to better monitor this leafing-out variability than NDVI (RMSE = 13 days, R^2 = 0.13, Bias = -5 days, N = 20). This result is attributed to the NDVI time series being noisier than GCC, as NDVI was more affected by changes in illumination conditions. It is concluded that UAV imagery has the potential to track leaf phenology at the individual tree level, but further studies are necessary to better understand this new level of information detected from UAVs.

KEYWORDS: Multispectral, phenology, calibration, forest

OBJECT BASED IMAGE ANALYSIS FOR MAPPING BUILT UP AREAS IN THOHOYANDOU, LIMPOPO PROVINCE (2007-2016)

Oliphant T[1]
[1]*SANSA*

RThe Earth's water resources are endangered pollution, inconsiderate use, and lack of conservation measures. Water scarcity is a major problem worldwide resulting in poor delivery of water which in turn results in severe socioeconomic and environmental problems. South Africa is a semi-arid country which makes a water scarce area and is experiencing significant water shortages due to climatic conditions. Monitoring these water reservoirs is very crucial for management and planning purposes however most of the reservoirs are located in remote regions which are not easily accessible and most are still maintained using only in-situ measurement systems Therefore the objective of this study is to provide a synopsis on the use of satellite images in managing and monitoring changes in surface area of water reservoirs. Multitemporal Landsat 8 images were used to investigate changes in the water surface area (extent) of Hazelmere dam. The result demonstrates a substantial decrease in the water surface area of the Hazelmere dam over the past (number) years.

KEYWORDS: Landsat8, remote sensing, temporal change and water reservoir

OBSERVATION OF GROUND SURFACE DEFORMATION IN COASTAL AREAS BY SAR INTERFEROMETRY

Raucoules D[1], Le Cozannet G[1], Poitevin C[2], de Michele M[1], Woppelmann G[2]
[1]BRGM (French Geol. Surv.), [2]LIENSS, Université de la Rochelle

Sea-level rise is one of the most unavoidable consequences of anthropogenic climate change. Coastal managers need to anticipate impacts and adaptation needs at local scale, which implies considering all components of sea-level changes, including vertical ground motions. Subsidence phenomena can exacerbate the local consequences of sea-level rise in terms of hazard (flooding, erosion) and coastal vulnerability.

In addition, studies attempting to reconstruct past-sea-level changes or to validate satellite altimetry also need to correct the local values obtained by tide gauges using vertical ground motions. Indeed, local surface deformation can affect the link between the global sea level evolution and its local expression measured by the tide gauges: subsidence enhances local sea level, uplift reduces it.

In all these cases, Differential SAR Interferometry – in combination with other in-situ measurements such as tide gauge, permanent GPS and DORIS Stations – provides useful complementary information on the spatial variability of vertical ground motions. Performances of DInSAR can be consistent (depending on the used processing technique) in terms of precision, with the observation of sea-level rise. In addition, using archived data (since the 1990's using past space-borne SAR missions) SAR interferometry allows to observe surface deformations on areas where no ground based geodetic network were available in the past.

We present results on coastal areas - mostly urbanized that are more suitable for DInSAR use – for illustrating the potential of the application of the SAR interferometric techniques to this research domain.

KEYWORDS: Sar Interferometry, Coastal Subsidence, Climate change

OBSERVING SURFACE CIRCULATION OF THE WESTERN MEDITERRANEAN BASIN WITH SATELLITE IMAGERY

Karimova S[1]
[1]*University of Liege*

Satellite oceanography provides multiple ways for observing surface circulation of marine basins. In the current study the benefits of using satellite imagery of different physical nature are being discussed. As a region of interest, we use the Western Mediterranean Basin, due to its prominent mesoscale eddy and frontal activity as well as an oligotrophic status.

In the first part of the study, we concentrate on a combined analysis of the fields of sea surface temperature (SST) and chlorophyll-a concentration (Chl-a), which is a widely-used proxy for bioproductivity of marine ecosystems. The aim of such study is to assess the correspondence between the two fields and to reveal at which extend the bioproductivity of surface waters is defined by the processes of vertical mixing and horizontal advection. The daily fields of SST and Chl-a were obtained by Aqua MODIS with a time coverage from 2003 to 2015 and a spatial resolution of 1 km.

The second part is devoted to the visual analysis of satellite imagery. This part is aimed at eddy detection and providing meso- and submesoscale eddy statistics for the region of interest. The daily fields of SST obtained from 2011 to 2013 were used for an analysis of spatial distribution of mesoscale eddies. Submesoscale eddies require a higher spatial resolution for their manifestation. Thus, a three-year-long dataset of Envisat Advanced Synthetic Aperture Radar (ASAR) imagery was processed for observing this type of water stirring.

Finally, the results of the first part were compared to those provided by the second part.
This research was supported by the University of Liege and the EU in the context of the FP7-PEOPLE-COFUND-BeIPD project. SAR imagery was obtained under the grant of the European Space Agency # 14120 "Spiral eddy statistical analyses for the Mediterranean Sea using Envisat ASAR Imagery (SESAMeSEA)".

OH BOUY: CLASSIFICATION OF OCEAN VESSELS USING SAR AND HISTORICAL AIS FEATURES

Meyer R[1], Kleynhans W[1], Schwegmann C[1]
[1]*CSIR*

It has been shown that SAR can detect non-cooperative, ocean going vessels regardless of weather or light conditions. This is an improvement on AIS, which is a transponder based system, that is be susceptible to spoofing and non-compliance.

There exists a trade-off between the resolution achievable in SAR imagery and the area that can be imaged. Course resolution, large swath SAR products can be used to monitor hundreds of thousands of square kilometers in a single image with the pixel resolution being several tens of meters.

While course resolution SAR is good for detecting vessels, due to the nature of the interaction between the SAR radar beam, ocean state and vessel structure, it is poor at extracting vessel characteristics such as length, width or gross-tonnage. Research has been done to improve this using machine learning algorithms but this research often ignores the most accurate feature that can be extracted from SAR data: the vessel's location.

AIS messages are received from ocean vessels using land based and satellite receivers. The messages are stored in a database giving a historical view of thousands of vessel's identity and movements. Historical parameters can be stacked in a spacial database giving information that describes the average size, bearing, speed or other parameter per vessel class over a vast area.

This research continues exploring existing classification techniques but adds features extracted from historical AIS messages that occur in the region of interest. By assuming that vessels follow predictable behaviors and patterns based on their economic activity, a classification algorithm can be built that uses historical AIS messages from multiple vessels performing the same, or similar activity, to classify a vessel detected on a SAR image or to detect AIS messages falling outside expected norms.

KEYWORDS: SAR, AIS, classification, ocean, vessels

OIL POLLUTION IN INDONESIAN WATERS: COMBINING STATISTICAL ANALYSES OF ENVISAT ASAR AND SENTINEL-1A C-SAR DATA WITH NUMERICAL TRACER MODELLING

Gade M[1], Mayer B[1], Meier C[1], Pohlmann T[1], Putri M[2], Setiawan A[3]
[1]*Universität Hamburg,* [2]*Institute Technology Bandung,* [3]*Agency for Marine and Fisheries Research and Development*

This Pilot Study aimed at improving the information on the state of the Indonesian marine environment that is gained from satellite data. More than 2000 historical and actual synthetic aperture radar (SAR) data from ENVISAT ASAR and Sentinel-1A C-SAR, respectively, were used to produce oil pollution density maps of two regions of interest (ROIs) in Indonesian waters. The normalised spill number and the normalised mean polluted area were calculated, and our findings indicate that in general, the marine oil pollution in both ROIs is of different origin: while ship traffic appears to be the main source in the Java Sea, oil production industry causes the highest pollution rates in the Strait of Makassar. In most cases hot spots of marine oil pollution were found in the open sea, and the largest number of oil spills in the Java Sea was found from March to May and from September to December, i.e., during the transition from the north-west monsoon to the south-east monsoon, and vice versa. This is when the overall wind and current patterns change, thereby making oil pollution detection with SAR sensors easier.

In support of our SAR image analyses high-resolution numerical forward and backward tracer experiments were performed. Using the previously gained information we identify strongly affected coastal areas (with most oil pollution being driven onshore), but also sensitive parts of major shipping lanes (where any oil pollution is likely to be driven into marine protected areas).

Our results demonstrate the feasibility of our approach, to combine numerical tracer modelling with (visual) SAR image analyses for an assessment of the marine environment in Indonesian waters, and they help in better understanding the observed seasonality.

ONLINE REMOTE SENSING EDUCATION: EXPERIENCE AT THE BRAZILIAN NATIONAL INSTITUTE FOR SPACE RESEARCH (INPE)

Ferreira H[1], Souza D[1], Rudorff N[1], Lucaccioni C
[1]*INPE*

The objective is to present an overview of Online Remote Sensing Technology Education Programs carried out by the Earth Observation Coordination (OBT) and Center for Weather Forecast and Climate Studies (CPTEC) at the Brazilian National Institute for Space Research (INPE), since the year 2004. Several online and hybrid courses, and webinars, have been provided so far. This long-standing experience with online courses and the information gathered from feedback surveys, after each course, were crucial for INPE professionals' responsible for the course to assess and determine the new paths for the educational outreach program in this area, investigating new teaching and learning strategies and pedagogies to suit students´ needs to improve their skills to succeed in their workplace. Information and communication technologies together with new teaching paradigms have been reshaping the whole learning environment of the courses.

KEYWORDS: Online Courses, Remote Sensing, Technology, Best Practices

ONLINE REMOTE SENSING TRAINING IN NORTH AND WESTERN AFRICA: A LEADING EXPERIENCE OF CASABLANCA VLAB CENTRE OF EXCELLENCY

Filali Boubrahmi N[1]
[1]Direction De La Météorologie Nationale

The Casablanca Centre of Excellency (COE) Belongs to WMO's Virtual Laboratory network. This COE was established in 2012 in order to consolidate the capacities of Francophone North and West African countries in weather remote sensing and related activities.

Given the training needs in the region, the challenges were high and many. It was not only a matter of organising training activities, but also meeting the needs of National Meteorological Services in the region.

With this goal in mind, Casablanca's CoE proposed the organising of training on "advanced remote sensing applications in the priority areas of meteorology" rather than remote sensing basics. At that time, the questions that were still in our minds were "What are these priority areas and what would be the most appropriate methodological approaches to follow?

To answer these, many online surveys were carried out that enabled us to highlight the urgent needs, priority sectors, barriers for online training in the regions and the optimal approach to carry out successful online training. These surveys also assisted us to start building a community of users.

This approach has helped to develop an action plan consistent with the needs of its users and to establish successful online training that have been meeting the needs of its users since then.

Examples of online training organised by Casablanca's CoE include the Aviation Week event (2013 and 2016), many Regional Focus Groups (RFG) that covered thematic such Marine Forecasting (2014), Numerical Weather Prediction (2013) and Modelling of low clouds and low visibility (2014).

This presentation aims to present this leading experience in online training in remote sensing, what's were the challenges and the keys of success.

OPEN SOURCE TECHNOLOGY FOR LARGE SCALE OPEN DATA SHARING

Bye B[1], De Lathouwer B[2]
[1]BLB, [2]Open Geospatial Consortium (OGC)

The Group on Earth Observations (GEO) coordinates and makes a multitude of remote sensing capabilities interoperable as it develops the Global Earth Observation System of Systems (GEOSS). GEO is a proponent of open data sharing. Using open source technology to implement the open data sharing principles is in line with the European Open ness policies, and the G8 Open Data Charter. H2020s NextGEOSS proposes to develop a hub for EOs, where users can connect to access data and deploy EO-based applications. The concept revolves around providing the data and resources to the users communities, together with Cloud resources, seamlessly connected to provide an integrated ecosystem for supporting applications.The Data Hub will be built on top of CKAN, an already powerful data management system that provides tools to streamline publishing, sharing, finding and using data with a system proven through successful mission operations. Data Hub harvesters will be demonstrated technology or in prototype that will be qualified for each EO data infrastructure to be harvested and provide links to the original source. This session will shed some light on different aspects of using the free and open technology CKAN as a scaleable solution for GEOSS as a big data infrastructure with illustrative examples from Africa. The goal of this session is to introduce the CKAN as technology supporting GEOSS, get input from all regions with respect to capability and challenges, demonstrate its potential through local examples. The input will be taken up by NextGEOSS and fed into its support of GEOSS ensuring relevance not only for Europe but GEO as a whole.

OPERATIONAL CROP AREA STATISTICS EXTRACTED FROM LANDSAT IMAGERY ON GOOGLE EARTH ENGINE

Reynolds C[1]
[1] *United States Department Of Agriculture*

Free Landsat imagery enables governments to monitor crop area on a monthly basis for improving monthly and annual crop area statistics. Landsat imagery at USDA/FAS is operationally processed so that 30-meter resolution crop classified products are available mid-season and after harvest to assist in crop area estimations. The Landsat imagery processing stream requires operational surface reflectance correction; removal of all clouds and shadows; generation of bi-monthly mosaics during the growing season, along with robust spectral classification algorithms. When producing bi-monthly Landsat mosaics, time series data gaps can also be filled with free Sentinel-2A imagery. With the availability of free Landsat and Sentinel-2A imagery on Google Earth Engine (GEE), USDA/FAS is making efforts to operationally improve monthly and annual crop area statistics derived from Landsat imagery available on GEE. Several operational Landsat imagery processing streams and final crop area products from USDA will be described.

OPERATIONAL GEOSTATIONARY SATELLITE AT THE SOUTH AFRICAN WEATHER SERVICE

Maseko B[1]
[1]*South African Weather Service*

Accurate forecasts of heavy rainfall or flooding is important for early warning, flooding can cause damage to human lives and property. A number of tools are used for rainfall observation such as, gauges, radar, and satellite. Used together they provide more accurate estimates of rainfall. However rainfall gauges are not available everywhere and provide point estimates, radar rainfall although provide better estimates are also not available in most of the African countries due to being expensive to buy and maintain. Hence large parts of Africa uses satellite based rainfall estimation. Although not as accurate as gauges or radar, it has high coverage and temporal resolution and can cover region where there are no radars or gauges.

OPERATIONAL HIGH RESOLUTION SOIL MOISTURE FOR AGRICULTURAL APPLICATIONS

Escorihuela M[1], Piou C[2], Merlin O[3], Abdallahi Ould Babah M[4], Ghaout S[5], Diakite F[6], Moumene K[7], Cressman K[8], Gao Q[1], Amazirh A[9], Er-raki S[9], Fontanet M[10], Ferrer F[10], Cissé S[6], Salem Benahi A[4], Chirhane J[5], Lazar M[7], Kerr Y[3]

[1]isardSAT, [2]CIRAD, [3]CESBIO , [4]Centre National Lutte Antiacridienne, [5]Centre National Lutte Anti Acridienne, [6]CNLCP, [7]INPV, [8]DLIS, FAO, [9]Université Cadi Ayyad, [10]LabFerrer

The purpose of this talk is to present the status and results from two projects that aim to use high resolution soil moisture operationally for agricultural applications.

The first one (SMELLS) aims to improve current desert locust early warning systems by introducing high resolution soil moisture (1km) based on SMOS disaggregated soil moisture.

Plagues of desert locusts, Schistocerca gregaria, have threatened agricultural production in Africa, the Middle East, and Asia for centuries. The livelihood of at least one-tenth of the world's human population can be affected by swarms of this insect. Currently, desert locust center use precipitation and NDVI index to forecast the areas with highest potential of desert locust breeding. To be able to take earlier the decision to send survey teams, one solution is to have timely information about soil moisture, which precedes vegetation. Our results show that soil moisture correlates well with desert locust presence and can explain current situation in Mauritania.

The other project (REC) proposes a solution to the need of root-zone soil moisture at the crop scale for irrigation management.

REC is based on an innovative operational algorithm that will allow for the first time to: 1) to map root zone soil moisture on a daily basis at the field scale and 2) to quantitatively evaluate the different components of the water budget at the field scale from readily available remote sensing data.

The methodology relies on the coupling between a surface model representing the water fluxes at the land surface-atmosphere interface (infiltration, evaporation, transpiration) and in the soil (drainage), and remote sensing data composed of land surface temperature, and near-surface soil moisture retrieved from microwave radiometers and radars.
These estimates will be integrated in an irrigation management system that will be used to trigger irrigation.

OPTIMAL PHENOLOGICAL PERIOD FOR ESTIMATING WOODY COVER USING MODIS IMAGERY

Cho M[1]
[1]*CSIR*

Savannas are predominantly grasslands with varying degree of tree cover. The various life forms in savannas differ in their phenology and this has important implications for remote sensing of vegetation characteristics e.g. percentage tree cover. Land surface phenology (LSP) derived from remote sensing data is determined by the abundances and condition of the various life forms within a pixel. Savanna trees generally show longer periods of growth than grasses. Our previous study showed that the start of growing season (SGS) of is highly influenced by the start of the rainy season (SRS) while, the end of rainy season (ERS) showed a much lower influence on the end of growing season (EGS), particularly on tree dominated pixels. Thus, the green signal of savannas towards the end of the growing season is more likely dominated by the reflectance of the trees than of the grasses. In this study, we sort to ascertain the superiority of the senescing phase of the savanna vegetation for estimating tree cover when compared to the growing phase and to determine the optimal dates of the year for tree cover assessment from remote sensing imagery. The tree cover map for the study region in South Africa was derived from Synthetic Aperture Radar imagery (2010 ALOS PALSAR data). MODIS 8-day reflectance data were downloaded from USGS explorer site. NDVI time series data was established from the MODIS imagery for the period of January 2001 to December 2015. Tree cover data were extracted for 100 random sample plots (2 by 2 pixels) and regressed against the NDVI data for each date. The results showed a mode of Julian day 161 (10 June), with a 95% confidence interval of 8 days as the optimal period of the year for estimating tree cover from MODIS. Clouds are less problematic during this period.

OPTIMISATION OF SAVANNAH FRACTIONAL WOODY VEGETATION COVER MAPPING USING OPTICAL AND RADAR DATA

Symeonakis E[1], Marqués-Mateu Á[2], Petroulaki K[1], Higginbottom T[1]
[1]*Manchester Metropolitan University,* [2]*Universitat Politècnica de València*

The fraction of woody vegetation plays an important role in natural and anthropogenic processes of savannah ecosystems. We investigate the optimal combination of Landsat optical and thermal bands as well as ALOS PALSAR L-band radar data from both wet and dry seasons for the mapping of fractional woody vegetation cover in southern African savannah environments. We employ colour aerial photography for sampling and validation and a random forest classification approach to map the fraction of woody cover in an area of 1200 km² in the Northwest Province of South Africa. Our results from random forests classifications show that the most accurate estimates are produced from the model that incorporates all parameters: Landsat optical and thermal bands and vegetation indices (NDVI and MSAVI) for the dry and wet seasons, and HH and HV polarised ALOS PALSAR L-band data. However, the combination of the six Landsat bands from either the wet or the dry season with either the HH or the HV PALSAR band, appears to be sufficient for achieving fractional woody cover balanced accuracies of >85%. Dry season optical bands alone are able to map fractional woody cover with more than 80% balanced accuracy. Our findings can provide much needed assistance to woody vegetation monitoring efforts in southern African savannahs where its observed expansion over the last decades is partly attributed to bush encroachment and land degradation brought about by recent climatic changes and/or land mismanagement.

KEYWORDS: Fractional woody vegetation cover, Landsat, ALOS PALSAR, Savannah, Vegetation index

OUTCOMES OF THE FOWLERS GAP UNMANNED AIRBORNE VEHICLE WORKSHOP (2016), NSW, AUSTRALIA

Lucas R[1], Lucieer A[2], Fisher A[1,4], Zhou Y[1], Hacker J[3], McGrath A[3]
[1]University of New South Wales, [2]University of Tasmania, [3]Airborne Research Australia, [4]University of Queensland

Over the past few years, unmanned aerial vehicles (UAVs) have become widely available to the ecosystem science community and are increasingly providing new avenues and opportunities for conducting and supporting environmental research. To demonstrate UAV platform and sensor capability, over 30 people from universities and government departments across Australia gathered at the University of New South Wales' (UNSW) Fowlers Gap Research Station in September 2016 to conduct Australia's first integrated UAV, aircraft and ground field campaign. Over 30 instruments, including RGB cameras, multispectral, hyperspectral, thermal sensors, and lidar, were carried on 17 airborne platforms. The participants addressed several challenges ranging from quantification of vegetation structure and species composition, identification and counting of large native mammals and stock and mapping of bird species distributions to discrimination of geological formations. TERN AusCover provided support for both ground and UAV operations, which also included measurements at the station's six TERN AusPlots sites, star transects for vegetation cover and animal surveys. For all sites flown, Airborne Research Australia (ARA) acquired concurrent aircraft observations using comparative sensors, including a UAV compatible survey-grade lidar. The campaign provided a collaborative platform to develop, test, and verify operating procedures, data standards and data processing techniques, which are needed if UAVs are to become a reliable and trusted tool for ecosystem science and management. Key outcomes included the establishment of optimal approaches for generating digital terrain and canopy height models, new methods counting large native animals (e.g., kangaroos) and options for understanding bird distributions, ecology and behavior. The integrated airborne, UAV and field data acquired at the Fowler's Gap station, including those from the six TERN AusPlots sites, will be made publicly available for scientific research via TERN's data infrastructure. The dataset, which is the most comprehensive in Australia, will also be released as an educational resource.

PARAMETERIZING NET RADIATION FOR EVAPOTRANSPIRATION MODELING USING REMOTE SENSING DATA

Majozi NP [1,2], Mannaerts CM [2], Ramoelo A [1,3], Mathieu R [1,4], Mudau AE [5] and Verhoef W [2]
[1]Council for Scientific and Industrial Research, [2]University of Twente, [3]University of Limpopo, [4]University of Pretoria, [5]Council for Scientific and Industrial Research

Net radiation (Rnet) controls the energy and water exchange between the biosphere and the atmosphere, and is a key component of the energy balance. It is a major driving force behind biophysical processes, including evapotranspiration, photosynthesis and respiration. While the incoming shortwave and longwave components of the Rnet can be estimated from weather data using empirical equations, the outgoing shortwave and longwave components are estimated from remote sensing data in remote sensing based evapotranspiration (ET) models. It is, therefore, important to understand that remote sensing based estimates of Rnet are instantaneous, and are supposed to be upscaled to daily estimates, especially when modeling ET. The aim of this study was to estimate and upscale instantaneous Rnet to daily averages using the sinusoidal method, using MODerate Resolution Imaging Spectroradiometer (MODIS) data products and ground data as inputs. The results vary per season and location, with the summer giving more accurate estimations of Rnet compared to the winter season. An accurate estimation of Rnet is critical for improved modeling of ET.

KEYWORDS: net radiation, remote sensing, instantaneous, sinusoidal method, MODIS

PARTICIPATORY MAPPING OF FOREST PLANTATIONS IN THE SOUTHERN HIGHLANDS OF TANZANIA WITH OPEN SOURCE DATA AND TOOLS

Koskinen J[1], Mankinen U[1], Pekkarinen A[2], Käyhkö N[1]
[1]*University of Turku*, [2]*Food and Agriculture Organization of the United Nations*

Growing amount of the global demand on forest related services such as timber, wood fiber and fruits are produced in planted forests, especially in tropical regions where forest plantations have expanded during the last 25 years. As the plantations may form a substantial proportion of regional and local landscapes, we need spatially explicit monitoring on the dynamics of forest cover to estimate environmental and socio-economic impacts and to support sustainable forest management regimes at all scales. However, in many tropical countries such estimates are challenged due to data scarcity. In this study, we mapped the extent and species composition of forest plantations in Southern Highlands of Tanzania, a region experiencing rapid growth of planted forest area, using open image catalogues and cloud computing capacity of Google Earth Engine (GEE) and participatory reference data collection with FAO developed Collect Earth tool. A large training sample of forest plantations was collected in a 'Mapathon' event where 22 local experts interpreted coverage, species and age information from Google Earth and Bing map images. The collected points were used to classify a stack of Landsat-8 OLI best pixel mosaic (2013-2015), Sentinel 2 median mosaic (2015-2016), Sentinel-1A mean mosaic (2015), elevation and slope, preprocessed in GEE, with Random Forest classifier. Based on the tentative results the plantation area was estimated with high overall accuracy with majority of the plantations being small and fragmented outside the largest government and company-owned plantations. The results show that local knowledge provides good accuracy even on more specific land cover interpretation tasks and holds potential for regional level land cover mapping endeavors. Moreover, automated classification methods based on freely available geospatial data sets and tools allow repetition of the mapping without massive effort, important for monitoring land cover dynamics in data scarce areas.

PAYLOADS WITH LIVE INTERNET ACCESS AND COMPLEMENTARY SATELLITE FLEET

Eickhoff J[1], Helm A[1], Jochum M[1]
[1]*Airbus DS GmbH*

This paper presents the concept for real-time EO services realized by individual or a constellation of LEO satellites (alt. ~ 600km, mass ~150 kg) communicating with ground via relay satellites in MEO or GEO. This allows permanent ground access to the LEOs.

The innovation consists of the LEOs providing TCP/IP accessible payloads and the cited relay connection being realized via an internet relay constellation. Depending on the required LEO mission characteristics, diverse relay systems are usable, e.g. OneWeb, O3B, Inmarsat and others.

Diverse instruments and services can be imagined to benefit from this permanent and easy payload access such as for

- GNSS-reflectometric measurement of wind & waves, called InstaOcean service
- Thermal-Infrared wild-fire detection, InstaView,
- Transport monitoring, e.g. ships with AIS, InstaLogistics,
- Airplane tracking with ADS-B, InstaFlight,
- Animal migration with ICARUS Trackers, InstaResearch

The design under conceptuation takes benefit of low-cost satellite platforms and allows a step-wise growth of business and constellation, e.g. starting with individual satellites, continuing with a small constellation 4-6 low inclination satellites for tropical belt coverage, and optional later deployment toward global coverage (high inclination orbits).

Due to the direct internet connectivity from ground (Web browser with VPN) – through the relay constellation to the LEO satellite's payload, the effort for science data access is very reduced compared to classic approaches.
Airbus offers the platform as a service and intends itself to fly an InstaOcean payload. The partner I-GOS is interested in potential flight of an ICARUS wildlife tracker. Renowned US institutions are currently evaluating platform use.

Become a partner and feel invited to contribute a hosted payload, to contribute to the InstaServices ground infrastructure, data evaluation services etc. The program is also open for Universities.

KEYWORDS: Realtime EO services, Internet access to in-orbit payload, low-cost small satellite platform as a service

PERSPECTIVES INTO EO-SAT1 SATELLITE SENSOR SPECIFICATIONS: A REVIEW OF THE OPPORTUNITIES AND TRADE-OFFS IN THE DESIGN PROCESS

Mhangara P[1], Mapurisa W[1]
[1]*South African National Space Agency (SANSA)*

The South African National Space Agency (SANSA) is in the process of developing a new medium sized earth observation satellite codenamed EO-SAT1. EO-SAT1 is funded by the Department of Science and Technology and is South Africa's contribution to the African Resource Management Constellation (ARMC) which is comprised of Nigeria, Algeria, Kenya and South Africa. ARMC aims to bridge the need for frequent high resolution data for resource management applications in Africa. Food security, land use and land cover mapping and disaster monitoring were prioritized as the key the applications to be addressed by ARMC satellites. A detailed discussion of the goals and user requirements for ARMC is provided by Mostert and Jacob (2008). Whereas EO-SAT1 will inevitably contribute significantly to a number of applications of socio-economic benefit, food security was identified as the primary mission objective for EO-SAT1 following a comprehensive user requirements assessment process. In essence EO-SAT1 was primarily designed as a vegetation sensor for agricultural monitoring. Agriculture is the backbone of most African economies and satellite based remote sensing is increasingly becoming an indispensable tool for monitoring agricultural food production globally. EO-SAT1 is therefore expected to contribute significantly to crop and foliage resource monitoring. The objective of this paper is to present the proposed spectral, spatial, radiometric and temporal resolutions for EO-SAT1 and how they address the primary mission requirements. Secondly the paper also highlights the rationale behind the proposed specifications and the trade-offs that were made in defining the sensor specifications.

KEYWORDS: satellite sensor specification, EO-SAT1, ARMC, food security, mission statement

PHENOLOGY OF THE ENKANGALA GRASSLANDS

Moyo M[1]
[1]*University Of The Witwatersrand*

Phenology is the study of the timing within the year of life history events in plants and animals. The phenology of plants is usually cued to climate; therefore climate change is likely to have an effect on the date of events such as greening and browning and thus the length of the growing season. Since the growth duration, the rainfall, and the temperature all control primary productivity and transpiration, phenological change will lead to changes in the ecosystem services of forage provision and water yield. The aim of this study is to gain a predictive understanding of the phenology of the Enkangala moist, high altitude grassland. Moderate spatial resolution, high time-resolution multi-temporal satellite-derived datasets are used to describe the phenology of the high-altitude Enkangala grasslands of South Africa by assigning attribute values to a minimal phenometric model which is related to climatic conditions. A long-term daily climate data record is then used to establish the climatic determinants (soil moisture and air temperature) and detect changes in phenometric attributes over the past century. Finally, projections of future phenological trends are made based on climate change projections for the region.

KEYWORDS: Phenology, Climate Change, Remote Sensing, MISR, Growing Season Length

PHENOLOGY-BASED DETECTION OF CONFLICT LOCATIONS DURING THE CIVIL WARS WITHIN SOUTH SUDAN

Sosnowski A[1], Ghoneim E[2], Crews K[1]
[1]University Of Texas at Austin, [2]University of North Carolina Wilmington

South Sudan has experienced many years of civil war, first within the Sudan and now as an independent country. Estimates of the number of people killed and the number of internally displaced persons (IDPs) are not closely monitored because of the magnitude and distribution of violence in the country. We propose a remote sensing based methodology of changes in vegetation phenological metrics to determine areas that have been affected by large scale violence. Outside of major cities, livelihoods are primarily agriculture and livestock based, meaning that they take advantage of the presence of vegetation. When an attack on civilians occurs, those who are not killed often flee the village. The vegetation of the original area may then regenerate or a different vegetation regime may take over due to the decreased human pressures.

This study utilized MODIS Enhanced Vegetation Index data from 2000 to 2016 to detect changes in overall annual greenness, peak greenness, and timing of green-up over the country of South Sudan. Seasonal trend analysis and contextual Mann-Kendall statistics were employed to differentiate significant deviations from overall trends in each pixel. These locations were compared to settlement, water point, and IDP locations from UN OCHA datasets. In cases where a settlement location corresponded with a significant change in phenological characteristics, the destruction of a nearby village was verified using either Google Earth or the Landsat NIR baseline method that has previously been used to indicate the destruction of villages in Darfur. The resulting maps indicate not only the regions from which people have fled or been killed, but the approximate time period that this occurred as well. The spatio-temporal detection of the effects of civil war has implications for human rights monitoring and crisis intervention.

KEYWORDS: harmonic regression, human displacement, human rights, MODIS EVI

PILLAR-COLLAPSE IN SOUTH AFRICAN COAL MINES AND RELATED SURFACE DEFORMATION AS DETECTED USING SENTINEL-1 DINSAR APPROACHES

Engelbrecht J[1], Haupt S[1], Theron A[1]
[1]*CSIR*

The room-and-pillar technique of mining is a system in which mined material is extracted across a horizontal plane while leaving pillars of untouched material to support the overburden. The optimum pillar size is designed to be small enough to minimise the amount of ore that is left behind, thereby increasing the profitability of the mine. However, if pillars are too small, the mine will collapse, leading to subsidence of the surface. In South Africa, a legacy of over 120 years of mining has led to extensive areas that have undergone room-and-pillar style mining that has since been abandoned. Frequently, the pillars start to fail in areas that are not actively mined or monitored. This leads to subsidence in unexpected areas that can pose significant threats to infrastructure and the environment in addition to affecting human health and safety. Differential interferometry (dInSAR) techniques are well known for their ability to provide centimetre- to millimetre-scale deformation measurements. The maturity of dInSAR has, in principle, overcome the limitations associated with field-based techniques and has been extensively used for its ability to monitor deformation over large areas, remotely. With the launch of the Sentinel-1 satellites together with the open access data policy, the routine monitoring of extensive areas has become a possibility. This paper presents the results of almost two years of monitoring mining-induced deformation in operational and abandoned coalmines in South Africa using Sentinel-1 data. The focus is on identifying areas associated with pillar collapse in underground environments and monitoring the expression of subsidence at the surface. The results suggest that surface deformation associated with pillar collapse can be detected and measured with a high degree of confidence. This means that long-term operational monitoring for early warning of hazardous conditions is a possibility.

KEYWORDS: Differential Interferometry, Surface deformation

PIXEL-BASED VS SEGMENTATION-BASED CLASSIFIERS USING SPOT-6 AND HEIGHT DIFFERENTIATION OVER SOSHANGUVE TOWNSHIP, SOUTH AFRICA

Gxumisa A[1], Breytenbach A[1]
[1]*CSIR*

The extraction of land use/land cover information has over the years become important for various EO applications in fields such as urban planning, transport and civil engineering. One way of extracting useful LULC information is through advanced image processing techniques. Advances in remote sensing technology and the availability of high resolution satellite ortho-optic imagery offer the potential for more detailed classification of very heterogeneous urban landscapes. The classification of remotely sensed images is normally performed using object-oriented or pixel based approaches, or a combination or hybrid of the two. In this study a single SPOT-6 multi-spectral acquisition was used to compare an object orientated versus a maximum likelihood classifier in order to identify primary and secondary LULC classes over the Soshanguve Township in Tshwane, South Africa. Two analyses were performed for each classification approach: one with and one without height information – obtained from LiDAR data – in order to assess the effect of integrating height metrics in the classification. The integration of height data was found to have significantly improved the quality of both the classification products. Ultimately, the object-oriented approach outperformed the pixel-based approach, particularly for the identification and separation of secondary built-up classes.

PLANET'S SUBSCRIPTION APPROACH TO ACCESSING DAILY, HIGH RESOLUTION COVERAGE OF THE EARTH

Ahlrichs J[1]
[1]*Planet*

The satellite Imaging business is rapidly changing from an order-task-deliver process (pay per sq. km) to subscription models where all data collected by multiple satellites are available for one fixed fee. No more having to ask permission to buy more data. This presentation will discuss Planet's capability to image the world each day at high resolution, API access to our platform and our fixed fee, data access models. This scalable, automated approach to data access enables businesses, institutions and universities to develop local to global scale monitoring and response programs that were never before possible. Applications that will be highlighted include near real-time disaster and forest loss response, agricultural forecasting, quantifying forest degradation and illegal urban growth as well as university training of the next generation of entrepreneurs and decision makers.

PLANNED GROUND SEGMENT ARCHITECTURE FOR EO-SAT1

Mapurisa W[1], Mhangara P[1]
[1]*South African National Space Agency*

The ground segment is a critical component of any earth observation satellite engineering system. Its primary functions include mission control, data reception, data pre-processing, processing, archiving, product generation, quality control, data dissemination and service provision. The objective of this paper is to introduce the proposed ground segment architecture of the upcoming South African earth observation satellite code-named EO-SAT1. EO-SAT1 is South Africa's contribution to the African Resource Management Constellation and possesses ten multispectral bands imaging at high and medium spatial resolution in the Visible and Near-Infrared (VNIR) Region of the electromagnetic region. The satellite will serve a wide range of applications that include crop monitoring, human settlements mapping, water resource monitoring, vegetation monitoring and general land use and land cover mapping. A customized and highly responsive ground segment design is fundamental in meeting and sustaining the needs of the global earth observation community. The paper will outline the ground segment image processing architecture that will be used by EO-SAT 1 mission and any other follow up mission including third party missions. A generic processing system adaptable to most earth observation sensors amendable for automated automated image processing and archiving is presented. This system will ensure timeous and efficient dissemination of radiometric and geometrically corrected data and value added products to customers.

POLAR SCIENCE AND OPERATIONAL SERVICES SUPPORTED BY CRYOSAT-2

Shepherd A[1], Armitage T[2], Forester L[3], Gilbert L[2], Gourmelen N[3], Hogg A[1], Konrad H[1], McMillan M[1], Muir A[2], Ridout A[2], Slater T[1], Tilling R[1]

[1]University Of Leeds, [2]University College London, [3]University of Edinburgh

CryoSat-2 is ESA's first satellite mission dedicated to measuring changes in the polar land ice and sea ice cover.

Following its launch in April 2010, we have examined the performance of the instrument over the continental ice sheets, sea ice, and the global oceans. We have confirmed the engineering performance at system level of the interferometer; the range precision is 19 cm RMS at 20 Hz, and the contribution of the across track slope error is 0.4 mm. With the corrected data products, we are able to confirm that the system performance of CryoSat-2 meets or exceed its specification over the continental and marine ice sheets.

This presentation summarises the key outcomes of the mission performance, and presents a series of example case studies where CryoSat-2 data have been applied to study changes in Earth's land and sea ice cover and their impact on operational services. We show that in the six years since launch, CryoSat has been able to detect changes in the mass of the Antarctic and Greenland ice sheets with an accuracy comparable to that of the past 20 years of conventional satellite altimetry, that important changes have occurred in these regions, and that these measurements inform international assessments of global sea level rise. We show also that the mission has been able to detect changes in the volume of rugged, glaciated terrain, that were beyond the capability of past altimeter missions and that are in places extreme in comparison to past observations. Finally, we show that CryoSat has been able to quantify changes in the volume of sea ice across the entire northern hemisphere for the first time and that unexpected changes have occurred, and that these measurements inform maritime operations.

POTENTIAL HABITAT SUITABILITY FOR SUSTAINABLE MANAGEMENT OF STERCULIA SETIGERA DEL. IN TOGO (WEST AFRICA)

Atakpama W[1], WALA K[1], Gouwakinnou G[2], Polo-Akpisso A[1], DIMOBE K[1], BATAWILA K[1], AKPAGANA K[1]
[1]University Of Lome, Faculty Of Sciences, Departement Of Botany, Laboratory Of Botany And Plant Ecology, [2]University of Abomey-Calavi, Faculty of Agronomic Sciences, Laboratory of Applied Ecology, P.O. Box 526 Cotonou, Republic of Benin.

The setting up of an economic activity based on natural resources require the appraisal of its return on investment employed, which is linked to the availability and the durability of resources. Accordingly, the present study which purposed at promoting the gum yield of Sterculia setigera aims to forecast the potential current and future suitable habitat for domestication of the species in Togo. Two types of data were used namely occurrence and environmental data. Occurrence data were gathered from field works, herbarium records and scientific published papers. Environmental data were formed by 19 bioclimatic variables, altitude obtained from worldclim database and soil data obtained from the Harmonized World Soil Database. The model based on Maximum Entropy (MaxEnt) was used to forecast current suitable potential habitat. The suitable distribution model (SDM) was run using occurrence data and environmental data. Based on cross-correlations among variables, the variables' contribution and jackknife test of variables' importance, six bioclimatic variables were selected for model running. The two main variables that contributed towards predicting the SDM were the annual precipitation and the temperature seasonality. Results showed the habitat suitability of the species within eco-floristic zone I, II, and III. Further studies on the nursery, regeneration, best cultivar selection and assessment of future climate impact on S. setigera population will be a great asset for its use and domestication.

KEYWORDS: Sterculia setigera, MaxEnt, Potential habitat, domestication, Togo

POTENTIAL OF SENTINEL-1 C-BAND FOR RETRIEVING FIRE SCAR IN DECIDUOUS SOUTHERN AFRICAN SAVANNAHS

Mathieu R[1], Main R[1], Naidoo L[1], Yang H[2]
[1]CSIR-NRE, [2]Princeton University

Fires in southern African savannahs are an important disturbance factor contributing to shape vegetation dynamic. Hence, fires impact on the hydrological regime, productivity, and biodiversity of ecosystems. Information on fire size, frequency, season, and extent is mostly provided by optical data (e.g. Active Fire Information System - AFIS) where the mapping of fire scars is based on the contrast between dark burned surfaces relative to non-burned surroundings. The Sentinel-1 constellation provides an opportunity to complement optical data by generating dense SAR time series (up to revisit time of 5 days), and may alleviate the problem of systematic optical data acquisition due to e.g. clouds for early season fires or winter smoke / heavy haze resulting from fires. In open forests – continuous layer of grass and discontinuous layer of shrubs and trees – the removal of the grass layer due to fire is hypothesized to decrease the C-band backscatter. The objective of our research was to investigate if time series of Sentinel-1 C-band data are useful to detect fire scars in deciduous southern African savannahs. High resolution (10m) time series of 19 C-band Sentinel-1 images were acquired in the Kruger National Park from March 2015 (autumn, prior to fire season) to December 2015 (spring, late fire season). SAR data were radiometrically calibrated, terrain corrected, and converted to dB values. Fire scars routinely extracted from AFIS (MODIS-based) were used to analyse the variability of the backscatter pre- versus post-fire situations across the various phenological seasons. A tree cover map derived from L-band ALOS PALSAR was used to analyse how varying tree cover from sparse to dense influences the detectability of fire scars. Initial analysis shows that most fire events can be detected and lead to a relative decrease of backscatter of 2dB, with early season fire presenting higher dB difference. Generally fire scars in areas dominated by grass (e.g. gabbro / basalt soils) are better detected than fire scars in areas dominated by trees.
Keyword: savannahs, fire scars, SAR, C-band

PREDICTING AIR TEMPERATURE FROM REMOTE SENSING AND WEATHER STATION DATA AT MT. KILIMANJARO, TANZANIA

Staeps F[1], Appelhans T[1], Detsch F[1], Otte I[1], Nauß T[1]
[1]*Philipps-University Marburg, Geography, Environmental Informatics*

In the face of climate change, the prediction of ecological systems becomes a vital tool for the undertaking of conservation. While it is still not feasible to model entire ecosystems, key elements to ecosystems like abiotic environmental factors can be predicted via remote sensing. Using computer-based modeling, we predicted 8-day aggregates of air temperature (at 2 meter level) for 64 points along the southern slope of Mt. Kilimanjaro, Tanzania, from an elevation of 855 meters to a maximum of 4541 meters ASL. The prediction models interpolate NOAA GSOD air temperature data from six stations in Tanzania and Kenya, at elevations ranging from 557 to 1392 meters ASL, based on the SRTM digital elevation model, MODIS land surface temperature (LST), and the soil adjusted and normalized differenced vegetation indices (NDVI and SAVI). We tested five modeling algorithms (random forest, partial least squares, gamSplines, k-neural networks and linear regression) with and without forward feature selection (FFS), resulting in ten different prediction models. The model predictions were validated against station data from the southern slope of Mt. Kilimanjaro. Four of the ten prediction sets proved suitable for mountainous regions, as the RMSE did not significantly increase with elevation, despite the relatively low elevation range of the six GSOD stations used for model training. The best prediction performance was achieved using partial least square regression with in a mean RMSE of 1.60 at an adjusted R-squared of 0.54. Based on this modeling method, it is now possible to create area-wide 8-day air temperature predictions of the Kilimanjaro region at spatial resolutions as high as 250 meters.

PREDICTING FUTURE DROUGHT EVENTS USING THE RELATIONSHIP BETWEEN SOIL MOISTURE, RAINFALL AND VEGETATION INDICES TO HELP SMALL SCALE FARMERS

Mugwena T[1]
[1]*University Of Johannesburg*

Agriculture plays an important role in South Africa by contributing to 2% of the country's Gross Domestic Product in 2013. Given the current drought in South Africa, there has been speculation on how the government would have predicted and put measures in place to alleviate the impacts of the drought. Drought is a natural recurring climatic phenomenon that is responsible for damage to agricultural systems and other sectors around the world. In agriculture, soil moisture essentially determines the amount and quality of crop yield from a given farm land. Agricultural drought occurs when there is not enough soil moisture to support average crop production. Using the relationship between soil moisture, rainfall and NDVI the paper aims to assess whether agricultural drought risk can be predicted using the relationship between soil moisture, rainfall and NDVI. The drought risk prediction will be mainly aimed at helping small scale subsistence farmers. First soil moisture will be estimated using SAR data and then correlated with the NDVI and rainfall of the study area. The NDVI will be correlated with the lagged variables or soil moisture and rainfall. The results will then be evaluated by computing an optimal lag and corresponding optimal correlation coefficient for NDVI and Soil moisture, as well as NDVI and rainfall.

KEYWORDS: Drought prediction, soil moisture, rainfall, NDVI

PRESENT AND FUTURE SOUTH AFRICAN EARTH OBSERVATION PLATFORMS

Steenkamp N[1]
[1]*Denel Spaceteq*

South Africa has a long and active history of involvement in satellite and Earth observation missions that started as early as 1953 – before the launch of Sputnik. The first successful satellite designed and built in Africa was the student-built Sunsat, launched in 1999. This was followed by two more locally developed satellites – one for an international client and one for the South African government – launched in 2007 and 2009 respectively. Each of these satellites contained increasingly more capable Earth Observation (EO) payloads. Denel Spaceteq is currently busy developing the first of a next generation of EO satellites, called EO-SAT1, for the South African National Space Agency (SANSA). EO-SAT1 hosts two EO payload instruments – HiRes and MedRes – that each image the Earth in 10 spectral bands spanning the visible to near infra-red (VIS-NIR) spectrum. The HiRes instrument images at 2.5m and 10m GSD and has a swath of 30km, while the MedRes instrument images at 15m and 60m GSD and has a swath of 184km (from 700km). By providing both a narrow and a wide swath instrument imaging in the same spectral bands, a large spectrum of EO applications can be serviced with one mission. The EO-SAT1 bus and payload are specifically designed to be modular and scalable with the emphasis on creating a proven technology base from which future enhanced EO platforms can be created. Using the EO-SAT1 technology base, Spaceteq can provide a range of EO platforms with increased GSD (down to 1m), wider swath (larger than 300km) and increased spectral bands.

PRESERVATION THROUGH EDUCATION - GEOTECHNOLOGIES FOR TRAINING UNESCO SITE MANAGERS IN COPERNICUS DATA ANALYSIS FOR MONITORING AND SUSTAINABLE DEVELOPMENT OF CULTURAL AND NATURAL WORLD HERITAGE

Wolf N[1], Siegmund A[1,2], Riembauer G[1], Fuchsgruber V[1]
[1]*Institute for Natural Sciences and GeographyFor* , [2]*Heidelberg University*

The UNESCO has declared a total of 1,052 sites worldwide to World Heritage worth preserving. An increasing number of these sites are flagged as "in Danger", menaced by environmental processes, impacts of climate change and manmade destruction. Aggravating this situation, 90% of these endangered heritage sites are located in developing countries or unstable areas of recent conflicts, where site managers have limited funds and are without access to sophisticated technologies for site monitoring. The potential of Earth Observation for documenting and monitoring world heritage sites is widely recognized. However, site managers often are not aware of the potential of free and open data such as those from the ESA's Sentinel missions and lack the theoretical understanding of remote sensing as well as practical image analysis skills and the technical infrastructure. Therefore, the presented work aims at empowering UNESCO site managers and planning authorities to incorporate Earth Observation data in their daily work routines. An online training environment is provided featuring an easy-to-use web-based remote sensing software alongside accompanying learning material providing the necessary theoretical grounding. The software enables the user to perform satellite image analyses with a didactically prepared toolset and sound explanations of each processing step. The set of functions comprises image classification or change detection. Hands-on exercises impart the technical skills while demonstrating exemplary workflows for monitoring ongoing environmental , economic and social processes. Thereby, this work contributes in communicating the rich World Heritage of our planet and preventing UNESCO sites from becoming endangered while also promoting future pathways for their development with regards to the implementation of the "Sustainable Development Goals" (SDGs).

KEYWORDS: Copernicus, Earth Observation, Natural and Cultural heritage

PROBABILISTIC FORECASTING OF CROP YIELD ACROSS CANADA UNDER ENVIRONMENTAL UNCERTAINTY

Newlands N[1]
[1]*Science and Technology Branch, Agriculture and Agri-Food Canada*

There is increasing concern over the destructive impacts of climate change, extreme weather events and crop disease on global food security. Reliable crop forecasts have the potential to address this challenge by aiding agricultural stakeholders a way to identify potential risks and benefits, with sufficient lead-time to adapt, adjust and improve their crop production plans. This is especially needed during times where crop production is at risk or decisions are more uncertain.

I will provide an overview of the Integrated Canadian Crop Yield Forecaster (ICCYF) – a probabilistic approach that uses machine-learning to predict regional-scale yield distributions, integrating satellite remote-sensing, meteorological, and survey data. I will also showcase my current inter-linked research work that is seeking to improve our ability to forecast changes and losses in yield across complex agricultural landscapes and the multi-scale effects of slower, inter-annual shifts (e.g., ENSO teleconnection forcing), and faster, sudden shocks (extreme weather events and disease outbreaks).

KEYWORDS: Canada, Forecasting, Machine-learning, Probabilistic, Regional variability

PROGRAMS FOR ACCESS COSMO-SKYMED DATA FOR SCIENCE

Coletta A[1], Candela L[1]
[1]*Italian Space Agency*

This paper will outline the various mechanisms for having free or low cost access to COSMO-SkyMed data for science use.

COSMO-SkyMed is an Earth Observation satellite system owned by Italian Space Agency and the Italian Ministry of Defense, intended for both military and civilian use. The constellation includes four identical medium-sized satellites equipped with Synthetic Aperture Radar (SAR) sensors operating at X-Band with global coverage of the planet.

The constellation, operated by ASI, has been completed in 2010 and became fully operational in 2011. The data policy foresees the data access for 3 main user categories:

- owners of the System: the Italian Space Agency and the Italian Ministry of Defence;
- institutional (national and international) users, including also scientific users;
- generic (or commercial) users, including all other users.

The commercial exploitation rights are exclusively granted to the private company, e-Geos.
The institutional users, including also scientific users, can have access to COSMO-SKyMed data through ASI. COSMO-SkyMed data are provided to institutional users at no cost for scientific research and application development through the mechanism of Announcement of Opportunity, Open Call or embracing international initiatives like the Geohazard Supersite and Natural Laboratories (GSNL) or the Global Forest Observation Initiative (GFOI). In addition, institutional national users can access the COSMO-SkyMed data over the Italian territory free of charge through the Map Italy project, started in 2009 on the base of specific needs of the Italian national Department of Civil Protection (DPC).

Moreover ASI can provide data to institutional users at low cost for scientific research and application development through the signature of an Agreement.

PROGRESS ON INTERNATIONAL COORDINATION OF SATELLITE EARTH OBSERVATIONS

Kelly F[1], Ross J[1]
[1]*Committee on Earth Observation Satellites*

Observing the integrated Earth system requires active international cooperation and coordination. No nation can do this on its own.

Through the Committee on Earth Observation Satellites, 32 space agencies work together to ensure satellite Earth observation data is available to be applied to the biggest challenges facing the world including sustainable development, responding to climate change, and disaster risk reduction.

CEOS engagement with end-users of products derived from satellite Earth observation has grown significantly over recent years. Through its links with partners such as the Global Climate Observing System (GCOS), and its role as the 'space arm' of the Group on Earth Observation (GEO), CEOS is able to ensure the data that is collected is put to maximum use, and that end-user and science requirements feed in to future mission design.

Increasingly, CEOS is coordinating satellite observations on a thematic basis, to reflect an increased focus on end-user needs. Strategies for carbon, climate, water, agriculture, disasters and forestry themes have already been developed, and are being implemented.

Long-term satellite programmes, and the adoption of more open data policies, means that for many applications a 'lack of data' will no longer be a problem. Through its work on 'future data architecetures', CEOS is working to exploit new 'big data' technologies and approaches to ensure this flood of data can be put to best use.

CEOS is also exploring how to improve interoperability of data from different missions, enabling users to develop more 'omniverous' applications.

CEOS will provide an update on these key topics, as well as its other activities.

KEYWORDS: satellite; collaboration; international; big data

PROGRESS ON NEAR-REAL TIME FOREST DISTURBANCE MONITORING IN GABON FROM 2015 TO 2016 USING A COMBINATION OF LANDSAT AND SENTINEL 2 IMAGERY

Sannier C[1]
[1]*Sirs Sas*

Gabon is one of the most heavily forested country in the world and the Gabonese Agency for Space Studies and Observations (AGEOS) was set up in 2010 and its knowledge centre was inaugurated in 2015 with direct reception of Landsat 8. One of its aim is to develop a national forest monitoring capability. Forest cover maps and forest cover change maps for 1990, 2000 and 2010 were produced to provide a baseline and confirmed the generally low level of deforestation expected in Gabon. The methodology developed provides a rigorous assessment of area estimates and associated uncertainties and was successfully transfered to AGEOS. As a result, the 2015 update was produced entirely by AGEOS.

This recent update showed that deforestation has now increased in Gabon with the development of its infrastructure and agro-industry. This highlights the need for near-real time monitoring capability which is required to identify forest disturbance as it occurs to take early actions in case of illegal logging.

However, Gabon has one of the heaviest cloud cover in the world preventing the acquisition of cloud free imagery. Nevertheless, initial trials based on dense time series simulating the Sentinel 2 constellation illustrated that despite the heavy cloud cover, regular coverage of most areas and complete national coverage on an annual basis would be possible.

Subsequently, a methodology was developed to combine state of the art pre-processing of Landsat and Sentinel 2 imagery with advanced time series analysis to detect forest disturbance in near real time.

Preliminary conclusions show that the adequate pre-processing techniques are critical to avoid false positive and that relatively simple forest disturbance methods provides good results thanks to the nature of forest cover and deforestation drivers in Gabon.

PROTECTED AREA'S: ISLANDS IN THE LANDSCAPE?

Jewitt D[1]
[1]*Ezemvelo Kzn Wildlife*

Natural habitats are being lost at an alarming rate worldwide, resulting in a loss of biodiversity and significant declines in species populations. Protected areas are becoming islands in a sea of anthropogenic transformation, limiting the ecological and evolutionary processes that create and maintain biodiversity. The landscapes outside of protected areas may be hostile to the survival of many species and different land-uses may limit the ability of species to naturally track changing climates. In this context, protected areas cannot be assumed to be safe havens for biodiversity into the future. Understanding the nature of change in the landscape is essential in order to direct appropriate conservation action. Using a suite of land cover products developed from Landsat and SPOT imagery, the amounts, rates and drivers of habitat loss are determined for KwaZulu-Natal, South Africa, between 1994 and 2011. The land cover information is further used to inform landscape-scale conservation of floristic diversity and natural habitats. Linkages between protected areas are identified and a vulnerability framework developed to guide conservation in the face of land cover change and climate change.

KEYWORDS: Land cover change, climate change, evolutionary and ecological processes, connectivity, landscapes

PROVING THE CAPABILITY FOR LARGE SCALE REGIONAL LAND-COVER DATA PRODUCTION BY SELF-FUNDED COMMERCIAL OPERATORS

Thompson M[1], Hiestermann J[1], Moyo L[1]
[1]*Geoterraimage Pty Ltd*

For service providers developing commercial value-added data content based on remote sensing technologies, the focus is to typically create commercially appropriate geospatial information which has downstream business value. The primary aim being to link locational intelligence with business intelligence in order to better make informed decisions. From a geospatial perspective this locational information must be relevant, informative, and most importantly current; with the ability to maintain the information timeously into the future for change detection purposes. Aligned with this, GeoTerraImage has successfully embarked on the production of land-cover / land-use content over southern Africa. The ability for a private company to successfully implement and complete such an exercise has been the capability to leverage the combined advantages of cutting edge data processing technologies and methodologies, with emphasis on processing repeatability and speed, and the use of a wide range of readily available imagery. These production workflows utilise a wide range of integrated procedures including machine learning algorithms, innovative use of non-specialists for sourcing of reference data, and conventional pixel and object-based image classification routines, and experienced/expert landscape interpretation. This multi-faceted approach to data produce development demonstrates the capability for SMME level commercial entities such as GeoTerraImage to generate industry applicable large data content, in this case, wide area coverage land-cover and land-use data across the sub-continent. Within this development, the emphasis has been placed on the key land-use information, such as mining, human settlements, and agriculture, given the importance of this geo-spatial land-use information in business and socio-economic applications and decision making.

QUALITY ASSESSMENT AND CONTROL OF OUTPUTS OF A NATIONWIDE AGRICULTURAL LAND COVER MAPPING PROGRAM USING LIDAR: PHIL-LIDAR 2 PARMAP EXPERIENCE

Pagkalinawan H[1], Gatdula N[1], Jose R[1], Rollan T[1], Tañada E[1], Tan G[1], Aves J[1], Dela Cruz P[1], Apura R[1], Dela Torre D[1], Magtalas M[1], Dimalanta C[1], Blanco A[2]

[1]*Phil-LIDAR 2 Project 1 Agricultural Resources Extraction from LIDAR Surveys (PARMap),* [2]*UP Department of Geodetic Engineering/ Training Center for Applied Geodesy and Photogrammetry*

The Philippines launched its Nationwide Detailed Resources Assessment using LiDAR (Phil-LiDAR 2) Program in 2014 with the goal of providing accurate and detailed resource maps for national agencies and local government units. One of its components is the Agricultural Resources Extraction from LiDAR Surveys (PARMAP) tasked to produce detailed agricultural maps using LiDAR data. Using LiDAR point cloud with a density of 2 points per meter, LiDAR data derivative layers, and orthophoto with resolution of 0.5 meter, object based image analysis using Support Vector Machine classifier was implemented. Accuracy of the resulting land cover classification were assessed through collection of validation points from the field. An accuracy of at least 90% is required for land cover classification before proceeding to post-processing and map lay-out.

Through knowledge sharing and capacity development facilitated by UP Diliman, partner universities across the Philippines have been producing agricultural land cover maps for their assigned region. Starting 2015, LiDAR blocks have been classified into detailed agricultural land cover at the crop level. Considering output layers are generated by multiple teams working on different landscape complexity with some degree of data quality variability, quality checking is crucial to ensure accuracy standards are met. UP Diliman PARMap devised a centralized and end-to-end scheme divided into four steps – land classification, GIS post-processing, schema application, and map lay-out. At each step, a block is reviewed and, subsequently, either approved or returned with required revisions specified. Turnaround time of review is at least one block (area ranging from 10 to 580 sq. km.) per day. For coastal municipalities, an additional integration process to incorporate coastal features mapped was applied and checked.

Common problems observed during quality checking include misclassifications, gaps between features, incomplete attributes and missing map elements. Some issues are particular to specific blocks such as problematic LiDAR derivatives. UP Diliman addressed these problems through discussion and mentoring visits to partner universities. As of December 2016, a total of 200 municipal agricultural maps have been turned-over. For the remaining six months of the program, an additional 317 maps are expected to be distributed.

KEYWORDS: agricultural land cover, nationwide mapping, quality assessment and control, LiDAR, object based image analysis

QUALITY OF SERVICE

Schreier G

NextGEOSS aims at meeting a high service quality, to adequately foster the private and public sector. This presentation will introduce the NextGEOSS approach to ensure Quality of Service (QoS). Daily operational investigations of the Quality of Service (QoS) of NextGEOSS' distributed infrastructure will be made. During low access periods, all relevant parts of the infrastructure will be verified on a random-choice basis. Probed products will be taken randomly out of the Data Catalogue, which contains metadata on all registered NextGEOSS products. Through this method necessary metadata will be extracted to provide a consistent description of where (at which URL) a product is accessible by which service (e. g. WMS, WMS-T, WCS, WFS, etc.) and what product type (image, NetCDF dataset, JSON) is expected as answers. Identification and correction of failures and bottlenecks of operational data providers will be done through various tests. The NextGEOSS approach to QoS, including a description of various types of tests-checks using examples from use cases, will be presented along with some requirements for CKAN.

RADAR INTERFEROMETRY USING SENTINEL-1 SAR DATA FOR IDENTIFYING SURFACE DEFORMATION IN KWAZULU-NATAL PROVINCE ASSOCIATED WITH THE DURBAN EARTHQUAKE OF 6 FEBRUARY 2016

Thomas A[1]
[1]*Council For Geoscience*

This paper describes the results obtained from radar interferometry using Sentinel-1 SAR data for identifying surface deformation and vertical displacement associated with the 3.7 magnitude Durban earthquake of 6 February 2016 in Durban and surrounding regions of KwaZulu-Natal Province (South Africa), using the differential interferometric synthetic aperture radar (DInSAR) technique. The two pass differential interferometric analysis performed using SNAP software on three Interferometric Wide format scenes (acquired on 31 January 2016 and 12 February 2016) involved the following steps: Slice assembly & coregistration of two SAR images, interferogram formation, topographic phase removal, phase filtering, phase unwrapping, orthorectification and calculation of vertical displacement. The unwrapped phase interferogram obtained from SAR images show that the surface deformation associated with the earthquake is around Durban and along the coastal region north and south of Durban with some minor surface deformation in the north-western regions especially west of Howick and Pietermaritzburg. The vertical displacement calculated using the full scene unwrapped phase interferogram ranged from -34 cm to +34.8 cm. The areas surrounding Durban area shows an uplift of about 14 cm whereas the coastal region north of Durban shows higher uplift values (ranging from 15 to 34.8cm). The vertical displacement calculated for areas of high coherence (<= 0.6) ranged from -27.9cm to +32.1cm; the Durban region shows uplift of about 13 to 15cm whereas the coastal region north of Durban shows uplift of 18 to 20cm. Comparison of surface deformation with simplified geology shows that the extent of the noticed surface deformation is confined mainly in areas underlain by Dwyka Group tillite & Ecca Group shale and sandstone (Karoo Supergroup) and Natal Group sandstone & Natal Metamorphic Province granite and gneiss regions. The shale and sandstone of Durban showed higher deformation as compared to other rocks.

KEYWORDS: Interferometry, SAR, surface deformation, displacement, earthquake.

RADARSAT CONSTELLATION MISSION

Iris S[1]
[1]*Canadian Space Agency*

The RADARSAT Constellation is the next step in evolution of the RADARSAT Program with the objective of ensuring data continuity, improved operational use of synthetic aperture radar (SAR) data and improved system reliability.

The mission consist of three identical C-band SAR satellites flying in a constellation which will provide complete coverage of Canada's land and oceans offering an average daily revisit, as well as a potential daily access to 90% of any location on the globe.

The main objective of the RADARSAT Constellation Mission is on meeting Government of Canada User Department's needs and requirements in Core Use Areas such as Maritime Surveillance, Disaster Management, Ecosystem Monitoring and Northern Development.

The constellation is designed primarily as a wide area monitoring system, offering medium resolution data, but it will also offers high resolution imaging capabilities, including a Spotlight Mode, as well as multiple polarization including Compact Polarimetry.

The greatly enhanced temporal revisit combined with accurate orbital control will enable advanced interferometric applications in between satellites on a four-day cycle that will allow the generation of very accurate coherent change maps. RCM frequent revisit capability, near real-time SAR data availability and vessel identification capabilities (through an AIS payload) will provide the capability to identify and monitor ships up before they enter national waters or ports.

The RADARSAT Constellation Mission is currently under construction with satellite launches planned for 2018.

This presentation will describe the RCM space and ground segments, provide an overall project status and discuss activities surrounding application development as well as Canadian government users' operational readiness.

KEYWORDS: SAR, RADARSAT, satellite, applications

RADIATIVE TRANSFER AND FALSE ALARM MODELLING FOR REMOTE DETECTION OF VEGETATION FIRES USING POTASSIUM SPECTRAL LINE EMISSION

Magidimisha E[1], Griffiths D[1]
[1]*Csir*

Detection and monitoring of wildfires is important in the prevention of loss of life and property. Many countries around the world have implemented various techniques to sense wildfire threats, ranging from forest surveillance from watch-towers through to remote sensing using aircraft and satellites furnished with sophisticated detectors i.e. cooled infrared sensors. A more recent approach for fire detection is by using trace element spectral emissions from burning biomass at fire temperatures. A commonly used trace element in this approach is Potassium (K), due to its low excitation energy and relatively high percentage biomass concentration. Emission intensity at the 770 nm spectral K-line is typically compared to the intensity at a nearby reference wavelength to discriminate fire incidence. However, there are potentially both natural and artificial materials that can cause false alarms in this scenario. For example, false alarms can be caused by highly reflective materials or surfaces such as water, glass, snow and also roofs of buildings. This study investigates the possible causes of false alarms in wildfire detection using K-line emissions. False alarm rates of occurrence from various natural and artificial materials are estimated through comparison of the radiation strength reflected from those materials to that of the fire. The false alarm rate is strongly related to solar elevation during the day. Image data from the Sentinel 2 satellite is used as a sensor data surrogate for statistical purposes. In addition, spectral measurements of Na lamps is conducted at close range to investigate the presence of K in high pressure Na lamps and their potential for false alarms at night.

KEYWORDS: Potassium, False alarm, Wildfires, radiative transfer

RAIN FOR AFRICA – DYNAMIC DECISION SUPPORT DATA FOR AGRICULTURALISTS

Newby T[1], Kaempffer C[1], Kroese N[2], Richard L[3]
[1]Agricultural Research Council, [2]South African Weather Service, [3]HydroLogic BV

Natural agricultural resources globally are finite. In order to ensure national and global food security for a growing global population, agriculturalists will increasingly need to rely on dynamic near real time information such as weather observations and forecasts. Modern information and communications technology facilitate the dissemination of relevant geo-located near real-time data directly to the agriculturalists at their geographic location via mobile devices.

The R4A platform based on the Digital Delta concept connects datasets, tools and customised applications so that users can access current, historical and near real-time information made relevant for specific decisions. Users of the R4A applications include Small holder farmers, Large commercial farmers, Extension services, input suppliers, Financial and Insurance service providers and national weather services in Africa.

This paper describes the operational R4A platform and operational applications for agriculturalists, meteorologists and other specific users through use case pathways.

Data from South African Weather Services, Agricultural Research Council and other partners are accessible via the applications running on the system.

KEYWORDS: Mobile Applications, Small Holder Farmers, Weather information.

REAL-TIME DYNAMIC SENSOR WEB BASED GIS FOR MONITORING AND MANAGEMENT OF ECOLOGICAL ENVIRONMENT IN THE YANGTZE RIVER, CHINA

Li D[1], Chen N[1], Zhu Y[1], Chen Z[1]
[1]*Wuhan University*

The contradiction between flood control, navigation, and generating electricity is serious in the Yangtze River. It needs air-space-ground collaborative sensing method to real-timely monitor the rain, water level, buoy, sediment, and soil moisture etc. information of the Yangtze River Basin. Meanwhile, there are over 5000 navigation marks, 1000 observation stations, and 100 observation satellites monitoring the Yangtze River every day, and it needs advanced GIS technology to manage and analyze the real-time dynamic observed data. To meet these needs, a real-time dynamic Web GIS based on Sensor Web including three main functions as collaborative observation, fused process, and real-time service are proposed. A platform called GeoSensor is implemented based on the Web GIS, and the GeoSensor manages the sensor resources in the Yangtze River. Based on the GeoSensor, a flood control system named "information management and analysis system of cascade hydropower stations in the lower reaches of Jinsha River", a navigation management system named "information sharing system of 21 reservoirs in the upper reaches of the Yangtze River", and a generating electricity management system named "auxiliary analysis system for the maintenance of the main channel of the Yangtze River" are developed respectively. The generating electricity management system realizing real-time dynamically managing more than 300 hydrological and meteorological stations as well as hydrology, sediment, meteorology, remote sensing observed data, hydrology and sediment evolution analysis, and river forecast, those meet the basic needs of water power generation and operation of hydropower station. The flood control system realizes real-time dynamically managing of the water level, rainfall, flow and flow rate of the data of the 8000 stations of the reservoir. The navigation management system realizes real-time dynamically managing 5034 sets of navigation mark monitoring systems and 132 sets of water level monitoring systems.

Water resource management, Real-time dynamic GIS, Sensor Web, Yangtze River

RECENT ADVANCES IN GLOBAL MAPPING OF HUMAN SETTLEMENT WITH SENTINEL DATA: ADDRESSING THE BIG DATA PARADIGM SHIFT THROUGH THE SYMBOLIC MACHINE LEARNING TECHNOLOGY

Corbane C[1], Pesaresi M[1], Syrris V[1], Kemper T[1], Politis P[1], Maffenini L[1], Soille P[1], J. Florczyk A[1], Ferri S[1], Rodriguez D[1], Sabo F[1]

[1]*European Commission, Joint Research Centre*

In the wake of the post-2015 sustainable development goals, the challenge of global mapping of human settlements is gaining momentum. Consistent and accurate information on the location and size of human settlements is essential for supporting the policy frameworks for sustainable cities. The Global Human Settlement Layer derived from Landsat satellite imagery is the first open and free information layer describing the spatial evolution of human settlements in the past 40 years. The recent availability of Sentinel-1 and Sentinel-2 data is expected to bring urban mapping and monitoring to an unprecedented level. With the great advantage of being free and immediately available for the users, Sentinel data can provide up-to-date global information on the status and evolution of human settlements. With the shift to Sentinel imagery, regular updates and incremental improvements of the GHSL will become more feasible and reliable. This study presents the recent developments in global mapping of human settlements with Sentinel data. It puts emphasis on the challenges posed by the processing and analytics of the Sentinel global coverage in the framework of the big data paradigm shift. To cope with those challenges, we exploit the capabilities of the Symbolic Machine Learning together with the functionalities of JRC Big Data infrastructure. The results show that noticeable improvements could be gained from the increased spatial detail and from the thematic contents of Sentinel-2 compared to the Landsat derived product as well as from the complementarity between Sentinel-1 and Sentinel-2 images. Additionally, the extension of SML workflow to a multi-class classification framework is demonstrated with examples on the assessment of the amount of greenness and open spaces in the built-up areas as derived from Sentinel-2. These indicators, underpinning the European regional Urban Policy are regarded as essential contributions to the sustainable development of cities.

RECENT RESULTS OF THE FIREBIRD MISSION

Lorenz E[1], Halle W[1]
[1]*Dlr*

Two years ago the German Aerospace Center (DLR) reported on the ISRSE36 in Berlin about the FireBird Mission. FireBird is a constellation of two small satellites equipped with a unique Bi- Spectral Infrared Instrument. Whereas this instrumentation is mainly dedicated to the investigation of high temperature events a much wider application field was meanwhile examined.

The first of these satellites- TET-1 - was launched on June 22nd 2012. On the ISRSE36 could be presented the results of two year operation. The second satellite- BIROS- was launched on June 22nd 2016.

The outstanding feature of the Infrared Instruments is their higher ground sample resolution and dynamic range compared to systems such as MODIS. This allows the detection of smaller fire events and improves the quality of the quantitative analysis. The detailed analysis of the large number of data sets acquired by TET in the last two years caused significant methodically improvements in the data processing.

Whereas BIROS has the same instrumentation as TET a number of additional technological features implemented in the satellite bus expand the application field of the instruments remarkably.

New High-Torque-Wheels will allow new scanning modes and generating with this new data products to be discussed. A Giga Bit Laser downlink terminal will enable near real time downlinks of large data volumes reducing the response time to disaster events. With an advanced on board processing unit, it is possible to reduce the data stream to a dedicated list of desired resulting parameters to be sent to users by an OrbCom modem. This technology can serve such Online Information Portals like the Advanced Fire Information System (AFIS) in South Africa.

The paper will focus on these new items of the FireBird mission.

KEYWORDS: Small Satellite Constellation, Infrared Instruments, High Temperature Events, Advanced On Board Processing

REDUCING THE RISK OF CYANOBACTERIA USING SATELLITE-BASED SERVICES: THE EARTH OBSERVATION NATIONAL EUTROPHICATION MONITORING PROGRAMME (EONEMP) EXAMPLE

Matthews M[1]
[1]CyanoLakes, [2]University of Cape Town

Cyanobacteria pose a serious health threat to recreational water users due to their potent toxins. Information services based on satellite earth observation may be the only way to effectively warn the public regarding the risks they pose to human health. This talk demonstrates a live web-based public information service developed through a Water Research Commission funded project entitled "The integration of earth observation into the national eutrophication monitoring programme (EONEMP)". The service makes it possible to warn the public regarding the risk from cyanobacteria at more than 100 South African dams using data from the Sentinel-3 Copernicus mission. This, it is hoped, will enable the public to informs their choices regarding recreational activities using simple indicators of health risk to partial and full contact recreational activities. It also charts the way forward for satellite earth observation based services aimed at the protection of public health, and how these can be integrated into the long-term monitoring programs of governments around the world.

REMOTE SENSING BASED ANALYSIS OF MASS MOVEMENTS AND SOIL EROSION IN THE ETHIOPIAN HIGHLANDS

Hochschild V[1], Kropacek J[2], Schillaci C[3], Maerker M[4]
[1]University of Tuebingen (Germany), [2]Czech University of Life Sciences, [3]Department of Agricultural and Environmental Science, [4]Department of Earth and Environmental Science

The Ethiopian Highlands are suffering from tremendous soil erosion processes, induced by an enormous population pressure causing land use changes with land degradation, overgrazing as well as unsuitable agricultural practices. Additionally, the geological setting and the high relief energy are causing high geomorphologic dynamics with mass wasting processes like huge rock falls or landslides amplified by unprofessional road construction. The overall objective of the presented study is the assessment of the present day geomorphic processes and the analysis of the complex system of the Ethiopian Highlands in terms of its erosion sensitivity. The project results provide a comprehensive method assessing erosion and mass wasting features by multisensoral high resolution remote sensing systems applicable in remote and data sparse regions. For the analysis there were IKONOS, Kompsat-2, WorldView-1 and -2, QuickBird as well as Geoeye-1 data used from the optical and ENVISAT data from the microwave range. The presentation will give an introduction to the different test sites at the contact zone of the Ethiopian Plateau with the African Rift Valley and the methods applied for the monitoring and mapping of the natural hazards. Two major landslides (Dessie, Debre Sina) as well as the soil erosion prone area of Andit Tid are subject of the sophisticated remote sensing analysis in order to transfer the approaches to larger Ethiopian areas. By using optical and microwave data combined with an analysis of Digital Elevation Models (DEMs), the objective was to map surface features in order to derive horizontal and vertical displacements. The horizontal gradients were calculated by a subtraction of two DEMs before and after the event in 2007 (one from 1986 derived from aerial photography using a Structure from Motion approach and the other from ALOS/PRISM data from 2008), the vertical component of the movement was validated with ICESat data.

REMOTE SENSING BASED CROP YIELD ESTIMATION: TOWARDS IMPROVING INFORMATION AND DECISION MAKING IN DEVELOPING COUNTRIES

__Forkuor G__[1], Thiel M[2], Tondoh J[1]
[1]Wascal, [2]Department of Remote Sensing, University of Wuerzburg

Remote Sensing Based Crop Yield Estimation: Towards Improving Information and Decision Making in Developing Countries.

Gerald Forkuor1, Michael Thiel2, Jerome Tondoh1
1 West African Science Service Center on Climate Change and Adapted Land Use
2 Department of Remote Sensing, University of Wuerzburg, Germany

In recent years, climatic changes, especially in rainfed dominated smallholder agricultural systems, have often led to crop failure and subsequent production loses. In order to reduce the effects of such changes on livelihoods decision makers would have to be equipped with tools that enable them to estimate potential yields of crops prior to harvest.

Previous studies have shown the potential of remote sensing data in estimating crop yields prior to harvesting. Remote sensing based vegetation indices have been shown to correlate well with crop yields reported by national agricultural ministries. Although these approaches are widely used in the developed world, it has been less explored in developing countries such as Ghana. This study explored the possibilities of developing a model to predict rice yields at district level in northern Ghana. Landsat, MODIS and agricultural statistics data from Ghana's ministry of food and agriculture between 2000 and 2015 were analyzed to: (1) develop a rice yield estimation model by regressing multiple vegetation indices with reported rice yields, (2) determine the optimal period within the season when vegetation indices are most correlated with reported yields and (3) determine which of the tested vegetation indices is most suitable for estimating rice yields. Results show that remote sensing data can be used efficiently in developing countries to improve decision making and enhance efforts at ensuring a food-secure world. With the recent launch of new high resolution and freely available satellite sensors, developing countries are encouraged to adopt such approaches in improving decision making.

KEYWORDS: Yield estimation, Vegetation indices, Rice, Landsat, MODIS

REMOTE SENSING DATA FOR MAPPING AND MONITORING THE AFRICAN SAVANNA WOODLANDS, CASE OF LIWALE, TANZANIA

Mabaso S[1], Bunting P[2], Hardy A[2], Brown S[3], Lucas R[2], Naesset E[4], Gobakken T[4], Kaniki N[5]
[1]The University Of Swaziland, [2]Aberystwyth University, [3]Winrock International, [4]Norwegian University of Life Sciences, [5]Tanzania Forestry Research Institute

Remote sensing data provide unprecedented opportunities for detecting and monitoring forest disturbance and losses. Disturbance and loss have been successfully mapped where the cleared land is of sufficient extent to provide discrimination within an image. However, methodologies for mapping and monitoring forest degradation are still lacking, primarily because these features are small in size. In Tanzania, the size and extent of degradation is currently unknown, yet there has been an increase in shifting cultivation, infrastructural and settlements developments, logging and charcoal burning of indigenous trees. The study sought to first establish a forest baseline, and then develop a method for detecting forest disturbance, using RapidEye and Landsat data. For each sensor, a baseline was derived using Land Cover Classification System (LCCS) approach. The baseline was then used to perform a feature-based change detection analysis on an image of a different date. Resultant possible features of change were further quantified into either real change or seasonality using Random Forests classifier. The results show that both RapidEye and Landsat are capable of being used as the primary sensors for forest mapping and forest monitoring in the savanna woodlands of Tanzania. Major causes of disturbance were shifting cultivation, fuelwood and timber harvesting, and wild fires. As means to reduce the cost associated with high-resolution imagery at a large (national) scale, it is recommended that these two sensors be used interchangeably, depending on whether an area has human interaction or not. Major conclusions included that components of a simple monitoring system (forest baseline establishment and change detection), applicable to Tanzanian REDD+ initiative were produced. Moreover, it was concluded that the developed methods were robust, and thus could be scaled up to a broader national scale and similar environments within the Southern African Development Community.

KEYWORDS: Forest disturbance, forest baseline, monitoring, RapidEye, Landsat

REMOTE SENSING FOR ESSENTIAL BIODIVERSITY VARIABLES (RS4EBV): APPLICATIONS TO PROTECTED AREAS

O'Connor B[1], Skidmore A[2], Darvishzadeh R[2], Wang T[2], McOwen C[1], Vrieling A[2], Harfoot M[1], Paganini M[3]
[1]UNEP-WCMC, [2]University of Twente, Faculty ITC, [3]ESA-ESRIN

Satellite remote sensing is gaining increased recognition as an efficient way to monitor biophysical changes in and around protected areas. Yet a lack of standardised approaches and harmonised observation systems are still hindering progress in its more widespread use. In response the Group on Earth Observation Biodiversity Observation Network (GEO BON) has developed a candidate set of Essential Biodiversity Variables (EBVs), which provide guidance to observation systems as to what and how to measure key aspects of biodiversity status and trends. Satellite remote sensing is the best tool to monitor a subset of these, known as the Remote Sensing Essential Biodiversity Variables (RS-EBVs).

This paper will showcase how the RS4EBV project aims to advance the conceptual development of RS-EBVs on terrestrial ecosystem structure and function by mapping plant functional diversity (FD); the value, range and relative abundance of plant traits in an ecosystem.

RS4EBV is trialling the methodology over two protected areas in Europe: the Bavarian National Park in Germany and the Schiermonnikoog National Park in the Netherlands. The RS-EBVs have been derived from some of the first acquisitions of Sentinel-2 imagery over Europe in 2016 and have been integrated with plot-level metrics of plant FD taken from within the park boundaries. Using environmental co-variates of ecological function such as air temperature and precipitation we have then extrapolated a model to derive a continuous layer of FD. This represents the very first attempts by an inter-disciplinary team of ecologists and remote sensing experts to indirectly derive a canopy-level estimate of plant FD from high resolution space borne imagery.

The initial results and limitations of the method will be discussed as will the policy relevance of the outputs. By monitoring the integrity of ecosystems in protected areas it will be possible to evaluate their resilience in the face of ongoing disturbance.

REMOTE SENSING OF NO2 OVER THE HIGHVELD: COMPARING AIRCRAFT AND SATELLITE PLATFORMS

Broccardo S[1], Heue K[2], Kokhanovsky A[3], Piketh S[1], Platt U[4]
[1]*North-west University,* [2]*DLR Earth Observation Centre,* [3]*EUMETSAT,* [4]*Institut fur Umweldphysik*

Aircraft remote sensing measurements of NO2 using a high-resolution imaging differential optical absorption spectrometer over the South African Highveld are presented. Uncertainty in the air-mass factor due to variation in the vertical profile of NO2 and aerosols is constrained by means of a radiative-transfer modelling sensitivity study. These high resolution aircraft measurements are compared with co-located satellite measurements at several positions downwind from known point sources.

KEYWORDS: DOAS, NO2, aircraft remote sensing, radiative transfer modelling

REMOTE SENSING OF SAVANNA WOODLANDS AT MULTIPLE SCALES: AN AUSTRALIAN EXAMPLE

Lucas R[1], Bunting P[2], Armston J[3], Scarth P[4]
[1]University Of New South Wales, [2]Aberystwyth University, [3]University of Maryland, [4]University of Queensland

For centuries, human activities have continually and gradually eroded the World's forests, with rates of clearance accelerating from the mid 20th century and continuing to this day. Recent (post 1970s) transitions (from forest to non-forest) have been quantified through time-series of remote sensing data but there is increasing interest in understanding and quantifying the events occurring within intact forests that have steadily been degrading over time as a result of natural or human-induced causes (including climate) or are regenerating from previous disturbances. Indeed, such change processes have not been sufficiently quantified even though the consequences include substantial exchanges of greenhouse gases (carbon dioxide, methane) with the atmosphere and losses or recovery of biodiversity and other ecosystem values.

Recognising the need for information on the changing state of intact forests and focusing on savanna woodlands in Queensland, Australia, this study aimed to establish, through integration of temporal airborne and field data, how tree species and communities have responded to natural and human-induced drivers of change, including vegetation management, fire and climatic fluctuations. The project also aimed to establish the efficacy of spaceborne radar and optical data, either singularly or in combination, for routinely detecting these responses. By combining these remote sensing datasets, algorithms for tracking changes within intact forests (e.g., disturbance and degradation, regeneration succession and woody thickening) have been developed which have application across Australia, and other regions globally with structurally similar vegetation.

REMOTE SENSING TRAINING IN AFRICAN CONSERVATION AND ECOLOGY

De Klerk H[1], Buchanan G
[1]*Stellenbosch University*

The potential of remote sensing (RS) to assist with conservation planning, implementation and monitoring is well described, and particularly relevant in African areas that are inaccessible due to terrain, finances or politics. We provide an African perspective on remote sensing (RS) training for conservation and ecology through investigating (i) recent use of RS in African conservation literature, (ii) use of RS in African conservation agencies, (iii) RS training by African institutions and (iv) RS capacity development by ad hoc events. Africa does not produce most of the research using RS in conservation and ecological studies conducted on Africa, with authors with correspondence addresses from the USA predominating (33% of a bibliometric analysis), although South African based authors constituted 20%, Kenya 6% and Tanzania and Ethiopia 4% each. Ideally research should be conducted as close to the point of use to ensure relevance and data residence in the country concerned. This is a point for attention, possibly through international funding to increase the capacity of African academic institutions. Part of this will need to include attention data costs and software costs, internet speeds and human capacity. Data costs have been alleviated by free Landsat and MODIS data, and the Copernicus programs, but there is need for higher resolution imagery to be freely available for certain conservation projects. Open Source software may well offer a long-term solution to software costs. Out of the 72 academic institutions surveyed a number of conservation programs supplied either tailored RS teaching or used 'service modules' to provide RS skills to young graduating conservation professionals, showing a recognition of the importance of RS in conservation in Africa. This paper highlights the success of capacity development in Africa, and the increasing use of remote sensing for conservation in Africa.

REMOTE SENSING-BASED TIME SERIES MODELS FOR MALARIA EARLY WARNING IN MADAGASCAR

Girond F[1,2,3], Herbreteau V[2], Mangeas M[2], Rakotomanana F[1], Brou T[2], Mwendera N[3], Dlamini B[3], Morris N[4], Makomva K[3], Piola P[1]

[1]*Institut Pasteur de Madagascar,* [2]*UMR 228 ESPACE-DEV (IRD, UAG, UM, UR),* [3]*Elimination 8 (E8),* [4]*Health GIS Centre, South African Medical Research Council*

In Madagascar, malaria incidence has decreased in recent decades mainly due to successful malaria control interventions. However, an upsurge of malaria outbreaks in recent years stressed the need for a Malaria Early Warning System (MEWS) adapted to Malagasy resources. Satellite remote sensing has been identified as a key source of meteorological and environmental data that can support early warning health monitoring by providing consistent, repeatable measurements across nearly the entire surface of the Earth.

Despite the importance of such data in the development of a MEWS, its implementation faces scientific and technical challenges. Many health scientists and public health professionals lack the necessary time, technical skills, and computational tools required to (i) identify proper data in terms of spatial and temporal resolution with sufficient archive but also to (ii) acquire (iii) process and (iv) analyze satellite remote sensing data in a way that is relevant to MEWS purpose.

We present here the construction of a MEWS for Madagascar. It relies on the epidemiological surveillance of malaria at 34 sentinel sites throughout the country, which have been recording information since 2007. Epidemiological records are transmitted by Short Message Service (SMS). We developed a computer application for concomitantly and automatically acquiring and processing remote sensing data (temperature, rainfall, Normalized Difference Vegetation Index). This site-specific satellite weather data allows the detection of malaria trends and malaria outbreak alerts based on various epidemic thresholds and a forecasting component.
Such web-based surveillance systems are fundamental monitoring tools for better control of malaria in real time. In an environment with limited resources, we had to optimize the system by limiting the downloading of data. We will discuss here how this system could be reproduced and optimized and our expectations in new satellite data.

RETRIEVAL AND VALIDATION OF GRASS TRAITS USING SENTINEL- 2

Darvishzadeh R[1], Azong Cho M[2], Ramoelo A[2], Wang T[1], Skidmore A[1]
[1]*Uinversity of Twente, ITC Faculty,* [2]*Council for Scientific and Industrial Research (CSIR)*

Assessment of grassland productivity and health require knowledge on the spatial and temporal distribution of grass traits. Savanna grasslands are characterized with high diversity and highly heterogeneous canopies, and therefore, present a challenge for remote sensing applications. Launch of the new Satellites, in particular the Sentinel-2 series have increased the potential and recognition of remote sensing as a reliable alternative to monitor grassland traits. Numerous efforts to estimate and quantify different grassland traits using Radiative Transfer Models (RTM) and statistical models have been carried out. Yet lack of high temporal and spectral imageries has hindered the widespread use of it. Sentinel-2 provide spectral information over the VIS NIR spectrum with high spatial resolution and offer opportunities for continuous monitoring. Therefore, as part of the training course: Mapping and monitoring vegetation using Sentinel-2 data, in this study, the retrieval of LAI, fraction of vegetation cover and chlorophyll are investigated utilizing Sentinel-2 images and RTM. Sentinel-2 images were acquired from Pretoria, South Africa during the course organization. In situ measurements of LAI, vegetation fraction and chlorophyll were performed by a large group of trainees during a field campaign concomitant with the time of image acquisitions in 2016. The widely used canopy radiative transfer model: PROSAIL was investigated for retrieval of traits. The RTM was first parameterized based on the spectral band settings of Sentinel-2. Consequently a large look-up table (LUT) was generated for the study site accounting for the available prior information related to the vegetation characteristics in the study site. To assess the performance of the model inversion and analyze the suitability of the image data and model, the normalized RMSE and R^2 between in situ measurements and estimated traits were used. Our results demonstrate the potential of model inversion using Sentinel-2 data for estimating vegetation traits in African savanna.

ISRSE37
TSHWANE 2017

RICE MONITORING USING SAR DATA: TOWARDS OPERATIONAL DATA EXPLOITATION

Le Toan T[1], Bouvet A[1], Phan T[1], Lam-Dao N[2]
[1]CESBIO, [2]VNSC/STAC

Rica is the staple food for more than half of the world population. Paddy rice cropland distributions, management intensity, and climate impacts will undergo changes over the coming decades. For global food security, it is important to have consistent methods to monitor the predicted changes in rice production, e.g. by using Earth Observation.

Among the remote sensing systems, satellite SARs are essential in tropical regions where much of the world rice is grown and where optical satellite data are severely hampered by cloud cover. Secondly, radar backscatter has been found to have a specific temporal variation during the growth cycle and to be sensitive to key parameters of the rice plant, including phenological stage, height, age, and biomass. Up to now, SAR data used for rice monitoring have been demonstrated only in research works . With the launch of Sentinel-1 which can acquire data every 12 days or 6 days, dense time series of C-band data are available at a number of rice grown regions. These systematic and open data are now subject to development of robust and generalized methods towards their long term exploitation.

This paper presents the works conducted in the Mekong Delta, Vietnam, within the Asia-RiCE Initiatives , and in the frame of the ESA DUE Innovators GEORICE project. Specifically, the paper presents a rice monitoring framework integrating the latest technologies available. For the first time, mapping of rice area and rice phenological stage has been obtained nearly every 12 days for the Mekong delta and test has also been done at national scale (Vietnam, Cambodge). The paper will also discuss the utility of such rice monitoring system to assist decision-makers for more informed decisions that will mitigate the impact of variability in climate and in rice production systems.

KEYWORDS: Rice monitoring, GEOGLAM, Asia-Rice, Sentinel-1, GEORICE

RIGOROUS COMPARISON OF DIVERSE CLASSIFICATION ALGORITHMS: THE ACMAC PROJECT REVISITED

Lawrence R[1]
[1]*Montana State University*

The AmericaView Classification Methods Accuracy Comparison Project (ACMAC) provided a framework for rigorously comparing classification algorithms using 30 moderate-resolution satellite datasets and applied it to six classification methods. ACMAC found that the best method was dataset dependent and there was no single best method. We have updated this approach by (1) increasing the number of datasets to 45, (2) increasing the algorithms to 14, including many popularly used ones, and (3) optimizing algorithm parameters using cross validation techniques. Tree-based methods, especially RandomForest and C5.0, performed best most often, followed by support vector machines, whose performance varied on the kernel selected. K-nearest neighbors also performed well with many datasets, while maximum likelihood and neural networks were comparatively poor. This research further supports an agnostic approach to image classification, where multiple methods are compared to determine which results in the highest accuracy. RandomForest was the best method most often (44% of the time), but was also close to the best method in all other cases, indicating that it was the preferred method if only one was used.

SAOCOM MISSION OVERVIEW

Frulla L
[1]Argentina National Commission of Space Activities (CONAE)

The SAOCOM (Satélite Argentino de Observación COn Microondas) mission is a constellation of two fully polarimetric L-band SAR satellites under development by the National Commission for Space Activities of Argentina (CONAE). The system will provide global coverage with a repeat cycle of 16 days per satellite and 8 days for the constellation. SAOCOM-1 is part of an agreement (SIASGE) between the Italian Space Agency (ASI) and CONAE. SAOCOM-1A is anticipated for launch by the end of 2017, and SAOCOM-1B one year later.

The SAOCOM-1 main driver soil moisture in support to agricultural and hydrological applications, particularly floods. SAOCOM also aims to meet emergency requirements within the framework of the Argentinian National Space Plan, including interferometric observations. SAOCOM can be thought as an end-to-end Earth Observation System dedicated to the remote sensing and data exploitation for enhancing socioeconomic activities and scientific studies.

SAOCOM-1 mission scenario involves observations from Argentina and the rest of the world, and is composed by three categories. One is dedicated to fixed observations related to the main driver (soil moisture) and calibration. The second category is dedicated to user requests and the third category to a background mission which is subject to a pre-defined acquisition strategy - the Integrated Mission Acquisition Strategy, IMAS - which is designed for the systematic collection of interferometric and polarimetric time-series data over Argentina and the world. From this mission scenario SAOCOM-1 mission aims to support international environmental coventions as part of Committee of Earth Observation Satellites (CEOS) and compliant with CEOS Space Data Coordination group recommendations for SAR missions.

The presentation will provide a mission review including description of the present status, its capabilities, the mission scenario and anticipated products from SAOCOM-1.

KEYWORDS: SAOCOM, L-band SAR, Earth Observation

SAR-EDU - THE EDUCATION PORTAL FOR RADAR REMOTE SENSING

Eckardt R[1], **Schmullius C[1]**, Thiel C[1], Pathe C[1]
[1]*Friedrich-schiller University Of Jena*

Radar remote sensing has a long and prosper tradition in the Earth observation sciences. Past, present and future satellite missions provide vast amounts of data for the analysis of the condition and development of the Earth surface as well as natural and anthropogenic habitats. The application of radar data is rife in nearly all fields of geoscientific research, decision making and the creation of commercial geographic products.

The project SAR-EDU is a joint education initiative of the Friedrich-Schiller-University of Jena (FSU), the German Aerospace Center (DLR) and numerous partners in radar-related scientific institutions. In a previous project phase two main cornerstones for education in the field of applied radar remote sensing were established. Since 2013 the FSU is hosting a yearly summer school on applied radar remote sensing. Furthermore DLR and FSU published the SAR-EDU learning portal in late 2014 (https://saredu.dlr.de). This web portal is designed to provide access to a vast range of teaching material regarding the basics, methods and applications for radar remote sensing.

In the recent project phase it is planned to equip the existing web portal with further interactive functionality in order to create a vital online community for radar remote sensing education. The teaching material is available under a creative commons license (CC BY-SA 4.0) allowing for the usage, adaption and distribution of the material.

Future visions for this education platform include the creation of Massive Open Online Courses (MOOC) and innovative ways to provide, share and communicate application oriented SAR knowledge.

SAR-MULTISPECTRAL FUSION FOR IMPROVED MAPPING OF CULTIVATED CROP FIELDS

Ngie A[2], Tesfamichael S[2], Ahmed F[1]
[1]*University Of The Witwatersrand,* [2]*University of Johannesburg*

The application of remotely sensed data in interrogating vegetation coverage on earth has evolved so much with different technologies with varying capabilities and inabilities. Optical remote sensing suffers from inability to differentiate structural information such as the difference between grass and crop, while radar is unable to differentiate albedo or spectral variations such as short vegetation and bare ground. The combination of data sets from both technologies provides structural and optical information that can be used to identify features more easily than individual data sets. Therefore, this paper sought to evaluate the potential of such combined data sets (optical and radar) in mapping summer agricultural activities in the north eastern part of the Free State province of South Africa. The fused data sets of multispectral bands (Sentinel -2) and radar bands (Sentinel -1) then underwent standard classification methodology for extraction of features. Through statistical comparisons with reference data sets collected from Google Earth high resolution images, fields of various summer agricultural activities were delineated. The findings showed the potential for improved reliability of mapping cultivation fields in areas where crop types exhibit a great deal of similarity with grasses.

KEYWORDS: Optical radar, remote sensing, mapping cultivation fields

SATELLITE AND AIRBORNE OBSERVATIONS OF SPECIATED AIRBORNE PARTICULATE MATTER

Ge C[3], Xu R[3], Wang J[3], Garay M[1], Wu L[1], Diner D[1], Kalashnikova O[1], Xu F[1], Franklin M[2]
[1]California Institute of Technology, [2]University of South California, [3]University of Nebraska

Ambient particulate matter (PM) is the top environmental risk factor worldwide, responsible for nearly 3 million premature deaths per year. Although there is a scientific consensus that exposure to PM increases the risks of death and disease, the relative toxicity of specific PM types—components having different size and chemical composition—is poorly understood. We demonstrate how data from the Multi-angle Imaging SpectroRadiometer (MISR) instrument that has been flying on NASA's Terra Earth Observing System satellite since 1999 can be used to provide estimates of surface PM concentrations. As a proof-of-concept, MISR high-resolution 4.4 km aerosol retrievals were combined with surface air pollution and meteorological measurements over Southern California. Spatio-temporal statistical models were built and cross-validated upon these combined data, resulting in an approach to reliably retrieve near-surface PM. Although MISR retrievals worked well for total PM2.5, we found limited sensitivity to PM10 or to PM2.5 components.

To improve particle type sensitivity, we introduce and evaluate a new approach of retrieving near-surface and atmospheric airborne PM types using information contained in multi-angle, spectropolarimetric remote sensing imagery constrained with a first-guess on aerosol chemical type and optical properties from a chemical transport model (CTM). The CTM vertical profile is used to convert total column aerosol properties differentiated by particle type to near-surface speciated PM. We tested the approach using Airborne Multiangle SpectroPolarimetric Imager (AirMSPI) data collected over the California Central Valley during recent field campaigns, and from a high-resolution (4 km) Weather Research & Forecasting (WRF)-Chem aerosol model over the same domains at the time of AirMSPI overpasses. We find that WRF-Chem-constrained AirMSPI retrievals of speciated PM show better agreement with ground-based observations (order of 30% difference) than WRF-Chem model predictions alone (order of 5-10 underestimation).

We conclude with a discussion of plans for future field campaigns and introduce the Multi-Angle Imager for Aerosols instrument that was recently selected under NASA's third Earth Venture Instrument call. MAIA will provide new information that enables estimates of speciated (size- and particle type- stratified) surface PM from space over major cities around the globe, and enable improved associations between particulate air pollution and human health.

ISRSE37
TSHWANE 2017

SATELLITE AND UAV DATA FOR THE CLASSIFICATION OF SAND DUNE VEGETATION

De Giglio M[1], Goffo F[1], Greggio N[1], Merloni N[2], Dubbini M[1], Barbarella M[1]
[1]University Of Bologna, [2]Ministero dell'Istruzione, dell'Università e della Ricerca

Within coastal systems, sand dunes are the only natural barriers able to counteract erosive processes, playing a significant role in coastline dynamics. Since their equilibrium is often threatened by both human activities and high vulnerability of the coastal environment, dunes require increasing attention and specific monitoring. Located between the mainland and the sea, dunes are unique residue habitats for some preserved plant and animal species. In particular, the vegetation is important for its function of consolidation that improves the vertical dune development. Generally, dune vegetation is organized into several formations along a profile according to different environmental and ecological gradients. In the study area (Emilia-Romagna, Italy), natural and anthropogenic subsidence, tourism and extreme events are reducing dune extension.

This work aims to determine the evolutionary status of the dunes through a vegetation analysis. A geolocation classification allows for the analysis of dune progradation or retreat, usually based on the geometric reconstruction. High spectral resolution satellite images (WorldView-2) and a multispectral orthophoto, obtained from high spatial resolution data acquired by UAV, were used. Objects and pixel classification algorithms were applied. Results were statistically compared to verify the validity of various combinations of data and methods. Using the same bands, results show that both data are usable but for the UAV orthophoto classification the object approach is more accurate.

Considering the availability of many bands of the WorlView-2, further details on the type and the plant phenological state were analyzed. Moreover, from UAV data, it has been possible to both identify some species by photo-interpretation, unrecognizable with automated methods, as well as to extract an accurate DSM of the dune.

Finally, all results were integrated to obtain a 3D reconstruction of the vegetation distribution that is useful for a reliable and easily achievable monitoring of dune spatial evolution.

KEYWORDS: Vegetation dunes, Classification, UAV/WorldView-2

SATELLITE BASED MONITORING OF INTRA-ANNUAL RESERVOIR STORAGE VOLUMES AND IT'S IMPLICATIONS FOR WATER AND FOOD SECURITY: A CASE STUDY OF LAKE BAM IN BURKINA FASO

Forkuor G[1], Becht R[2], Ibrahim B[3], Idriss S[4]
[1]Wascal, [2]ITC Faculty, University of Twente, [3]Abdou Moumouni University, [4]INERA/CNRST/MRSI

Climate change and variability is negatively affecting food security in most countries in West Africa. Unreliable rainfall patterns in recent years has led to crop failure in the rainfed agriculture dominated region. This situation has brought to the fore the need to improve irrigation facilities and expand irrigable land. But expanding irrigable land requires availability and efficient management of water resources, especially considering the potential reduction of rainfall totals as predicted by some climate models.

Small reservoirs and natural lakes are common in arid regions of West Africa where a 7 month dry season often lead to poor water availability for agricultural (irrigation) and domestic water usage. In this regard, a better understanding of irrigation water demand, vis-à-vis available water resources in the dry season is critical for efficiently managing the scarce resource. But such analysis have been hampered in the past due to poor data availability. The proliferation of earth observation (EO) data in recent years, especially freely available data such as that of the European Space Agency's Sentinel satellites, present an excellent opportunity for data scarce regions to monitor available water resources from space and evaluate its sustainability in light of irrigation water demand during the dry season. This paper presents preliminary results of a study that uses Sentinel-1, 2 and other freely available EO data to monitor the temporal evolution of water availability in a natural lake in Burkina Faso and the corresponding crop water requirements during the 2016/2017 dry season. Results indicate the potential of the Sentinels to support water resource managers in efficiently managing the resource for increased agricultural productivity.

KEYWORDS: Sentinel, Water demand, Irrigation, Earth observation, West Africa

SATELLITE DRIVEN DISTRIBUTION MODELS OF ENDANGERED ATLANTIC STURGEON OCCURRENCE IN THE WESTERN MID-ATLANTIC TO REDUCE HARMFUL INTERACTIONS

Breece M[1], Fox D[2], Haulsee D[1], Oliver M[1]
[1]University Of Delaware, [2]Delaware State University

Unintentional interactions with endangered species can severely limit recovery even after directed harvests have been eliminated. Estimating when and where these species occur can facilitate the reduction of harmful interactions. Daily movements and seasonal migrations require habitats to be measured on similar scales to that of the animal observations and management. If scales or timing are not aligned the researcher could be measuring properties that are not being experienced or even avoided by the organism. Polar orbiting satellites like MODIS-Aqua have the spatial (~ daily) and temporal (~ 1km) resolution, and footprint needed to estimate the habitat of marine organism at scales relevant to their daily movement. Atlantic sturgeon (Acipenser oxyrinchus oxyrinchus) are an endangered species that migrate through, and occupy the coastal waters of the Mid-Atlantic where they interact with anthropogenic activities. Measures to understand and avoid Atlantic sturgeon that take into consideration the dynamic nature of their habitat may reduce harmful interactions. In this study we match fisheries independent biotelemetry observations of Atlantic sturgeon with daily satellite remote sensing observations to construct a time resolved spatial distribution model of Atlantic sturgeon. We find that depth, day-of-year, sea surface temperature, and ocean color are the most important predictors of Atlantic sturgeon occurrence. Demographic factors, such as sex and river of origin are of secondary importance. We find strong spatial differences in Spring and Fall migration patterns, when anthropogenic interactions are their greatest. Our cross-validated models correctly identify > 88% of biotelemetry observations and 67% of fisheries dependent observations throughout the year. However, during their migrations, when harmful interactions are highest, our models correctly identify 91% of fisheries dependent observations. We suggest that this model can be used for guidance to practitioners and stakeholders to reduce interactions with this endangered species.

SATELLITE MONITORING OF DUST STORMS OVER SOUTHWEST ASIA

Rashki A[1]
[1]*Ferdowsi University Of Mashhad*

Satellite remote sensing provides important observational constraints for monitoring dust life cycle and improving the understanding of its effects on local to global scales. The present work analyzes the dust aerosol patterns over the arid environment of southwest Asia particularly in southeastern Iran, by means of multiple satellite platforms aiming to reveal dynamic, spatio-temporal distribution and trends. This work also investigates the modulation in dust activity over southwest (SW) Asia attributed to changes in the mean sea level pressure (MSLP) between the Caspian Sea (CS) and Hindu Kush (HK) during the summer months (June-July-August-September, JJAS) of the period 2000-2014.satellite observations were includes TOMS, OMI, METEOSAT, MODIS, MISR and HYSPLIT forward trajectories. The results indicated, several dust's hot points in the study area. It is found that in few cases the dust storms from Sistan in south east Iran affect central/south Arabian Sea and India, while they control the aerosol loading over northernmost Arabian Sea. The Infrared Difference Dust Index (IDDI) images confirming the main pathways of the dust plumes and illustrating the importance of the region as one of the most active dust sources in southwest Asia.

SATELLITE OBSERVED WATER QUALITY CHANGES IN THE LAURENTIAN GREAT LAKES DUE TO INVASIVE SPECIES, ANTHROPOGENIC FORCING AND CLIMATE CHANGE

Shuchman R[1], Sayers M[1], Fahnenstiel G[1], Leshkevich G[2], Bosse K[1]
[1]Michigan Tech Research Institute, Michigan Technological University, [2]NOAA GLERL

Long time series ocean color satellite data from sensors such as SeaWiFS, MODIS, MERIS and VIIRS can be used to measure Laurentian Great Lakes water quality parameters that include Chlorophyll, dissolved organic carbon, suspended minerals, harmful algal blooms (HABs), attenuation coefficients (Kd), photic zone depth, and primary productivity on weekly, monthly and annual observational intervals. The observed changes in these Great Lakes water quality parameters over time are a direct result of the introduction of invasive species such as the Dreissena mussels as well as anthropogenic forcing and climate change. Time series of individual lake wide, nearshore and open lake averages of the above mentioned water quality parameters have been generated starting in 1998 with SeaWiFS, and to the present using MODIS, MERIS and VIIRS. These time series have documented the affect the mussels have had on water clarity improvement by decreasing the chlorophyll and mineral concentrations. Primary productivity has decreased in several of the Lakes due to the decrease in phytoplankton abundance even though the photic zone has increased by greater than a factor of 2 during this 18 year observation period. Comparing monthly and annual water quality values of Lake Superior to the lower lakes is insightful because Lake Superior, the largest and most northern of the five Great Lakes, has not to date been affected by the invasive mussels and thus can be considered a control. In contrast Lake Erie, the most southern and shallow of the five Laurentian Great Lakes, is heavily influenced by agricultural practices (i.e. nutrient runoff) which directly influence the annual extent of HABS in the Western Basin of that Lake.

SCIENTIFIC ACHIEVEMENTS BY GREENHOUSE GASES OBSERVING SATELLITE (GOSAT) AND FUTURE PROSPECTS FOR GOSAT-2

Imasu R[4], Nakajima M[2], Matsunaga T[1], Yokota T[1], Inoue G[3]
[1]National Institute for Environmental Studies, [2]Japan Aerospace Exploration Agency, [3]University of Tsukuba, [4]The University of Tokyo

Greenhouse Gases Observing Satellite (GOSAT) and its successor, GOSAT-2, are Japanese earth observing satellites for greenhouse gases measurements from space. Both satellite projects are joint efforts among Ministry of the Environment (MOE), Japan Aerospace Exploration Agency (JAXA), and National Institute for Environmental Studies (NIES).

GOSAT was launched in January 2009, already finished its design lifetime (five years), and is currently in its extended operation period. It has a Fourier transform spectrometer (FTS) for the measurements of columnar abundances of greenhouse and other gases and a UV-VIS-NIR-SWIR imager (CAI) for cloud and aerosol detection. Its data have been used to calculate whole-atmosphere monthly mean carbon dioxide concentration and to identify locations with large anthoropogenic emissions of CO2 and methane.

GOSAT-2 will be launched in FY2018. GOSAT-2 instruments (FTS-2 and CAI-2) will be modified or improved based on the experiences of GOSAT instruments. FTS-2 will have the extended spectral coverage for carbon monoxide measurement and the intelligent pointing capability to avoid cloud contamination. CAI-2 will have multiple UV bands for more precise land aerosol monitoring and the forward/backward viewing capability to avoid sun glint over oceans. Critical design reviews of GOSAT-2 spacecraft, earth observing instruments, and ground systems have been completed and manufacturing of them is ongoing.

In this presentation, scientific achievements by GOSAT and future prospects for GOSAT-2 will be discussed.

SEA SURFACE SALINITY FROM SMOS SATELLITE MISSION: MAJOR ACHIEVEMENTS AFTER 7 YEARS IN ORBIT (2010-2017)

Boutin J[1], Reul N[2], Delcroix T[3]
[1]LOCEAN, [2]IFREMER, [3]LEGOS

Ocean salinity is an important driver of ocean circulation and represents a key indicator of changes in the global water cycle. Recent advances in observing sea surface salinity (SSS) from space have provided an unprecedented capability to study the influence of salinity on ocean circulation and its relations to climate variability and the water cycle. The SMOS satellite mission is the first satellite mission that provides a global monitoring of SSS from space and a quasi-synoptic monitoring of the global ocean every 3 to 5 days. In tropical and subtropical areas, the precision of monthly SSS anomalies is close to 0.2 pss. Here we review main scientific results achieved thanks to these novel measurements.

Large scale interannual SSS anomalies in the tropical Pacific Ocean, linked to El Niño/La Niña events (2010-2011 and 2015-2016) and in the Indian Ocean, linked to the Indian Ocean Dipole, have been detected by SMOS with a spatio-temporal resolution very complementary to Argo and ships of opportunity. In the cold tongue region of the equatorial Atlantic Ocean, SSS seasonal cycle has been shown to be lagged with respect to the SST. In addition, tropical instability waves in the equatorial Pacific and Atlantic Ocean have been observed to vary interannually. SMOS data have also allowed to conduct unprecedented studies on the small scales of SSS at 50 to 100 km. This concerns in particular the propagation of eddies in the Gulf Stream, North Equatorial Counter Currents, rings in the north Brazil current, the seasonal variability of the surface horizontal thermohaline structure in the subtropical Atlantic Ocean. The influence on the SSS variability of the freshwater flux relative to the ocean dynamics has been studied both in main river plumes and in rainy regions of the open ocean (fresh pool of the eastern Pacific Ocean, Intertropical Convergence Zones).

SEMANTICS AND INTEROPERABILITY WITH CKAN

Maso J

Sharing Earth observation data requires exchange of geoscientific information across discipline boundaries. This information can be both rich and complex, and content is not always readily interpretable by either humans or machines. Difficulties arise through differing exchange formats, lack of common semantics, divergent access mechanisms, etc. Recent developments in distributed, service-oriented, information systems using web-based (W3C, OASIS, ISO, OGC) standards are leading to advances in data interoperability. Furthermore, work is underway to understand how meaning may be represented using ontologies and other semantic mechanisms. In this presentation an overview of developments in interoperable data sharing will be given including the potential represented by CKAN. It will show how representation of semantic meaning can enable interpretation of geoscientific information.

SEN2-AGRI NATIONAL DEMONSTRATION IN UKRAINE

Kussul N[1], Kolotii A[1], Bellemans N[4], Shelestov A[2], Bontemps S[4], Lavreniuk M[2], Yailymov B[1], Yailymova H[5], Koetz B[3], Defourny P[4]

[1]*Space Research Institute,* [2]*National Technical University of Ukraine "Igor Sikorsky Kyiv Polytechnic Institute,* [3]*European Space Agency,* [4]*Université catholique de Louvain,* [5]*Taras Shevchenko National University of Kyiv*

ESA's project "Sentinel-2 for Agriculture" is started in 2014 as a major contribution to the GEOGLAM initiative and dedicated to Sentinel-2 data usage intensification as well as new satellite products benchmarking [1].

This project is coordinated by Universite catholique de Louvain (UCL). During demonstration stage 3 countries were selected for country level demonstration. Among them Ukraine was selected due to wide range of main crops (both winter and summer), big fields and high climate variability over the territory (3 main climatic zones).

Within the project Space Research Institute (as sub-contractor of UCL) was responsible for ground data collection along the roads of main agro-climatic zones for land cover and crop type mapping [2]. For biophysical parameters validation Leaf Area Index (LAI) was estimated with use of non-destructive method based on DHP-imagery and VALERI protocol.

For Sentinel-2 data processing an automated information system was deployed and launched in Ukraine. This system was developed within Sen2-Agri project and should provide the effective Sentinel-2 data processing for big territories. Within Ukrainian national demonstration for the first time satellite products were obtained for the whole Ukraine at 10 m spatial resolution and good accuracy. In particular, crop land mask overall accuracy is more than 96%.

KEYWORDS: crop type, classification, land cover, mapping, Sentinel-2

SENSING THE PAST FROM SPACE: REMOTE SENSING FOR DOCUMENTATION, MONITORING AND PRESERVATION OF CULTURAL HERITAGE

Lasaponara R[1]
[1]*CNR-IMAA*

The use of EO technologies in archaeology is stepping in its golden age characterized by an increasing growth of both classical and emerging technologies and multidisciplinary methodologies, addressed to the study and conservation of natural and cultural heritage. The availability of the new technologies have opened new infinite possibilities, unthinkable only a few years ago especially for archaeology and cultural landscape that is an integral part of our archaeological heritage being that it preserves the main features that identity the evolutionary history of civilization over time.

A wide availability of 3D technologies, from active and passive satellite, aerial and ground sensors, including laser scanning, GIS mapping tools, virtual 4D modeling, augmented reality etc enables us to address manifold strategic challenges. Thus opening a new horizon in Archaeology. The impact of digital technologies for archaeology regards researchers, professionals as well as end-users. This is clearly evident thinking about, for example, the new portable devices, as tablets and smart-phones, nowadays equipped with integrated GPS, very powerful processors and video cards, which permit us to enjoy virtual reconstructions as well as an increasing amount of information available "exactly on site and on time".

Moreover, we already live in an age of a growing availability of free data and open access software tools that enhance a powerful link between in situ investigations and computer-based analysis thus offering a new opportunity for the exploitation of scientific outcomes. Therefore, the new and emerging challenges are the dissemination of data, the interoperability, the costs, simplicity in use and speed of applications, to make them open and understandable to everybody.

The lecture will be focused on:

--An Overview on active and passive satellite remote sensing technologies for documentation, monitoring and preservation of natural and cultural heritage
--Remarkable case studies selected from Europe, Africa, Asia and Southern America

KEYWORDS: Archaeology and Landscape, water systems, looting- monitoring, Spectral signatures, Classifications.

SENTINEL 2 FOR AGRICULTURE: PROVISIONAL RESULTS FROM THE SOUTH AFRICAN NATIONAL VALIDATION SITE

Newby T[1], Chirima G[1], Nyamugama A[1], Ferreira S[2], Defourny P[4], Du Preeze E[3], Bellemans N[4], Bontemps S[4]
[1]Agricultural Research Council, [2]GeoTerraImage (Pty) Ltd., [3]SIQ (Pty) Ltd., [4]Université catholique de Louvain

The Sentinel-2 for Agriculture system was designed to deliver Earth observation based products dedicated to agriculture monitoring in an operational way. Developed as an open source system, it allows users to generate at their own premises products useful for crop production monitoring.

During the 2016 – 2017 crop production season in South Africa, the system is being validated for both the winter and summer grain crops. Four main products are being produced, namely cropped area maps, crop type maps produced twice during the growing season, Leaf area index(LAI) and NDVI product produced every 10 days, as well as a monthly cloud free image composite. These products are based on Sentinel 2 and Landsat 8 image data. The products are being generated by the consortium of the Sentinel-2 for Agriculture project and in parallel, by national partners having the system installed at their premises. National partners are also in charge of in-situ data collection at national scale. This close collaboration with teams working on the field has the additional objective to transfer the system to their operations after the end of the project.

The products are being validated by comparison with the current operational Producer Independent Crop Estimation System (PICES) a component of the national crop estimation system. Observers in light aircraft collect point observations of crop type, non-crop points and other ancillary data. Some of the data is used as training data for the systems classification algorithms. LAI validation data is also being collected by means of photographs captured from the light aircraft and in field.
Successful validation of the system will result in its incorporation into the current operational national crop estimation system resulting in a cost effective and technical improvement of the already cutting edge system used in South Africa.

KEYWORDS: Crop monitorng, Sentinel 2, South Africa,

SENTINEL HUB - NEXT GENERATION PLATFORM FOR SATELLITE IMAGERY APPLICATIONS

Milcinski G[1], Sernek B[1], Kadunc M[1], Mocnik R[1], Kolaric P[1], Batic M[1], Repse M[1], Sovdat B[1]
[1]*Sinergise*

Sentinel-2 data are being distributed since November 2015. However, there are still not many publicly available applications, based on these data. The reason probably lies in technical complexity of using S-2 data, especially if one wants to use full potential of multi-temporal and multi-spectral imaging. Vast volume of data to download, store and process is technically challenging, even more so using "current-generation" principles, such as building pyramids of final products, resulting in billions of small files, which have to be managed efficiently.

At Sinergise, we have approached the problem from a different angle - thinking about which products/services we can offer without significant pre-processing, tiling or even manual effort. We have set-up the copy of global S-2 archive at AWS and implemented a processing chain, which taps into the data in real-time, when requests are coming in. We were able to establish WMS/WMTS service, which gets request with specific parameters (e.g. AOI, maximum cloud coverage, time range), queries the database for matching scenes, downloads the relevant data from AWS, creates a composite, a mosaic and then the final result, based on set rendering parameters (e.g. true color, false color, NDVI, EVI, LAI, etc.). All of these actions are being done on full, worldwide, S-2 archive (multi-temporal and multi-spectral) in just a few seconds. To demonstrate the technology, we created a simple web application, called "Sentinel Playground", which makes it possible to query Sentinel-2 data anywhere in the world.

Sentinel-2 data are only as useful as the applications built on top of it. We would like people to not bother too much with basic processing and storing of data but rather to focus on value added services. This is why we are looking for partners, who would bring their remote sensing expertise and create interesting new services.

SENTINEL-1 AND -2 DATA FOR OPTIMIZED FOREST COVER DETECTION IN EUROPEAN TEMPERATE FORESTS AND SOUTH AFRICAN SAVANNA

Heckel K[1], Schmullius C[1]
[1]*Friedrich-Schiller-University Jena*

Reliable information about the extent of tree canopy cover (TCC) is essential for knowledge-based policy-makers and different research fields such as environmental modelling. Latter are interested in this topic, due to the crucial importance of forests as co-determinators for the regulation of the global carbon cycle. Hence, optimal forest cover detection requires, input data that provides both high spatial and temporal resolution, allowing a more realistic spatial representation of forests, particularly in sparsely forested regions. These criteria are met by the Sentinel-2 satellite which is controlled by the European Space Agency (ESA). In order to derive accurate maps of forest cover in different latitudes preprocessing and classification was firstly carried out in this study to derive TCC from optical Sentinel-2 data only. Following, a comparison with results from microwave data (Sentinel-1) alone and from a fused classification approach was accomplished, to rate the beneficial effect of data fusion for forest cover derivation. As study sites, Thuringia (Germany) and Kruger National Park (South Africa) were chosen, to assess the robustness of the classification. Therefore, the combination of spectral bands and vegetation parameters such as the Tree Canopy Index (TCI) and Fraction of Absorbed Photosynthetically Active Radiation (fAPAR) from Sentinel-2 were used in two empirical non-parametric approaches, namely Random Forest (RF) and Support Vector Machine (SVM), leading to considerably high classification accuracies above 85 %. In order to evaluate their performance, an additional knowledge-based approach featuring user-defined thresholds based on prior separability analysis was carried out. Analysis of variable importance revealed that TCI and fAPAR/LAI where the most valuable vegetation indices while bands 8 (NIR), 7 (red-edge) and 11/12 (SWIR) where the most important sources of spectral information. The results indicate that Sentinel-1 and –2 provide valuable information for forest cover derivation in varying latitude/ecosystems.

KEYWORDS: Forest Cover, Data Fusion, Non-parametric

SENTINEL-1 DATA PROCESSING AT THE EODC EARTH OBSERVATION DATA CENTRE FOR WATER RESOURCES MONITORING

Wagner W[1], Elefante S[1], Naeimi V[1], Briese C[2]
[1]*Vienna University of Technology (TU Wien)*, [2]*EODC Earth Observation Data Centre for Water Resources Monitoring*

The EODC Earth Observation Data Centre for Water Resources Monitoring has been founded in 2014 as a public-private partnership for fostering the use of earth observation data for global land and water monitoring. One of its centrals goals is to build up a collaborative cloud infrastructure capable of storing large volumes of satellite data and processing them on a supercomputer and a dedicated EO processing cluster in a highly parallel fashion. EODC provides to its expert users access to virtualized resources, parallel processing, and Petabyte-scale storage containing amongst other data sets an almost complete archive of Sentinel data. In addition to optical satellite data (Sentinel-2, Landsat) EODC holds a complete global archive of Sentinel-1 Synthetic Aperture Radar (SAR) which can be processed with a supercomputer for continental to global scale mapping of soil moisture, water bodies and other hydrological parameters. In this presentation we will present the technical capabilities of EODC for processing of Sentinel-1 data and compare them with the service offerings and specifications of other cloud platforms offering Sentinel data.

SENTINEL-2 FOR AGRICULTURE: SENTINEL-2 OPERATIONAL EXPLOITATION FOR SUPPORTING NATIONAL AND GLOBAL CROP MONITORING

Bontemps S[1], Defourny P[1], Bellemans N[1], Cara C[2], Dedieu G[3], Grosu A[2], Guzzonato E[4], Hagolle O[3], Inglada J[3], Morin D[3], Nicola L[2], Rabaute T[4], Savinaud M[4], Udroiu C[2], Valero S[3], Koetz B[5]
[1]Earth and Life Institute, Université cath. de Louvain, [2]CS-Romania, [3]CESBIO, [4]CS Système d'Information, [5]ESA-ESRIN

Developing better agricultural monitoring capabilities based on Earth Observation (EO) data is critical for strengthening food production information. Since 2011, support for agricultural monitoring using satellite data has formal institutional support, objectives and timelines, e.g. with the Global Agricultural Monitoring Initiative (GEOGLAM) building on GEO's Agricultural Community of Practice and the Joint Experiment of Crop Assessment and Monitoring.

Launched in 2014, the ESA Sentinel-2 for Agriculture project aims at developing an open-source system to process Sentinel-2 (S2) data in an operational way into relevant EO agricultural products for major worldwide agriculture systems. The S2 mission has indeed the optimal capacity for agriculture monitoring in terms of resolution (10-20 meter), revisit frequency (5 days) and coverage (global). The products generated by the system consist of (i) monthly cloud-free composites, (ii) monthly cropland masks, (iii) cultivated crop type map for main crops and (iv) LAI and NDVI indicators describing on a 10-day basis the vegetative development of crops.

The first phases of the project focused on algorithms selection, system design and implementation. They were achieved using S2-like time series (SPOT-Take5, Landsat-8) complemented with in-situ data shared by local teams over 12 globally distributed sites. Started in March 2016, the last phase aims at demonstrating the developed system to deliver the products over the 2016-2017 growing seasons in near real-time and with S2 data. This demonstration is done at national scale over Ukraine, Mali and South Africa and at local scale over eight sites spread over Europe, Africa and Asia. It is carried out in interactions with local teams, with the objective to transfer the system to their operations. Doing so, the project contributes to filling the gap between state-of-the-art remote sensing practices and operational systems, thus providing a strong scientific contribution to the GEOGLAM initiative.

KEYWORDS: Sentinel-2, agriculture monitoring, operational, GEOGLAM

SENTINEL-2 MSI DATA FOR MAPPING AND MONITORING URBAN GREEN SPACES

Furberg D[1], Ban Y[1]
[1]*KTH Royal Institute Of Technology*

Urban green spaces such as parks, woodlands and gardens provide a variety of benefits to urban populations. By improving physical fitness and reducing depression, the presence of green spaces can enhance the health and wellbeing of people living and working in cities. Green spaces also indirectly impact our health by improving air quality and reducing urban temperatures thus limiting the impact of heat waves. In addition, urban vegetation stores carbon, helping to mitigate climate change, and reduces the likelihood of flooding by storing excess rain water. Green spaces also contributes to the preservation of biodiversity, promotes the state of ecosystems and their resilience, and thereby strengthens ecosystem services that are important for society in general. The objective of this research is map urban green spaces and to assess the impact of urbanization on green spaces. The study area is the Stockholm County in Sweden, which are built on islands with a dense core of high density built-up areas surrounded by several areas of low density residential built-up in the outskirts. 2016 Sentinel-2A MSI data as well as 1986 and 2006 SPOT data over Stockholm are used in this research. Land cover classifications were performed using an object-based Support Vector Machines. Based on the classifications, the environemtal impact of urbanization are being quantified through several spatial attributes of green and blue patches and their configuration, including area (CA), connectivity (COHESION), core area (TCA), diversity (SHDI), edge effects (CWED), percentage of land cover (PLAND) and a proximity measure. The preliminary results show that Sentinel-2A MSI data is suitable for mapping urban land cover and green spaces with an overall accuracy at 80%. Urban green space changes and the environmental impact of urbanization during the past 10 and 30 years are being analyzed and will be presented at ISRSE2017.

SERVICING THE MARKET WITH HIGH RESOLUTION DATA - A PLUG FOR SERVICES & INFORMATION - NOT IMAGES

Hausknecht P[1], Blain R[1]
[1]*Earth-i Ltd.*

The last years have seen an increased provision of very high resolution satellite data* (VHR) by a number of providers in the commercial satellite data supply world. Since 2015 satellites produced by SSTL^ in the UK are part of this family, with the launch of three identical DMC3 satellites providing multispectral images with a panchromatic channel offering better than 1 metre spatial resolution. The high agility and revisit rate provide for imaging opportunities of every place on Earth at least once a day – cloud cover depending. In the context of all other suppliers, the availability of high resolution satellite data is better than ever.

But do we really need all that additional data? And who is paying the commercial rates for such data given the wide availability of the new 10m Sentinel-2 data? The answer lies in the needs of the end-user, which is often not the data itself but rather the information contained in the satellite data, extracted and delivered as a value-added service linked directly to their decision-making systems. Very high spatial resolution satellite data has its place in that overall value proposition and future business opportunities will demand services more than raw data, a need which Earth-i seeks to address.

This presentation will look at some ongoing EO activities and compare the information content from different services. Demonstrating the DMC3/TripleSat high resolution data capability we show some image examples amongst others from different areas across the globe, servicing customers in Africa, Australia, Middle East and Europe.

*1 metre GSD and better is the definition; ^ Surrey Satellite Technology Ltd

SHARING AND STRENGTHENING AGRICULTURAL MONITORING KNOWLEDGE IN AFRICA, MARS TOOLS FOR CROP ANALYSTS

Rembold F, **Baruth B**[1], Urbano F, Kerdiles H
[1]*European Commission - JRC*

Monitoring crop and vegetation conditions is highly relevant, particularly in the food insecure areas of the world. Data from remote sensing at high temporal and medium to low spatial resolution can assist this monitoring by providing key information in near real-time over large areas. For more than 10 years and in collaboration with numerous international and national partners, MARS has invested in the development of crop monitoring and yield forecasting tools for an international cooperation and development context.

In particular 3 examples can be mentioned here:
1.) In 2014 together an E-learning course was produced for FAO´s E-learning catalogue in Food and Nutrition Security, entitled: Remotely Sensed Information for Crop Monitoring and Food Security - Techniques and methods for arid and semi-arid areas.

2.) The Software for the Processing and Interpretation of Remotely sensed Image Time Series (SPIRITS) developed together with VITO, is a stand-alone flexible analysis environment to facilitate the processing and analysis of large image time series and for providing information about vegetation status. A selection of global data are distributed directly in SPIRITS format on the website: http://spirits.jrc.ec.europa.eu

3.) CGMS Statistical Tool (CST), a dedicated tool for yield forecasting, helps to understand the inter-annual variability of crop yield and to determine if this variability can be explained by one or more indicators. These indicators can then be used to predict crop yield early in the growing season.

SPIRITS and CST are also part of a larger collection of environmental analysis tools which is managed by the EU funded MESA (Monitoring for Environment and Security in Africa) project. Over the years an important number of trainings based to a large extent on the 3 tools mentioned above has been organized either directly by the JRC or in collaboration with other partners.

KEYWORDS: agriculture monitoring, food security, remote sensing

SIGMA, STIMULATING INNOVATION IN GLOBAL MONITORING OF AGRICULTURE

Gilliams S[1]
[1]*VITO*

SIGMA is part of Europe's contribution to GEOGLAM, actively networking expert organizations world-wide, in a common effort to enhance current remote sensing based agricultural monitoring techniques. Its aim is to develop innovative methods and indicators to monitor and assess progress towards "sustainable agriculture", focused on the assessment of longer term impact of agricultural dynamics on the environment and vice versa. Throughout the past years in SIGMA a number of data fusion techniques have been developed to create daily cloud free simulated data sets to increase the temporal resolution of EO data for agricultural monitoring. Fusion between proba-v and sentinal2 data or S1 and S2 data for ag mapping as an example. The results of those studies along with the results of the general reasearch of SIGMA in Agricultural Mapping, agricultural production levels and the impact on the environment will be presented.

KEYWORDS: agricultural monitoring, data fusion, production

SIMOCEAN: COLLABORATIVE SYSTEM FOR THE CREATION OF MARINE SERVICES

Almeida N[1], Grosso N[1], Catarino N[1], Almeida S[2], Lamas L[2], Alves M[2], Deus R[3], Oliveira P[3]
[1]*DEIMOS,* [2]*Instituto Hidrográfico,* [3]*Instituto Português do Mar e da Atmosfera*

The creation of services based on the combination of products from different sources, from satellite EO imagery to models and ground observations, generated and managed by different institutions must rely on the interoperability of the different information systems and tools.

The European landscape for service creation systems is rapidly evolving based in: a) the proliferation of open datasets such as the Copernicus services and; b) the efforts made to harmonize those data streams with the definition of new standards for data generation, catalogue and publication, through initiatives such as the Working Groups promoted by OGC and the European directive INSPIRE.

These global initiatives are then translated into national efforts to implement value added services and information systems based on the harmonized datasets coming from different institutions. It's in this context that SIMOcean appears, a Big Data platform for the creation of marine EO-based services based on datasets managed by the Portuguese Hydrographic and Meterological and Atmospheric institutes. This platform was developed to ensure interoperability with other National, European and International information systems and allow scalability by applying industry lead technologies, standards and lessons learned from the research projects FP7 SenSyF, H2020 Co-ReSyF and ESA TEP for Hydrology.

Three flagship services are already deployed to validate the key principles of the platform: Characterisation of Fishing Areas; Sea State Index for Harbour Approaches and Diagnostic of Meteo-Oceanographic Fields. These services are ready to be exploited by Civil Protection Authorities, Port Authorities and Fishing Associations, where these new products have a significant positive impact in their operations. The full experience on the development, setup and operations of such a platform will be shared demonstrating the challenges and lessons learned across the different areas, from data access up to dissemination of value added information to different communities.

KEYWORDS: Big-data, marine services, interoperability, earth observation

SMALL UNMANNED AIRCRAFT SYSTEMS IN CORAL REEF ENVIRONMENTS: FROM PROMISE TO REALITY

JOYCE K[1], Leahy S[1], Maier S[2], Roelfsema C[3], Kovacs E[3], Phinn S[3], Leon J[4]
[1]James Cook University, [2]maitec, [3]University of Queensland, [4]University of the Sunshine Coast

Remote sensing plays a critical role in understanding reef spatial patterns related to habitat distribution, productivity, geomorphic zonation, bathymetry, water quality, and water temperature at scales of square meters to square kilometres. Exciting advances in consumer grade small unmanned aircraft systems (UAS) are now progressing the discipline from initially providing highly detailed pictures in different environments, to producing highly detailed maps. Quality sensor miniaturisation also means that UAS are becoming increasingly capable of carrying thermal and even hyperspectral sensors, providing a complete image acquisition package. Here we detail a case study using UAS to map benthic habitats and sea surface temperature (SST) on Heron Reef, Great Barrier Reef. Using the Aeronavics Bot with interchangeable payloads of a Sony a7R (DSLR) and FLIR A65 (thermal imager), we present optical data that is being used to map and monitor coral reef benthos and structure at centimetre-level spatial resolutions. In addition, we present quantitative, remotely-sensed SST data collected at unprecedented temporal and spatial resolutions. With a thermal pixel size of just 6 cm and multiple repeat image acquisitions during the day and night over the course of a week, we were able to detect fine scale temperatures differences and tidal fluctuations. By combining the thermal and optical data sets we hope to better understand the thermodynamic processes occurring on the reef flat.

KEYWORDS: Coral reef, UAS, drones, thermal, SST

SMARTER THAN YOUR AVERAGE SENSOR: AIS SENSOR THAT INTELLIGENTLY RE-TRANSMITS MEANINGFUL INFORMATION DERIVED FROM RAW AIS DATA IN NETWORK LIMITED AREAS

Meyer R[1], Kleynhans W[1], Swanepoel D[1]
[1]*CSIR*

AIS is a transponder based, anti-collision system used by the majority of ocean traffic. Ships regularly transmit their identity, position and speed. The information used to populate the AIS fields come from ship based sensors, such as GPS, and user populated fields, such as the vessels name and MMSI number.

These fields are susceptible to spoofing and can be changed to hide the identity or location of a vessel. This is often done to disguise the vessel as a different class to avoid inspections or to enter a protected area without raising alarms.

The AIS system has found great utility in monitoring global shipping trends and traffic but this was never intended when the protocol was designed. Docked vessels still transmit messages regularly that contain no new information. These, and other redundant messages, are still transmitted and stored. In situations where a sensor is remote and has limited access to the Internet this can become costly.

The Smart-AIS sensor records all incoming messages locally and makes a decision on whether the message is of special interest or not. Messages of interest are re-transmitted to an AIS database. The Smart-AIS sensor also builds up an expected message reception area where the bulk of messages have been received. AIS messages that fall far outside this radius are local vessels spoofing their location while vessels that are expected to be within this radius and are not transmitting are marked as "dark targets". The expected distribution of other parameters is also learnt by the sensor; speed, bearing, vessel turn-rate etc. Vessels performing unexpected maneuvers or travelling too fast (calculated from successive locations or reported speeds) also cause alerts. This reduces the load on the sensors network connection.

KEYWORDS: AIS, spacial database, smart sensor

SMOS AFTER 7 YEARS IN ORBIT: LESSONS LEARNED, ACHIEVEMENTS AND WAY FORWARD

Kerr Y[1], Mecklenburg S[2]
[1]Cesbio Cnes, [2]ESA ESRIN

The Soil Moisture and Ocean Salinity (SMOS) mission, launched in November 2009, is the European Space Agency's (ESA) second Earth Explorer Opportunity mission with CNES and CDTI. Significant progress has been made over the course of the now 7-year life time of the SMOS mission in improving the different products while a large number of new science products emerged. The paper will provide an overview on the mission status and a performance assessment of the SMOS core data products. The key scientific findings with a special emphasis on soil moisture and ocean salinity will be summarised.

The main emphasis of this paper will be though on providing i) an overview and performance assessment of newly developed SMOS data products, namely sea ice thickness, soil moisture in NRT and high wind speeds over ocean, ii) give an outlook on future data products currently in preparation (e.g. vegetation optical depth for agricultural applications, freeze/thaw etc) and ii) summarise the potential use of SMOS data for operational applications, either currently available or planned for the near future.

After more than 7 years in orbit, SMOS provides a valuable source of data for observing and understanding longer-term processes and phenomena that were not necessarily targeted in the original mission design (e.g. El Nino and El Nina signals in sea surface salinity, drought pattern monitoring). This paper will give a first indication of how SMOS can contribute to this aspect.

Finally, we will discuss evolving user requirements based on feedback from the user community. An attempt will be made to identify the driving requirements for a future mission concept and commonalities across the various application areas. Based on this analysis, the aims and objectives for potential future mission concepts will be formulated independently of any technical or programmatic implementation routes.

SMOS AND CLIMATE APPLICATIONS

Kerr Y[1], Mahmoodi A[1], Rodriguez-Fernadez N[1], Al Bitar A[1], Parrens M[1], Mialon A[1], Richaume P[1], Molero B[1], Wigneron J[2]
[1]Cesbio Cnes, [2]INRA ISPA, [3]LTHE

The SMOS (Soil Moisture and Ocean Salinity) satellite was successfully launched in November 2009. This ESA led mission for Earth Observation is dedicated to provide soil moisture over continental surfaces (with an accuracy goal of 0.04 m3/m3), vegetation water content over land, and ocean salinity. These geophysical features are important as they control the energy balance between the surface and the atmosphere. Their knowledge at a global scale is of interest for climatic and weather researches, and in particular in improving model forecasts.

The purpose of this communication is to present the mission results after more than seven years in orbit in a climatic trend perspective, as through such a period anomalies can be detected. Thereby we benefit from consistent data sets provided through the latest reprocessing using most recent algorithm enhancements.We have also designed data fusion schemes enabling to extend the temporal coverage using other microwave missions (15+ years currently).

Using the above mentioned products it is possible to follow large events such as the evolution of the droughts in North America, or water fraction evolution over the Amazonian basin. In this occasion we will focus on the analysis of SMOS and ancillary products anomalies to reveal two climatic trends, the temporal evolution of water storage over the Indian continent in relation to rainfall anomalies, and the global impact of El Nino types of events on the general water storage distribution.

SOCIOECONOMIC IMPACTS ON CROPLAND TRANSITION AND ADAPTATIONS IN BEIJING, CHINA

Zhang L[1]
[1]*Institute Of Remote Sensing And Digital Earth*

Over the last several decades, the cropland in China experiences the rapid loss as a result of nationwide urbanization, industrialization, socio-economic development and ecological project implement which driven by continuous economic reform, environment protection policies. Cropland is particularly precious land resource with pressure of overpopulation in China and loss of cropland threaten the national food security. The paper aims at the analyses of spatial pattern of cropland changes, driving forces and its adaptation. The object based approach was performed to accurately measure cropland changes using remote sensing data. Conversion matrix and statistical regression model were used to explore the impacts of socioeconomic factors on cropland changes. The case study in Beijing concludes that cropland in Beijing decreased by 47% in last 35 years. The primary driving forces to cropland loss are contributed by urban expansion, ecological projects, water environment protection, and orchard plantation. The increased migration particularly resulted in urban expansion. These changes caused cropland fragmentation, landscape diversity and low ecological value. For sustainable development, to meet urban and environment protection, the cropland adaptation were performed by increasing value added commercial crop plantation, improving the crop yield and decreasing cultivation intensity.

KEYWORDS: Cropland transition, Adaptation, Socioeconomic impacts, Urban expansion

SOIL EROSION MODELLING USING SATELLITE-BASED DIGITAL ELEVATION MODELS (DEMS) – A COMPARISON FROM THE THURINGIAN BASIN, GERMANY

Baade J[1]

[1]*Department of Geography, Physical Geography, Friedrich-Schiller-University Jena*

Soil erosion, i.e. the accelerated displacement of soil or parent material by natural drivers following human interference with the landscape, is considered a major driver of soil degradation. Despite some limitations, the Universal Soil Loss Equation (USLE) is still used in many countries to assess soil erosion and to identify hotspots of erosion. The USLE is appealing because of its limited data demand. Soil erosion risk might be calculated from the rainfall erosivity (R), the erodibility of the soils (K) and the slope length and slope angle (LS). Within this group of model inputs, the predicted soil loss is most sensitive to the slope angle.

In countries where high resolution Digital Terrain Models (DTMs) derived from airborne LiDAR acquisitions are missing satellite-based Digital Elevation Models (DEMs) are often used to represent the relief and calculate slope length and slope angle. However, most satellite-based DEMs have to be considered Digital Surface Models (DSMs) representing not only buildings and other infrastructure, but as well the canopy cover, e.g. forests. This might result in unrealistic high slope angles at the field-forest boundaries. Furthermore, there are considerable differences in the posting and accuracy of the different available DEMs in open terrain.

This contribution presents the result of an evaluation of different satellite-based DEMs, including the ASTER-GDEM2 (1 arcsec posting), the two SRTM editions (1 and 3 arcsec posting) as well as the new global TanDEM-X DEM (0.4 and 1 arcsec posting), used to assess the soil erosion risk in a 260 km^2 catchment in the Thuringian Basin, Germany. High resolution airborne LiDAR based DTMs (5 m and 25 m posting) are used to set the benchmark.

KEYWORDS: USLE, ASTER-GDEM, SRTM, TanDEM-X, Model comparison

SOIL MOISTURE VALIDATION USING IN SITU MEASUREMENTS: THE SMAP APPROACH

Colliander A[1], Jackson T[2], Bindlish R[2], Chan S[1], Das N[1], Kim S[1], Cosh M[2], Dunbar R[1], Asanuma J[3], Aida K[3], Berg A[4], Rowlandson T[4], Bosch D[2], Caldwell T[5], Caylor K[6], Holifield Collins C[2], Al Jassar H[7], Lopez-Baeza E[8], Martinez-Fernandez J[9], Gonzalez-Zamora A[9], Livingston S[2], McNairn H[10], Pacheco A[10], Moghaddam M[11], Montzka C[12], Notarnicola C[13], Niedrist G[13], Pellarin T[14], Prueger J[2], Pulliainen J[15], Rautiainen K[15], Ramos J[16], Seyfried M[2], Starks P[2], Su Z[17], Zeng Y[17], van der Velde R[17], Thibeault M[18], Dorigo W[19], Vreugdenhil M[19], Walker J[20], Wu X[20], O'Neill P[21], Entekhabi D[22], Yueh S[1]

[1]Jet Propulstion Laboratory, California Institute Of Technology, [2]USDA Agriculture Research Service , [3]University of Tsukuba, [4]University of Guelph, [5]University of Texas at Austin, [6]Princeton University, [7]Kuwait University, [8]University of Valencia, [9]University of Salamanca, [10]Agriculture and Agri-food Canada, [11]University of Southern California, [12]Research Center Juelich, [13]European Academy of Bozen/Bolzano, [14]University of Grenoble, [15]Finnsih Meteorological Institute, [16]Universidad Nacional Autonoma de Mexico, [17]University of Twente, [18]Comisión Nacional de Actividades Espaciales (CONAE), [19]Vienna University of Technology, [20]Monash University , [21]NASA Goddard Space Flight Center, [22]Massachusetts Institute of Technology

NASA's Soil Moisture Active Passive (SMAP) Mission was launched in January 2015. The objective of the mission is global mapping of soil moisture and landscape freeze/thaw state. The primary validation reference of the data products are ground-based measurements. Well characterized sites with multiple calibrated in situ sensors within the SMAP 40-km footprint have been used to determine the quality of the data products; these sites are designated as core validation sites. The mission success has been evaluated with respect to these core site comparisons. Sparse networks of soil moisture stations and other remote sensing and model-based products have been used as complimentary resources to expand the spatial and temporal scope of the evaluation.

In an effort to ensure the geographic distribution and diversity of conditions captured by the core validation sites, SMAP partnered with investigators across the globe. Because different SMAP Level 2 and 3 soil moisture products have different spatial scales, the suitability of the various sites for validation of the different products was assessed considering several factors. The main factors were the availability of a geographically distributed in-situ soil moisture network consisting of multiple points, gravimetric calibration of the in situ sensors within a site, determination of a spatial scaling function of the sensor measurements up to the SMAP resolution scales, and timely accessibility of the data.

The mission has been able to utilize the core site measurements since the launch of the satellite because the infrastructure for data transmission and processing was established well before the launch. The validated SMAP soil moisture products were released in May 2016. This presentation will show the performance of the latest radiometer-based surface soil moisture data products using 2 years of data, and discuss the lessons learned during the validation process.

SOLID EARTH ASPECTS OF ESA'S SWARM MISSION

Jackson A[1]
[1]*ETH Zurich, Institute for Geophysics*

The Swarm mission is one of ESA's Earth Explorer missions that is currently in orbit. Consisting of three identical spacecraft flying in formation, it is tasked with exploring near-Earth's magnetic and electrical environment with unprecedented precision. There are a multitude of scientific aspects to the mission, however, I will report on the aspects of the mission that pertain to the interior of the Earth. The constellation aspect of the mission has allowed us to perform separation of the different magnetic effects that the magnetometers record. One part of the signal pertains to the tiny magnetic field induced by the tidal motion of seawater at around 12 hours. Measurement of this signal allows us to perform a depth-sounding of the electrical conductivity structure of the Earth, since this is the quantity that affects the amplitude and phase of the tiny electrical currents that are induced within the Earth. New results have been obtained that show an abrupt electrical conductivity change at the base of the lithosphere. Work is underway to examine how this jump changes as a function of the age of the oceanic crust.

I will also report on the new results that are forthcoming concerning the main magnetic field of the Earth. The main field is generated in the Earth's fluid core 2900km beneath Earth's surface by a self-sustaining dynamo process. On top of the slowly-changing background field can be a host of shorter-time scale dynamical processes associated with the magnetohydrodynamics of the core. Our constellation of satellites constrain time changes in this field with previously unobtainable levels of accuracy. I will report on the rapid changes in the field that are discernible in the observations and their relevance to the quest to understand core dynamics.

SPATIAL ANALYSIS OF LANDUSE DYNAMICS OF FOREST PART SOUTHWESTERN NIGERIA

Francis A[2], Ojo A[1]
[1]African Regional Centre For Space Science And Technology Education, [2]Obafemi Awolowo University

The study examines the landuse and landcover dynamics of the cocoa area of the southwestern Nigeria between 1986 and 2014.

Cocoa is one of the most important cash crops grown largely (more than 80%) by the small-scale farmers of the region. Thus, its extensive pattern of cultivation has brought about opening up of newly uncultivated forest areas, settlement expansion and environmental degradation. This study used geo-spatial techniques based on multi-source imageries to enhance the utilization of images with coarser resolutions in landuse analysis of southwestern Nigeria. The objective of the study is to evaluate the variations in landuse characterization with multispectral images. The remotely sensed data sets used included Landsat TM 1986, 1991, 2002, 2007 and 2014 imageries. The tonal values recorded in the images with the features on the ground were validated by ground truthing. The data from ground truthing were combined with visual image interpretation for "supervised" classification. The classes defined and analyzed included "built-up area", "bare rock", "farmland", secondary forest regrowth" and "waterbody". The results show that no concerted effort has been directed towards quantifying and assessing the spatial pattern extent of deterioration components of the environment as a result of mismanagement of landuse in the cocoa belt region. The study confirms the relevance of the growing interest in the use of geospatial techniques for landuse analysis.

KEYWORDS: spatial, remote sensing images, environmental monitoring, change detection and land use/ cover.

SPATIAL AND TEMPORAL VARIATION IN AEROSOLS ACROSS NAMIBIA OVER A YEAR. IMPLICATIONS FOR REMOTE SENSING ATMOSPHERIC CORRECTION MODELS

Knox N[1], Holben B[2]
[1]*Namibia University Of Science And Technology,* [2]*NASA - GSFC*

Climate change studies making use of time-series optical earth observation data should require that the data has been atmospherically corrected. Atmospheric correction will ensure that when the analysis of the imagery is conducted the reflectance values are comparable over the study time frame. For Namibian time series studies that have included atmospheric correction typically make use of dark object subtraction (DOS), or models such as the S6 model based on the subtropical atmospheric correction model. These models do not account for temporal and spatial variation of the aerosol across the country. In this study we make use of the recent AERONET network that has been installed across Namibia. Seven instruments have been installed across Namibia five of these in 2015. These instruments provide the opportunity to analyse the spatial and temporal variations of aerosols derived from marine, burning biomass and aeolian sources. In this study we have conducted a spatial and temporal study, over the year 2015, to determine the variation of these aerosol sources across the country. The variation observed during this time frame shows that the aspect of scale of temporal analysis is important as there are both variations diurnally and across seasons. The input will become an important contribution to fine tune a regional atmospheric correction model that will capture the variation that has been observed in the region, taking into account the times typical for sensor over passes over the region.

SPATIAL DATA INFRASTRUCTURE TO SUPPORT THE NATIONAL OCEANS AND COASTAL INFORMATION MANAGEMENT SYSTEM: A DATA GOVERNANCE VIEW

Fourie N[1], Hankel L[1]
[1]*CSIR*

The development of the National Oceans and Coastal Information Management System (O&C IMS) and the implementation of national development initiatives, such as Operation Phakisa highlighted the requirement for legislative and policy structures to create an enabling environment for the required spatial data. The initiatives will contribute to South African national objectives of economic growth and sustainable development as outlined in the South African National Development Plan (NDP) and the successful implementation and maintenance is imperative to delivering on the NDP.

The South African Spatial Data Infrastructure (SDI) Act (Act 54 of 2003) was established as the technical, institutional and policy framework to facilitate the capture, management, maintenance, integration, distribution and use of spatial information. The objective of the SDI Act is to establish the South African Spatial Data Infrastructure (SASDI) and the Committee for Spatial Information (CSI) as convening structure. Furthermore the SDI Act aspires to establish an electronic metadata catalogue to avoid the duplication of data capture initiatives. This will be done by the appointment of base data custodians and provision of policies, standards and prescriptions to facilitate the sharing of geospatial information

South Africa, as is the case in the majority of SDI initiatives around the world had an initial terrestrial focus. Initiatives such as National O&C IMS highlighted the gap in the SASDI approach. This presentation will provide an overview of the processes followed to establish a formal structure for the development of SDI initiatives to address coastal and marine data requirements thru the existing SDI legislative and policy frameworks. Lastly data governance elements such as input data audits and the collection, processing, analysis, dissemination and archiving of data products will be reviewed. This review process is an operational measure to ensure sustainability of the National O&C IMS.

KEYWORDS: data governance, Spatial Data Infrastructure, legislative frameworks

SPATIAL MODELING OF MOSQUITO POPULATION DYNAMICS: AN OPERATIONAL TOOL FOR THE SURVEILLANCE OF VECTOR-BORNE DISEASES

Tran A[1], Herbreteau V[2], Demarchi M[3], Mangeas M[2], Roux E[2], Degenne P[1], Dehecq J[4]
[1]Cirad, [2]IRD, [3]Marie Demarchi, [4]Agence de Santé Océan Indien

Context. Mosquitoes are vectors of major pathogens worldwide, such as the pathogens of Malaria, Chikungunya, dengue, Rift Valley or West Nile fevers. Accurate understanding and prediction of mosquito population dynamics are needed to optimize surveillance and control of mosquito-borne diseases.

Objectives. This study had two main objectives i) understanding the relationships between environmental conditions and mosquito population dynamics to predict mosquito densities and ii) developing operational tools for surveillance and control of vector-borne diseases, taking the example of Aedes albopictus in Reunion Island.

Methods. We developed different models using respectively process-based and data-based approaches to study the relationships between meteorological variables (daily temperature and rainfall), land cover classification derived from SPOT-6 imagery, and entomological collections of Aedes albopictus larvae from 9 sites located around the Island. The best models were implemented with Ocelet, a domain specific language and simulation tool for modelling changes in geographical landscapes, and a user-friendly interface was developed.

Results. The observed and predicted abundances of Aedes albopictus tallied very well. Spearman's correlation coefficients ranged between 0.45 and 0.90. Higher correlations were obtained in the sites with a higher seasonality of the mosquito population dynamics.

Conclusions. We developed a flexible and efficient tool that predicts mosquito abundance based on local environmental and meteorological factors. It is operational with a simplified, user-friendly interface and used by vector-control agencies to target surveillance areas in Reunion Island.

KEYWORDS: Land cover, classification, mosquito dynamics, modeling

SPATIOTEMPORAL ANALYSIS OF URBAN LAND COVER CHANGE IN KIGALI, RWANDA USING MULTITEMPORAL LANDSAT IMAGERY AND LANDSCAPE METRICS

Mugiraneza T[1,2], Haas J[1], Ban Y[1]
[1]*KTH, Royal Institute Of Technology,* [2]*University of Rwanda-Centre for GIS and Remote Sensing*

Urbanization mapping and environmental impact assessment using space-borne data and landscape metrics has recently been a hot research topic. However, few studies were found on monitoring urban landscape dynamics using satellite images and landscape metrics in Sub-Saharan African cities and none exist in case of Kigali in Rwanda. The objectives of this research are to analyze spatio-temporal patterns of urbanization in Kigali, Rwanda during the last three decades (from 1984 to 2015) using multitemporal Landsat data and to assess the urbanization environment impact using landscape metrics. Landsat-5 TM data acquired in 1984, 2001 and Landsat-8 OLI-TIRS data acquired in 2015 were selected as input data. The pre-processing step consisted of texture analysis using Gray-Level Co-Occurrence Matrix (GLCM) and Normalized Difference Vegetation Index (NDVI). The aforementioned indices, a Digital Elevation Model (DEM), and the R/NIR/SWIR bands were used for supervised Support Vector Machine (SVM) classification. Seven land cover classes were generated and transformed into ecosystem services supply and demand including high- and low-density built-up areas, agriculture, forest, bare land, water and wetland. Eight landscape indices were computed and evaluated for characterizing the spatial evolution of composition and configuration of landscape patterns. The landscape in the study area was found highly fragmented from 1984 to 2015. Agriculture was the most fragmented landscape entity and forest was found the most reduced class with a decrease evaluated at 18.4%. Built-up areas were found to have expanded to 382.6% of their size since 1984. The change in landscape composition and configuration affected some ecosystem services, especially flood and climate regulation services due to combined side-effects of forest clearing and fragmentation, fragmentation of agriculture patches and increase in built-up space and impervious surfaces. Satellite based change detection analysis and quantification of urbanization using landscape metrics are found to be cost effective method for urban information extraction.

SPATIOTEMPORAL VARIATIONS IN THE IMPACTS OF URBAN LAND USE TYPES ON URBAN HEAT ISLAND EFFECTS: THE CASE OF RIYADH, SAUDI ARABIA

Aina Y[1,2], **Adam E[2]**, Ahmed F[2]
[1]Yanbu Industrial College, [2]University of the Witwatersrand

Urban heat island (UHI) effect is one of the important indicators of the impacts of urbanization on the environment especially since the effects could be linked to climate change. Thus, the growing interest in studying the impacts of urbanization on changes in land surface temperature (LST). The literature on LST indicates the need for more studies on the relationship between changes in LST and land use types especially in the arid environment. This paper, through a case study of Riyadh, Saudi Arabia, examines the spatial and temporal changes in land surface temperature influenced by land use type. Multi-temporal Landsat images of the study area, 1985 to 2015, were processed to derive land surface temperatures. UHI index was computed for the different land use types (high density residential, medium density residential, low density residential, industrial, vegetation, desert) in the study area. A population data of the study area was used to analyze the exposure of the populace to high temperatures. The results indicate a trend of rising temperatures in all the land use types in the study area. This is probably due to climate change. The industrial area has the highest temperatures among the land use types. The lowest temperatures are found in the vegetation area as expected. There is a need to implement mitigating measures to reduce the effects of rising temperatures in the study area.

SRTM30 VERSUS ASTER GDEM2 FOR HYDROLOGICAL ANALYSIS

Mashimbye Z[1], De Clercq W[1], Van Niekerk A[1]
[1]*Stellenbosch University*

Digital elevation models (DEMs) are essential for deriving hydrological parameters which play crucial roles in hydrological and hydraulic modeling. DEMs invariably contain errors which are mostly attributed to the source of data, methods of DEM creation, topography complexity and spatial resolution. Also, the accuracy of a DEM is reliant on the application. This study aimed to compare the accuracy of streamlines and catchment boundaries derived from the second version of the 30-m advanced spaceborne thermal emission and reflection radiometer global digital elevation model (ASTER GDEM2) and the 30-m shuttle radar topography mission (SRTM30). Catchment boundaries and streamlines were extracted from these DEMs using the Arc Hydro module. A reference catchment boundary was generated from a very-high resolution DEM derived from GeoEye stereo imagery. Reference streamlines were digitized at a scale of 1:10 000 from 1-m orthorectified GeoEye images. The DEM derived catchment boundaries and streamlines were validated using visual inspection, and quantitative measures, namely correctness index (Cr), mean absolute error (MAE), root mean squares error (RMSE) and figure of merit index (FMI). The SRTM30 yielded lower RMSE and MAE, and higher Cr and FMI for the delineated catchment boundaries and streamlines in comparison to the ASTER GDEM2. Overall, the SRTM30 DEM demonstrated superiority to the ASTER GDEM2 for catchment boundary and streamlines delineation in this study. These findings lay the foundation for the choice of a DEM for hydrological analysis.

KEYWORDS: Hydrology, catchment delineation, correctness index, figure of merit index, Euclidean distance index.

STANDARDISED SATELLITE-BASED CLASSIFICATIONS OF LAND COVERS IN PROTECTED AREAS AND SURROUNDING LANDSCAPES

Lucas R[1], Mitchell A[1]
[1]*University Of New South Wales*

The Earth Observation Dynamic Habitat Mapping (EODHAM) system was developed as part of the EU funded BIOSOS project to support monitoring of protected landscapes and surrounds from airborne and/or spaceborne Very High (VHR) and Moderate Resolution (MR) remote sensing data. The system initially classified these landscapes according to the Food and Agricultural Organisation (FAO) Land Cover Classification System (LCCS-2) taxonomy but the requirement for habitat information by conservationists led to the development of standardized rules for translating the LCCS categories to General Habitat Categories (GHCs). To detect change, references were made to transitions in LCCS component codes (e.g., A2 to A4; grasslands to trees), GHCs and spectral indices and whether objects (e.g., field units) were split or merged.

Despite success in classifying land covers within and around Natura 2000 sites in Greece, Holland, Italy, Wales and the Netherlands, several limitations to the uptake of the technique by practitioners were noted, including a reluctance to adopt LCCS categories and GHCs, a preference to use other satellite-focused classification techniques (e.g., machine learning) and incomplete descriptions of habitats (e.g., because the LCCS system does not account for sub-pixel proportions).

To address these concerns, the EODHAM system has been modified through the EU-funded ECOPOTENTIAL project such that existing land cover and habitat maps can be routinely translated to LCCS categories, similar maps can be generated from classifications of satellite sensor data, and mixtures or gradients of vegetation communities can be described to better quantify the distribution of land covers and assist their conversion to habitat categories. The LCCS-3 (LCML) approach has also been considered. Classifications are demonstrated for a number of protected areas, primarily in Europe but also in Africa. The classifications are being used subsequently to quantify ecosystem services and changes in these as a function of human-induced and natural events and processes.

STATE OF LAND DEGRADATION AND TRENDS OF SOIL ORGANICS MATTER CONTENTS: AN EXPERIMENTATION OF SPECTRAL INDICES CROSSING TO ASSESS STATIC AND DYNAMIC ENVIRONMENTAL PROCESSES FOR SUSTAINABLE AGRICULTURE IMPROVEMENT

Ngandam Mfondoum A[1]
[1]*Universite Of Yaounde I*

Several spectral indices (SI) have been developed to improve agriculture practices. Some of them are applied to assess the land degradation, especially soil and vegetation indices. Others are created to assess the soil organic matter. In most cases, they are calculated and compared for these purposes.

Thence, the main objective of this paper is to highlight the relationship between samples of SI used in any case, and cross them to enhance the assessment of the said environmental processes.

It can then be hypothesized that SI developed and that are in relationship enhance the sensing of soil sub-surface phenomena.

The methodology to assess the state of land degradation uses the Second Modified Soil Adjusted Vegetation Index (MSAVI2), Normalized Difference Bare Soil Index (NDBSI), Texture Index (NDTeI), Crust Index (CI), Top Soil Grain Size Index (GSI), Normalized Difference Sand Dune Index (NDSDI) and first Specific Principal Component of the red, near infrared, shortwave infrared bands stacking (SPC1R-NIR-SWIR1-SWIR2). While the trends of soil organics matter contents uses Albedo, Organic Matter Index (OMI), Potential Evapotranspiration (PET), soil conductivity, Color Index (CI) and Normalized Difference Vegetation Index (NDVI), Normalized Difference Built-up Index (NDBI), Soil Adjusted Vegetation Index and Modified Normalized Difference Water Index (MNDWI). In both cases, SI are calculated, compared and crossed using Landsat 8 and Digital Elevation Model of Shuttle Radar Topography Mission (DEM-SRTM) images.

The results are enhanced maps of the state of land degradation and annual trends in soil organics matter contents. The first result can help to control the process of land degradation at a specific moment or season; while the second is helpful to build an agricultural calendar for farmers between seasons.

KEYWORDS: Spectral indices; State of land degradation; Trends of soil organics matter contents; environmental processes; Sustainable agriculture.

STRUCTURAL CLASSIFICATION OF AUSTRALIA'S FORESTS AND DETERMINATION OF RELATIVE GROWTH STATES

Scarth P[1], Lucas R[1], Armston J[1]
[1]*Joint Remote Sensing Research Program*

Estimates of vegetation structure and growth stage across large areas from single source optical or radar remote sensing data has proven difficult due to the need to extimate growth form, cover and vertical structure through association with broad floristic formations and subformations. More detailed maps of vegetation structure have also been generated for many areas using lidar data but, due to the cost and processing overheads, mapping is discontinuous across the landscape. To address these limitations, this research focused on investigating the benefits of using Landsat, ALOS PALSAR and ICESAT GLAS data to retrieve vegetation structural information continuously across the landscape and at a national scale.

The approach adopted used fused 25m resolution ALOS PALSAR and Landsat-derived persistent green cover to segment the Australian continent into more than 33 million segments. K-means clustering then grouped the segments with similar cover and L-band HH and HV backscatter into similar clusters. Where GLAS-ICESat footprints intersected these clusters, canopy profiles were extracted and, once adjusted for the different laser period intensities and footprints, were aggregated to produce a mean vegetation profile for each cluster that was used to derive mean canopy and understorey height, depth and density. Due to the large number of returns, these retrievals are near continuous across the landscape, enabling them to be used for inventory and modeling applications.

The derived height and cover were classified into national vegetaion information system (NVIS) classes to produce a consistent national vegetation structure classification. A numerical decomposition of the vegetaion profiles within each structure class was then undertaken to assess the variance in the foliage profiles, and to assign them to relative growth stages. The integration of the three datasets provided options for future operational monitoring of structure and AGB across large areas with potential benefits for quantifying carbon dynamics and biodiversity.

SUITABILITY OF COSMO-SKYMED DATA FOR DETECTING FOREST COVER CHANGE

Mahmood A, Watt P, Donoghue D, Bholanath P
[1]*Research Associate, Geography Department, Durham University,* [2]*Head of Resource Mapping and Climate Change, Indufor Asia Pacific Ltd.,* [3]*Professor, Geography Department, Durham University,* [4]*Head of Planning and Development Division, Guyana Forestry Commission*

Efforts to quantify forest resources at both the national and global scale is a long standing and nontrivial problem to estimate the level of CO2 emissions and removals through forest loss and gain. Reported estimates of CO2 emissions and removals show large differences and high levels of uncertainty. The UNFCCC REDD+ programme envisages that satellite data will provide detailed and accurate estimates of forest cover change. Guyana has a well-developed MRV system that uses 5-m RapidEye to map the entire country and systematically detect and record all land cover change events with an annual census since 2011, that identify and record all deforestation events greater than 1-ha and less than 1-ha for forest degradation. While the quality of the mapping in the Guyana-MRV system is verified by an independent probability-based accuracy assessment using independent reference data based largely on 0.25-m digital aerial imagery and some independent re-analysis of RapidEye imagery. The aim of the project is to evaluate the potential of COSMO-SkyMed 1-m Spotlight and 5-m Stripmap data to detect deforestation and forest degradation. With COSMO-SkyMed 1-m Spotlight and 5-m Stripmap data there is an opportunity to assess the suitability of high resolution radar data for validation 5-m resolution forest change maps against 0.25-m digital air photos and RapidEye. Therefore, our first objective is to assess and compare the accuracy of the RapidEye-based Forest Change Maps using three different validation data sets; reinterpretation of RapidEye imagery, COSMO-SkyMed data sets, and 0.25-m digital air-photographs. The second objective is to evaluate the capability of Cosmo-SkyMed Spotlight and Stripmap data to identify change in forest cover (deforestation and forest degradation) along with associated drivers of the changes. The results look to assess the applicability of COSMO-SkyMed data for detecting tropical forest changes to assist with UNFCCC-REDD+ verification process.

KEYWORDS: REDD+, MRV, and COSMO-SkyMed

SURFACE WATER AND OCEAN TOPOGRAPHY (SWOT) MISSION APPLICATIONS USER NEEDS ASSESSMENT

Srinivasan M[1], Peterson C[2], Hossain F[3], Beighley E[4], Andral A[5]
[1]JPL Caltech, [2]MSU Science & Technology Center, NASA Stennis, [3]University of Washington, [4]Northeastern University, [5]Centre National d'Etudes Spatiales (CNES)

An international collaboration between the U.S. and French space agencies, with contributions from the Canadian and United Kingdom space agencies, is developing innovative wide swath altimetry technology that will cover most of the world's ocean and surface water. It will provide measurements of the height of lakes, rivers, and wetlands, as well as of the ocean surface topography with unprecedented resolution compared to existing technologies. The Surface Water and Ocean Topography (SWOT) mission will build an important time series of measurements of changes in these water bodies over time. These data will allow scientists and operational users to monitor the hydrologic cycle, flooding, small scale features of ocean circulation, and the climate impacts of a changing environment.

The applied science community is an important element in optimizing the benefits of the SWOT mission by demonstrating the high value of the science and data products to address societal issues and needs. SWOT will inform a broad range of applications derived from the science and engineering advances of this new technology and the resulting data products.

In order to understand and effectively address the applications requirements of SWOT data users the SWOT Applications Working Group has developed the 1st SWOT User Survey. Results from this survey provide important insights to the latency, spatial scale, technical capabilities and assets, and other practicable considerations of using SWOT data. Results of the survey are presented, with implications on the future actionable outcomes from the mission's user communities.

SURVEILLANCE OF THE SOUTH AFRICAN EEZ WITH RADARSAT-2

Staples G[1]
[1]*MDA*

The South African Exclusive Economic Zone (EEZ) extends seaward from the South African coastline approximately 370 km and encompasses an area of about 1.5 million sq. km. This vast expanse of ocean is currently monitored by shore-based radars that have seaward coverage of about 25 km and are spatially distributed along approximately 29% of the coastline, thus providing coverage for about 3% of the EEZ.

The primary EEZ surveillance needs are ship detection and oil slick detection, both of which require different radar imaging parameters. In general, ship detection is better at large incidence angles using cross-polarized data and oil slick detection is better at small incidence angles using co-polarized data. To address these somewhat conflicting needs, the RADARSAT-2 Ocean Surveillance, Very wide swath, Near incidence (OSVN) mode was developed. The OSVN mode is dual-polarized, with a 500 km swath width and 50 m nominal pixel spacing. Analysis of the OSVN mode indicated that the oil slick detection performance (HH polarization) was similar to the RADARSAT-2 300-km swath-width ScanSAR Narrow mode. On the other hand, the ship detection performance (HV polarization) was slightly better than ScanSAR Narrow, with a nominal minimum detectable ship length of ~ 35 m at 45°incidence and a wind speed of 8 m/s.

Based on orbit prediction analysis, complete coverage of the EEZ using the OSVN required about five days. A concept of operations was developed that entailed the integration of RADARSAT-2 data, AIS data, and shore-based radar with the aim to provide comprehensive surveillance of the South African EEZ.

KEYWORDS: RADARSAT-2, ship detection, marine pollution, SAR, maritime surveillance

SURVEY ON THE HEALTH & WELL BEING OF THE AFRICAN EO COMPANIES

Woldai T[1]
[1]*School Of Geosciences, University Of Wits,* [2]-

Survey targeting the African private sector involved in EO and geospatial sciences specifically, was never done before. The current survey is the first of its kind on the continent and a beginning to understand how companies in Africa operate, capture their expertise and assess their state and health. The survey covers the private EO services industry across Africa defined as any company selling products or services which contain some data coming from EO satellites. The EO industry comprises satellite operators, data suppliers, value-adding companies or geo-information (GI) companies using derived products where the satellite data are not always visible.

The survey was conducted between February and April 2016. A total of 229 companies were contacted with useful responses coming from 78 of them via an on-line questionnaire. These respondent companies represent 21 out of 54 countries in Africa. The results of the survey show an industry in development. There has been steady growth in revenues in the past few years, accompanied by a good growth of employment in the private sector. The latter has also seen a lot of change in recent years, e. g. the launch of a number of African mini-satellites and other new commercial satellite systems as well as significant technology change with the development of Google Earth and the advent of cloud computing. These and other technologies such as Remotely Piloted Aircraft systems are arriving on the horizon and maybe there are other technologies which will emerge in the near future. The author, find an optimistic outlook, albeit cautious, on the near future.

SYSTEMATIC ACQUISITION OF SENTINEL 2 FOR BURNT AREA MAPPING

Vhengani L[1], Sibanda P[1], Wessels K[1]
[1]*Csir*

Wildfires can result as a natural process or can be human-induced. These fires, if left unmonitored and uncontrolled can burn thousands of hectares of land. The consequence of these, can be loss of life, atmospheric pollution and loss of biodiversity. Satellite data presents means for burnt area assessments. The assessments of burnt areas includes the delineation of the burnt area and estimating burn severity. The aim of this work is to show how Sentinel 2A scenes where automatically downloaded, pre-processed and repackaged to create a local image repository and database used for automated burnt area mapping.

TAILORING EARTH OBSERVATION TO RANCHERS FOR IMPROVED LAND MANAGEMENT AND PROFITABILITY

Scarth P[1], Trevithick B[1]
[1]*Joint Remote Sensing Research Program*

VegMachine Online and the NRM Spatial Hub are web applications that allows ranchers across Australia to view and interact with satellite derived ground cover state and change maps on their property and extract information in a graphical format using interactive tools. They rely on the web-time interrogation and summarization of a massive earth observation data set in an accessible, producer friendly way.

All available Landsat images and more than 2500 field sites across the Australian rangelands were used to derive endmembers used in a unmixing approach to estimate the per-pixel proportion of bare, green and non-green vegetation. A seasonal metoid compositing method was used to produce national fractional cover virtual mosaics for each three month period since 1988. The time series of green fraction is used to estimate the persistent green due to tree and shrub canopies, and this estimate is used to correct the fractional cover to ground cover for our mixed tree-grass rangeland systems. Finally, deciles are produced for key metrics every season to track a pixels relativity to the entire time series.

These data are delivered through time series enabled web mapping services and web processing services that enable the full time series over any spatial extent to be interrogated in seconds via a RESTful interface. These services interface with browser applications that provide product visualization for any date in the time series, tools to draw or import polygon boundaries, plot time series ground cover comparisons, look at the effect of historical rainfall and tools to run the revised universal soil loss equation in web time to assess the effect of proposed changes in cover retention.

These tools are being used by ranchers monitoring paddock condition, organisations supporting land management initiatives in Great Barrier Reef catchments and by students developing tools to understand land condition and degradation.

TANDEM X-BAND INTERFEROMETRIC AND POLARIMETRIC OBSERVATIONS OF SUB-ARCTIC SNOW

Brown I[1], Panagiotopoulou D[1]
[1]*Stockholm University*

Seasonal snowpacks deliver a range of ecosystem services including thermal insulation for vegetation, small mammals and insects; important seasonal contributions to stream flow, including during dry summers when high altitude snowpatches and glaciers are important fresh water sources; and in providing a high albedo surface reducing the radiation absorption at the Earth's surface. This study investigates the impact of snowpack metamorphosis on time series of TanDEM-X (TDX) polarimetric and interferometric products. We demonstrate the effects of snowpack changes on scattering processes at alpine heath and sub-arctic boreal forest locations in northern Sweden. Polarimetric scatterer decompositions snow the effects of snow depth changes on scattering contributions even in the absence of liquid water in the snow. Phase differences and eigenvector/eigenvalue dynamics are also analysed under changing snow and ground conditions. InSAR products are sensitive to temperature dynamics at the base of the snowpack and the freezing of water in the upper ground layer. The results show the complexity of the interaction between climate drivers, snowpack, vegetation and soil properties that result in changes in scattering and transmission through the winter dry snow period. Temperature effects on vegetation, snow and ground dielectric properties are shown to be a major control on backscatter and coherence. These effects are not reduced to linear relationships with near surface air temperature but rather the temperature history of the vegetation-snow-ground system. Finally, the quality of TanDEM-X data and utility of Tandem SAR missions is emphasised.

TEMPORAL MONITORING OF WATER LEVEL CHANGES IN HAZELMERE DAM USING REMOTE SENSING AND GIS TECHNIQUE

Oliphant T[1]
[1]*SANSA*

The Earth's water resources are endangered pollution, inconsiderate use, and lack of conservation measures. Water scarcity is a major problem worldwide resulting in poor delivery of water which in turn results in severe socioeconomic and environmental problems. South Africa is a semi-arid country which makes a water scarce area and is experiencing significant water shortages due to climatic conditions. Monitoring these water reservoirs is very crucial for management and planning purposes however most of the reservoirs are located in remote regions which are not easily accessible and most are still maintained using only in-situ measurement systems Therefore the objective of this study is to provide a synopsis on the use of satellite images in managing and monitoring changes in surface area of water reservoirs. Multitemporal Landsat 8 images were used to investigate changes in the water surface area (extent) of Hazelmere dam. The result demonstrates a substantial decrease in the water surface area of the Hazelmere dam over the past (number) years.

KEYWORDS: Landsat8, remote sensing, temporal change and water reservoir

TERRADUE CLOUD PLATFORM - ADVANCING EARTH SCIENCE COLLABORATIVELY

Gonçalves P[1], Caumont H[1]
[1]*Terradue*

Earth observations from satellites produce vast amounts of data. Especially, the new Copernicus Sentinel missions are playing an increasingly important role as a regular and reliable high-quality, free and open data source for scientific, public sector and commercial activities. The latest developments in Information and Communication Technology (ICT) facilitate the handling of large volumes of data. Most importantly, this has started to modify the expectations that organisations have on new service development and on support to Earth Observation (EO) data exploitation, now aiming at chains of value-creation where openness and accountability of data products play a key role. Nevertheless, research institutes and commercial companies dealing with EO are still adapting to the new different economic models provided by Cloud computing for big data processing and still need to understand the implications of having their business delivered through a Cloud platform.
Terradue Cloud Platform is addressing this issue by targeting solutions for the authoring and portability of EO processing algorithms to cloud infrastructures. It provides services to move the processing to where the data is, and it optimizes the connectivity services of the data centres with more integrated discovery and processing methods.

The concept of service integration is an evolution of previous work on European Space Agency (ESA) and European Commission (EC) projects that provided experience in integration and deployment APIs for Cloud operations.

In this presentation, we will describe the evolution of the Cloud Platform and its application to EO data processing. We will show Terradue Cloud Platform support for ongoing ESA exploitation platforms (e.g. Geohazards, Urban, Hydrology), for the new EC project NextGEOSS covering the evolution path of GEO, and for the support of new research communities within the EVER-EST, INTAROS and Co-ReSyF projects.

KEYWORDS: Thematic Exploitation Platforms Cloud ICT

THE GLOBAL EARTH OBSERVATION SYSTEM OF SYSTEMS COMMON INFRASTRUCTURE: COMPONENTS GEOSS PORTAL AND GEO DISCOVERY AND ACCESS BROKER (GEO-DAB).

Van Bemmelen J[1]
[1]*ESA-ESRIN*

Components GEOSS Portal and GEO Discovery and Access Broker (GEO-DAB) GEOSS, the Global Earth Observation System of Systems, is providing access to millions of diverse and heterogeneous Earth observation resources (satellite sensed, in-situ, buoys, airplane, etc.) at different scales (local, regional and local) to users with all kind of backgrounds (scientists, decision-makers, policy-makers, citizens) and disciplines (oceans, land, atmosphere, etc.) around the globe. It is built around a common infrastructure (GCI) with the main components being the GEOSS Portal and the GEOSS Discovery and Access Broker (GEO-DAB), the portal being the main user interface and the GEO-DAB being the main interface with the resource providers. The presentation will provide a general overview of how users can discover, access and benefit from the different resources available via GEOSS, and have a particular focus on the latest enhancements of the GCI-elements in function of improved (e.g. more intuitive and direct) uptake by the different user communities. There is the possibility of a demonstration depending on the Internet connection.

THE APPLICATION OF INTEGRATED TOPOGRAPHIC DATA FOR LAND USE MAPPING: A CASE STUDY OF EDEN DISTRICT, SOUTH AFRICA

Ngcofe L[1]

[1]*Department Of Rural Development And Land Reform*

One of the issues that have not been effectively resolved in existing classification schemes is the distinction between land use and land cover. Most schemes group these two geographic variables together since they are closely related. There are, however, key differences that create confusion when land cover and land use are blended together in a single classification scheme and map product. The distinction between land cover and land use is fundamental but often ignored or forgotten. Confusion and ambiguity between these two terms lead to practical problems, particularly when land cover and land use data needs to be matched, compared and / or combined. South Africa is no exception, having many land cover maps deemed as land cover and land use maps.

Land cover is defined as (bio) physical cover of the earth's surface, including various combinations of vegetation types, soils, exposed rocks and water bodies as well as anthropogenic elements, such as agriculture and built environment. On the other hand land use refers to the socioeconomic uses of the land such as agriculture, environmental and residential / industrial uses.

The Department of Rural Development and Land Reform (Spatial Planning and Land Use Management together with National Geospatial Information directorate of National Geomatics Management Services branch) have embarked on a national land use classes and definition standard development. The proposed hierarchical structure of land use classification provides back and forth aggregation of related land use classes, thus allowing capability of land use mapping from various scales (from national to micro scale). The National Geospatial Information (NGI) Directorate is responsible for land use and topographic mapping amongst some of its other responsibilities. In so doing this study investigates and assesses the potentials and limitations of topographic data for land use mapping in the Eden District.

KEYWORDS: Land use/land cover, topographic data

THE AUSTRALIAN GEOSCIENCE DATA CUBE, A PLATFORM FOR EARTH OBSERVATION SCIENCE AND INNOVATION

Ross J[1], Oliver S[1]
[1]*Geoscience Australia*

The Australian Geoscience Data Cube (AGDC) is enabling data-intensive science through the application of High Performance Data and High Performance Computing to analysis of Earth observation data collections. The Data Cube enables scientific examination of geophysical properties of the Earth by providing a common analytical framework for structuring gridded data for rapid, iterative, and parallel processing. The AGDC reduces the barriers to direct application of geoscience data by providing access to analysis ready, spatially and spectrally aligned, quality assured observations through a highly abstracted application programming interface. This provides increased capacity for development of information products by the Earth observation community, and an increased realization of the value of Earth observation information. The AGDC is currently being used in applications including the mapping of surface water from Landsat observations over the Australian continent since 1986. The application areas are rapidly expanding and currently include mineralogical, vicarious sensor calibration, vegetation health and condition assessment and land cover change detection. This paper provides a status update on the AGDC programme.

THE BENEFITS OF A LIGHTNING DETECTION NETWORK AND ITS APPLICATIONS

Gijben M[1]
[1]*South African Weather Service*

The South African Weather Service (SAWS) owns a Vaisala cloud-to-ground lightning detection network consisting of 25 sensors situated across the country, which became operational in 2006. This network can detect cloud-to-ground lightning strokes with a predicted detection efficiency of at least 90% and location accuracy of 0.5km or better over most of South Africa. The lightning data recorded by the SALDN has been used extensively to analyse the distribution of lightning over South Africa. A lightning climatology was developed for South Africa that currently covers the 11-year period from 2006 to 2016. These maps are particularly useful for identifying the lightning hot spot areas over the country where lightning safety is of paramount concern. Lightning data is used extensively for the validation of many thunderstorm nowcasting and forecasting products at SAWS since lightning is a good indicator of where convection is occurring. Recently a Lightning Threat Index was developed to forecast the probability of lightning occurrence. SAWS also has many external clients, which utilises lightning data. These clients range from the insurance sector that requires lightning data for lightning verification purposes, to the power utility of South Africa for the design and monitoring of their infrastructure. Operational weather forecasters at SAWS utilises lightning information to monitor thunderstorms. Lightning is a good indicator of the convectively active regions of the country. Lightning related deaths and damages pose a serious risk to South Africa, Africa and the rest of the world. In South Africa, many communities are severely at risk from lightning due to the lack of sufficient shelter and awareness of the dangers of lightning. Extensive work is needed to ensure the protection of people and animals, and to minimise the damages caused by lightning.

KEYWORDS: Lightning, Nowcasting, Climatology, Lightning Threat Index

THE BIOMASS MISSION: QUANTIFYING BIOMASS FOR GLOBAL CARBON ASSESSMENT

Le Toan T[1], Quegan S[2], Scipal K[3], The Biomass Mission Advisory Group[3]
[1]*CESBIO,* [2]*University of Sheffield,* [3]*ESA-ESTEC*

To determine the distribution of forest biomass at a global scale is the objective of the BIOMASS mission, selected in May 2013 as the 7th ESA Earth Explorer Mission, for a launch in 2021. Over the mission lifetime, BIOMASS will map the full range of the world's above-ground biomass with accuracy and spatial resolution compatible with the needs of national scale inventory and carbon flux calculations. BIOMASS is based on a P-band Synthetic Aperture Radar to provide continuous interferometric and polarimetric radar observations of forested areas. BIOMASS will measure and map forest above-ground biomass, as well as forest height, over tropical, temperate and boreal forests at a spatial resolution of around 200 m every 6 months . However, the particular focus is on the carbon-rich dense tropical forests which constitute by far the largest current stock of biomass, but also the largest proportion of carbon emissions from deforestation and forest degradation. By using a P-band SAR, BIOMASS allows high values of AGB in tropical forests to be measured. The combination of three mutually supporting measurement techniques, namely polarimetric SAR, polarimetric interferometric SAR (PolInSAR) and tomographic SAR (TomoSAR) will significantly reduce the uncertainties in biomass retrievals.

This paper will present an overview of the mission, comprising the scientific requirements, the system concept, and results of mission preparation activities.

ISRSE37 TSHWANE 2017

THE CASE FOR AFRIGAM (A CO-ORDINATION, DATA ACCESS, CAPACITATING AND EVIDENCE BASED DECISION MAKING SUPPORT INITIATIVE FOR AFRICAN AGRICULTURAL MONITORING)

Newby T[1], Mlisa A[2], Nakalembe C[3], Whitcraft A[4]
[1]Agricultural Research Council, [2]GEO Secretariat, [3]Department of Geographical Sciences, University of Maryland, [4]GEOGLAM Secretariat

The AfriGEOSS initiative, within the GEO framework, strengthens the links between GEO activities and existing capabilities and initiatives in Africa and provides the necessary framework for countries to access and leverage on-going EO-based initiatives across Africa. GEO's Global Agricultural Monitoring initiative(GEOGLAM) reinforces the international community's capacity to produce and disseminate relevant, timely and accurate forecasts of agricultural production at national, regional, and global scales using Earth Observation data. Furthermore, food security and agricultural monitoring in Africa have been identified as priority activities by the African Union Commission. Evidence-based decisions on food security management require current, understandable and accessible information. All five sub-regions of Africa list food and nutrition security as a sustainable development priority. This is manifested in the continents Sustainable Development Goals. The AfriGEOSS Agricultural Monitoring initiative (AfriGAM), is a joint initiative of AfriGEOSS and GEOGLAM and I contributes in facilitating the collection of and access to food security information allowing for evidence based decision-making by African governments, industry decision makers and other stakeholders.

The case for AfriGAM rests in its objectives:

1. In collaboration with GEOGLAM identify, develop and document best practices, methodologies and systems suitable for sustainable crop monitoring and production estimation in Africa and African countries.
2. To promote the uptake of best practices, methodologies and systems for the monitoring of agriculture and agricultural production in African countries for both cultivated crops, fodder and rangelands.
3. To co-ordinate training, support and capacity development for agricultural monitoring at the national level in African countries.
4. To encourage local research and development of sustainable best practice methodologies suitable for uptake by African countries considering available technologies, resources, skills and user needs. This will be achieved by promoting the establishment of JECAM sites in African countries.
5. To facilitate liaison, participation in, contribution to and interaction with GEOGLAM activities.

THE CEOS FEASIBILITY STUDY FOR AN AQUATIC ECOSYSTEM IMAGING SPECTROMETER

Dekker A[1], Gege P[2], Pinnel N[2], Briottet X[3], Peters S[4], Court A[5], Sterckx S[6], Botha E[1], Costa M[7], Bergeron M[8], Heege T[9], Turpie K[10], Giardino C[11], Brando V[11], Krasemann H[12]
[1]CSIRO, [2]DLR, [3]ONERA, [4]Water Insight, [5]TNO, [6]VITO, [7]University of Victoria, [8]Canadian Space Agency, [9]EOMAP, [10]NASA, [11]CNR, [12]HZG

The Committee on Earth Observation Satellites (CEOS) response to the Group on Earth Observations System of Systems (GEOSS) Water Strategy developed under the auspices of the Water Strategy Implementation Study Team was endorsed by CEOS at the 2015 Plenary. As one of the actions, CSIRO has taken the lead on recommendation C.10 : A feasibility assessment to determine the benefits and technological difficulties of designing a hyperspectral satellite mission focused on water quality measurements.

More specifically this report is a high-level feasibility assessment of the benefits and technological difficulties of designing a hyperspectral satellite mission focused on biogeochemistry of inland, estuarine, deltaic and near coastal waters - as well as mapping macrophytes, macro-algae , seagrasses and coral reefs - at significantly higher spatial resolution than 250 m, which is the maximum spatial resolution of dedicated current aquatic sensors such as Sentinel-3 and future planned aquatic sensors such as the Coastal Ocean Color Imager (COCI – 100 m res). Further, the GEO Community of Practice Aquawatch suggested that alternative approaches, involving augmenting designs of spaceborne sensors for terrestrial and ocean colour applications to allow improved inland, near coastal waters and benthic applications, could offer an alternative pathway to addressing the same underlying science questions. Accordingly, this study also analizes the benefits and technological difficulties of this option as part of the high-level feasibility study.

KEYWORDS: Imaging Spectrometry, Aquatic Ecosystems; Future Sensors; Water Quality; Coral Reefs and Seagrasses

THE COASTAL WATERS RESEARCH SYNERGY FRAMEWORK, FOR UNLOCKING OUR POTENTIAL FOR COASTAL INNOVATION GROWTH

Terra-Homem M[1], Catarino N[1], Grosso N[1], Scarrott R[2], Politi E[2], Cronin A[2]
[1]*Deimos Engenharia S.A.*, [2]*University College Cork*

Until recently, scientists had to deal with the daunting task of mining large datasets for suitable data, and often downloading EO information from various different sources. In addition, as the datasets increased in volume, the processing has become slower and demanding of better computing facilities. The Coastal Waters Research Synergy Framework (Co-ReSyF) project aims to tackle these issues, by developing a platform for combined data access, processing, visualisation and output in one place. The platform is based on cloud computing to maximise processing effort and task orchestration. The platform is to support researchers in the field of monitoring the economic and social coastal activities (e.g. fisheries, harbour operations, ship traffic monitoring, oil spill detection) in a changing world.

Co-ReSyF is a 3-year project (2016-2018) funded by the European Union, within the European Union's Horizon 2020 research and innovation programme under grant agreement No 687289. The project supports research applications using Earth Observation (EO) data for Coastal Water Research. Co-ReSyF will create a cloud platform, which simplifies integration of EO data use into multi-disciplinary research activities. This platform aims to be user friendly and accessible to inexperienced scientists as well as EO and coastal experts. We will reach a wide community of coastal and oceanic researchers, who are offered the opportunity to experience, test and guide the development of the platform, whilst using it as a tool for their own research. The platform will include a set of 5 core Research Applications, developed under the project, and additionally other potential applications added by the community. The set of core applications are:

- Bathymetry Determination from SAR Images
- Determination of bathymetry, benthic classification and water quality from optical sensors
- Vessel and oil spill detection
- Time-series processing for hyper-temporal optical data analysis
- Ocean coastal altimetry

THE CONTRIBUTION OF THE EARTH OBSERVATION PROGRAMS OF THE EUROPEAN SPACE AGENCY TO THE REALISATION OF THE 2030 AGENDA FOR SUSTAINABLE DEVELOPMENT

Paganini M[1]
[1]*European Space Agency*

With the adoption of the 2030 Agenda for Sustainable Development by the UN General Assembly, all countries agree to undertake some collective and transformative efforts to integrate the three pillars of sustainable development (economic growth, social development and environmental protection) into their national policies and programmes. The 17 Sustainable Development Goals (SDG) and the 169 SDG Targets will need to be incorporated into national processes. A Global Indicator Framework for monitoring and reporting on the SDG Targets was recently adopted by the UN Statistical Commission and is currently further developed and tested in a number of pilot countries, following a 3-tier system based on the level of methodological development and data availability. When adopting the 2030 Agenda for Sustainable Development, UNGA recognised the importance of satellite observations to ensure access to high quality, timely and disaggregated geospatial data, in particular for developing countries.

Through its EO programmes, ESA supports the development of innovative applications aimed at providing user communities at large with continuous and reliable information about the Earth Environment. Since the 2002 World Summit on Sustainable Development, the European Space Agency has conducted a number of projects related to environmental governance and sustainable development, in close partnership with major actors of Multilateral Environment Agreements like the UNFCCC, UNCCD, CBD or the Ramsar Convention on Wetlands. The recent emergence of operational satellite missions with open and free data policies, such as the Sentinels of the European Copernicus Program or the US Landsat series, offer an unprecedented ensemble of satellite observations that collectively enable the development of satellite-based monitoring systems, facilitating national ownership in monitoring and tracking progress towards the SDG targets. The presentation will showcase how ESA EO programs and projects can the support the collective realisation of the 2030 Agenda for Sustainable Development.

THE COPERNICUS CLIMATE CHANGE SERVICE (C3S) IN THE MAKING - THE C3S SERVICE AND PRODUCTS

Thépaut J[1], Dee D[1], Buontempo C[1], Armstrong D[1], Garces de Marcilla J[1]
[1]*ECMWF*

Copernicus is the European Commission's flagship Earth Observation programme that delivers freely accessible operational data and information services. ECMWF has been entrusted to operate two key parts of the Copernicus programme, which will bring a consistent standard to the measurement, forecasting and prediction of atmospheric conditions and climate change.

The Copernicus Climate Change Service (C3S) will routinely monitor and analyse more than 20 essential climate variables (ECVs) to build a global picture of our climate, from the past to the future, as well as developing customisable climate indicators for relevant economic sectors, such as energy, water management, agriculture, insurance, health, etc. This climate information will be made available to a wide variety of users via a Climate Data Store, which will also provide facilities and tools to further develop tailored information and generate downstream services.

C3S is now entering its pre-operational phase. The C3S portfolio includes earth observation based (in particular from satellites) ECV datasets, global and regional reanalyses, multi-model seasonal forecasts, as well as user-friendly access to climate projections products at global and regional level. An alpha version of the Climate Data Store technical infrastructure as well as a toolbox has been made available to a limited set of users. Meantime, climate monitoring information is routinely produced and available at climate.copernicus.eu. In addition, a number of proof-of-concept sectoral climate services have been initiated, including demonstrators for specific economic sectoral applications.

This paper will focus on (a) the most recent achievements of the Service, with a focus on the description and expected outcome of the proof-of-concept activities as well as the definition of a roadmap towards a fully operational European Climate Change Service and (b) a demonstration on how to make use of the already existing C3S portfolio will also be provided.

THE COPERNICUS GLOBAL LAND HIGH RESOLUTION HOT SPOT MONITORING PROGRAM - PROVIDING LAND COVER INFORMATION FOR PROTECTED AREAS AND BEYOND

Brink A[1]
[1]*Joint Research Centre - European Commission*

Environmental information is of crucial importance. It helps to understand how our planet and its climate are changing, the role played by human activities in these changes and how these will influence our daily lives. To take the right actions, decision makers, businesses and citizens must be provided with reliable and up-to-date information on how our planet and its climate are changing. The European Earth monitoring Copernicus program has been set up to provide this information. Users are provided with information through services dedicated to a systematic monitoring and forecasting of the state of the Earth's subsystems. The scope of the Copernicus Global Land – Hot Spot Monitoring program is to provide detailed inventory of land cover and land cover change over identified areas of interest, with a change assessment frequency of 1 to 20 years, and based on high or very high resolution satellite data. Initial priority areas have been identified as Key Landscape for Conservation areas (KLC) comprising several protected areas and their surroundings. First results will be presented highlighting land cover changes and related indicators such as loss of natural vegetation and agriculture encroachment over selected KLC's.

THE COPERNICUS SPACE SEGMENT IS GROWN-UP: STATUS AND FUTURE

Jutz S[1]

[1]*European Space Agency - ESA*

The Copernicus environment-monitoring programme with its fleet of Sentinel satellites forming the heart of the programme's space component, entered its operational phase in 2014 with the launch of the first dedicated satellite, Sentinel-1 A.

In the meantime three more spacecraft, Sentinel-2 A, Sentinel-3 A and Sentinel-1 B have been launched in 2015 and 2016.

The Copernicus programme, technical managed by ESA for the space component but overall led by the European Union, expects to launch in the following years, over 16 more Sentinel satellites or instruments, covering all environmental domains. The data will be distributed free of charge as part of the European policy seeking to stimulate downstream value-added Earth observation-related businesses.

Around 52000 users world-wide are accessing Sentinel data from several data access hubs developed by ESA. Over 12 PB of data have been already downloaded with an average of several TB of data products downloaded every day. These figures will grow as new satellites will be put in orbit.

Data continuity will remain an objective for Copernicus but it will not justify per se a follow-on programme. The evolution of Copernicus is very ambitious and oriented to a larger set of activities, e.g. enhancing the sustainability of EO in Europe, supporting many of the Sustainable Development Goals of UN, widening the spectrum of observations also in the light of technology advances, etc.

In the meantime, and thinking of the near term future, new priorities have been introduced in the EU policies arising from recent events in Europe and new societal needs.

This joint EU-ESA endeavour, which just started with a priority list of missions for new measuring capabilities such as CO2 monitoring, polar ice/ocean interferometric altimetry, thermal Infrared, hyper-spectral land imaging, and soil moisture applications, will enrich the already existing family of Sentinels.

ISRSE37
TSHWANE 2017

THE EAGLE CONCEPT – FRAMEWORK FOR A PARADIGM SHIFT IN LAND MONITORING

Bock M[1], Arnold S[1]
[1]*Federal Statistical Office (Destatis)*

The current environmental challenges require the interconnection of ecological, economic and social factors at local to global scales. There is therefore a fundamental need to monitor these factors, their impact on land, their spatial distribution and changes over time in the form of land cover (LC) and land use (LU) observations. To work effectively across the required temporal and spatial scales these observations need to be modeled in a consistent and machine readable way.

A broad variety of LC/LU classification systems have evolved over time in response to specific needs and available technology. Each application emphasizes different aspects of LC and LU and many have mixed LC and LU information. Incompatibility caused by variations in class definitions (semantic overlaps/gaps etc.) often hampers the exchange of data between different applications. The globalization of information on land requires harmonization, which so far was approached by spatial and thematic generalization resulting in coarsely aggregated data. Future tasks require a more differentiated and detailed description of landscape.

Meanwhile, progress in the development of remote sensing and database technology has increased the methodological capabilities in the land monitoring domain and opened the way for a more sophisticated approach in land description.

The EAGLE concept represents a data model that consistently separates LC and LU information by decomposing landscape into Land Cover Components, Land Use Attributes and further landscape Characteristics. The concept can be used (a) as a semantic translation tool, (b) for semantic ontological analysis of existing class definitions, (c) as a guideline for the design of classification systems or mapping activities. It can be considered as a conceptual proposal for a future European land monitoring framework building upon preceding achievements while integrating new possibilities of parameterized data storage and modelling.

KEYWORDS: Land Cover, Land Use, Data Model, Land Monitoring, Ontology, Semantics

THE EFFECT OF FIRE ON VEGETATION DYNAMICS OBSERVED THROUGH MULTIPLE REMOTE SENSING SENSOR SYSTEMS AND VEGETATION INDICES

Christiansen T[1], Crews K[1], Meyer T[1]
[1]*The University of Texas at Austin*

Fire is a key determinant of vegetation structure, distribution, and composition across spatio-temporal scales in savanna ecosystems. Remotely sense data have improved our environmental understanding of fire regime properties (extent, season, frequency, intensity, and severity) in these systems resulting in the development of land management recommendations. This paper presents both methodological and ecological findings from a fifteen hectare prescribed burn in the Ghanzi region of the Botswana Kalahari and leverages pre- and post- structural vegetation measurements (three-dimensional) with optical remote sensing. Four primary research objectives were addressed: (1) To what extent were different sensor systems (e.g., Sentinal-2, Landsat, MODIS) and different projects (e.g., raw imagery, MODIS burn area products) capable of detecting the burn? (2) What was the level of spatial heterogeneity of vegetation structure pre- and post- burn as detected by Sentinel-2 imagery? (3) How did the pre-burn vegetation species, structure, and biomass map onto regrowth in the first post-burn rainy season as measured by Sentinel-2 vegetation indices? (4) What are the implications of scale (extent) of both the vegetation plots and the burn itself for sensors with different scale (grain and spectral) sensor systems? The results will inform both remote sensing and savanna ecology as to best practices in leveraging in situ and remotely sensed measurements, modeling, and management.

THE EFFECT OF THE AGULHAS CURRENT ON ADVANCED SYNTHETIC APERTURE RADAR DERIVED WIND FIELDS

Shilperoort D[1]
[1]*Nansen Tutu Center For Marine Environmental Research*

Five years (987 swaths) of high resolution wind speeds, derived from Advanced Synthetic Aperture Radar data collected over the Agulhas Current region, are studied to investigate the effect of warm, high intensity currents on the ocean's surface roughness and resulting derived wind fields. Globcurrent ocean current velocity data is used to investigate the difference between the satellite derived wind speeds compared to surface velocities of the current and the true wind speed. The current-relative wind effect is investigated for different wind direction regimes, namely: upcurrent, downcurrent, crosscurrent west and crosscurrent east. Our analyses are conducted for 6 locations of interest, evenly spaced along the Northern Agulhas Current. MODIS, SEVIRI and OSTIA SST data are used as proxy for locating the core of the Agulhas and its temperature fronts, as well as to investigate wind speed modifications as a result of ocean atmosphere energy transfer. It is found that higher resolution SAR derived winds have a greater ability to represent higher intensity and smaller scale wind features in comparison to winds derived from Scatterometers. A combination of the current relative effect and SST atmospheric heating for upcurrent wind directions results in a sharp increase in mean wind speeds over the inshore boundary of the current of between 5m/s and 7m/s (50–60%). Individual events can reach as high as 15 m/s (100%) over 10's of kilometres. For downcurrent winds, the expected current relative effect is overridden by increased wind speeds of up to 5m/s (40%) across the entire current due to the influence of SSTs. The mean effect of SSTs on wind speeds has a stronger effect than the current relative effect on wind speed changes over the current. The wind speed differences are best represented under moderate wind speeds between 5–15 m/s.

THE EFFECTS OF EI NINO ON THE CO2 CONCENTRATION CHANGES IN OCEANIA REGION BY USING GREENHOUSE GAS MONITORING SATELLITES

Yin S[1], Wang X[2], Tani H[2], Zhong G[1], Sun Z[1]
[1]Hokkaido University, Graduate School of Agriculture, [2]Hokkaido University, Research Faculty of Agriculture

The Greenhouse Gases Observing Satellite (GOSAT) was launched successfully on January 23, 2009, in Japan, and it is the world's first spacecraft to measure the concentrations of carbon dioxide and methane, the two major greenhouse gases, from space. The other corresponded satellite, Orbiting Carbon Observatory 2 (OCO-2), was also successfully launched on 2 July 2014 by NASA. Because these two satellites have different spatial and temporal resolution, they will be able to provide sufficient data and information to ascertain the distribution of greenhouse gas and enhance scientific understanding on the causes of global warming. Here we use the CO2 data from two satellites to analyze the spatial-temporal distribution in Oceania region during the EI Nino event (2015-2016). The data of both GOSAT and OCO-2 are point data, thus the Kriging interpolation method has been used to get the monthly CO2 concentration distribution map of Oceania region. Since EI Nino is always characterized by a pattern of sea surface temperature anomalies and precipitation anomalies, the sea surface temperature data and precipitation data from NOAA are also used to analyze the changing pattern between climate factors and CO2 concentration during EI Nino event. The results indicate that the CO2 concentration presents a strong seasonal pattern and EI Nino affects the CO2 concentration in Oceania region. For this study, the integrated view of multiple date resources provides a powerful tool for characterizing the CO2 spatial distribution and EI Nino.

THE ESA ADM-AEOLUS DOPPLER WIND LIDAR MISSION: CONCEPT, STATUS AND VALIDATION STRATEGY

FLAMANT P[1]
[1]*Latmos-Ipsl*

The European Space Agency's wind mission, ADM-Aeolus, is getting ready for launch in December 2017. By providing timely and accurate profiles of the World's winds along with information on aerosols and clouds, the ADM-Aeolus mission will advance our understanding of atmospheric dynamics, and will provide information to improve weather forecasts and contribute to climate research and air quality issues. The satellite will fly in a polar dusk/dawn orbit, measuring at 6 am/pm local time. The global coverage is ~16 orbits per day. In clear air, the lidar shall deliver 2D horizontally projected single line-of-sight wind observations from ~30 km down to the Earth's surface. The vertical sampling will be 500 m in the PBL, 1 km in the troposphere, and 2 km in the stratosphere. The ALADIN lidar operates continuously at a 100 Hz PRF. The horizontal observation are consecutive blocks of ~85 km. The required accuracy of the wind measurements is 2 m/s in the PBL, 2-3 m/s in the free troposphere, and 3-5 m/s in the lower stratosphere. Spin-off products are aerosol layers and semi-transparent clouds extinction and backscatter coefficient profiles. The wind and optical properties products will be delivered near-real-time (NRT) to operational numerical weather prediction (NWP). The ADM-Aeolus Doppler wind lidar instrument, ALADIN, is a High Spectral Resolution (HSR) Lidar operating at 355 nm with a line of sight at 30° from nadir and perpendicular to the satellite vector direction. The UV illumination is provided by a powerful Nd-YAG laser. The HSR separates broad air molecules scattering from narrow particles scattering. The light is analyses by two separate interferometric units i.e. a Fabry-Perot etalon for molecular scattering and a Fizeau interferometer for particles scattering. ALADIN, successfully passed the Instrument Full Functional Performance test in April 2016 and was integrated on the satellite platform in autumn.

THE ESA EARTH EXPLORER BIOMASS MISSION

Le Toan T[1], Quegan S[2], Scipal K[3], Chave J[4], The Biomass MAG[3]
[1]CESBIO, [2]University of Sheffield, [3]ESA-ESTEC, [4]EDB

To determine the distribution of forest biomass at a global scale is the objective of the BIOMASS mission, selected in May 2013 as the 7th ESA Earth Explorer Mission, for a launch in 2021. Over the mission lifetime, BIOMASS will map the full range of the world's above-ground biomass with accuracy and spatial resolution compatible with the needs of national scale inventory and carbon flux calculations. BIOMASS is based on a P-band SAR to provide continuous radar observations of forested areas. BIOMASS will measure and map forest above-ground biomass, as well as forest height, over tropical, temperate and boreal forests at a spatial resolution of around 200 m every 6 months throughout the five years of the mission. However, the particular focus is on the carbon-rich tropical forests which constitute the largest stock of biomass, but also the largest proportion of carbon emissions from deforestation. By using a P-band SAR, BIOMASS allows high values of AGB in tropical forests to be measured. The combination of three mutually supporting measurement techniques, namely polarimetric SAR, polarimetric interferometric SAR (PolInSAR) and tomographic SAR (TomoSAR) will significantly reduce the uncertainties in biomass retrievals.

This paper will present an overview of the mission, comprising the scientific requirements, the system concept, and results of mission preparation activities.

KEYWORDS: BIOMASS mission, P-band SAR, global carbon assessment, REDD+

THE ESTABLISHMENT OF THE EXCELLENCE RESEARCH CENTRE FOR EARTH SURVEILLANCE AND SPACE-BASED MONITORING OF THE ENVIRONMENT (EXCELSIOR) IN CYPRUS

Hadjimitsis D[1], Kontoes H[2], Schreier G[3], Ansmann A[4], Komodromos G[5], Themistocleous K[1], Mamouri R[1], Michaelides S[1], Nisantzi A[1], Papoutsa C[1], Mettas C[1], Tzouvaras M[1]

[1]*Cyprus University of Technology,* [2]*National Observatory of Athens,* [3]*DLR,* [4]*Leibniz Institute for Tropospheric Research,* [5]*Ministry of Transport, Communications and Works*

The aim of this paper is to present our strategy and vision to upgrade the existing ERATOSTHENES Research Centre (ERC), established within Cyprus University of Technology (CUT), into a sustainable, viable and autonomous Centre of Excellence (CoE) for Earth Surveillance and Space-Based Monitoring of the Environment (EXCELSIOR), which will provide the highest quality of related services both on the National, European and International levels. The EXCELSIOR project is within the Horizon 2020 Teaming project, which addresses the reduction of substantial disparities in the European Union by supporting research and innovation activities and systems in low performing countries. It also aims at establishing long-term and strategic partnerships between the Teaming partners, thus reducing internal research and innovation disparities within European Research and Innovation landscape. The EXCELSIOR project envisions upgrading the existing ERC into an inspiring environment for conducting basic and applied research and innovation in the area of Earth Observation. The main objective of the EXCELSIOR project is to provide a clear vision for upgrading the existing ERC to a sustainable CoE. The upgrade will regard the expansion of this vision to systematic monitoring of environment using space- and ground- based cutting-edge technologies. Such an approach will lead to the thorough study of all three domains of the environment; Air, Land, and Water. Five partners have united to upgrade the existing ERC into a CoE, with the common vision to become a world-class innovation, research and education centre, actively contributing to the European Research Area (ERA). More specifically, the Teaming project is a team effort between the Cyprus University of Technology (CUT, acting as the coordinator), the German Aerospace Centre (DLR), the National Observatory of Athens (NOA), the German Leibniz Institute for Tropospheric Research (TROPOS) and the Cyprus' Department of Electronic Communications of the Ministry of Transport, Communications and Works (DEC-MTCW).

KEY WORDS: Center of Excellence, remote sensing, Cyprus, environmental monitoring

THE EU COPERNICUS GLOBAL LAND COMPONENT OF THE LAND SERVICE - THE GLOBAL LAND COMPONENT: BRINGING SPACE DOWN TO EARTH

Cherlet M[1]
[1]*European Commission*

The Copernicus European Earth monitoring program is the European Union's flagship programme on Earth Observation and entered full operation phase following the adoption of the Regulation by the European Parliament and Council in April 2014. The Global Land component, further defined in short as "Global Land" or through the acronym COP-GL is earmarked as a component of the Land Service, one of the six Services identified for operational implementation in the above-mentioned regulation.

The COP-GL has currently three components. The first component provides a set of bio-geophysical terrestrial variables in a systematic and quality controlled production and time dissemination. The set biophysical variables, at resolutions of 1km or 300m, are relevant for environmental monitoring of the earth's land surface, including products that are useful for biomass productivity monitoring, crop monitoring, crop production forecast, carbon budget, forest cover, biodiversity, cryosphere, water and climate change monitoring at worldwide level. The second component is the Global Land Hot spot monitoring that delivers high resolution products on land cover and land cover change, including derived indicators, for hot spot over specific areas of interest to Commission Services.. In the first phase these focused on protected areas in Africa. The third component underway will make available routine products such global mosaics of Sentinel 2 Level 2 and 3 Pre-Processed data layers; further on some thematic high-resolution global layers will be provided. These layers will be tailored to specific agriculture applications as well as to support to programmes such as REDD+.

A first presentation will focus on the technical aspects of the various products of this earth observation service.
A second presentation will focus on the users, applications and use cases.

THE EU COPERNICUS MARINE SERVICE - AN INTEGRATED OFFER BASED ON STATE-OF-THE-ART SCIENTIFIC KNOWLEDGE

Thomas-Courcoux C

The Copernicus programme places a world of insight about our planet at the disposal of citizens, public authorities and policy makers, scientists, entrepreneurs and businesses on a full, free and open basis. It consists of a complex set of systems which collect data from multiple sources: earth observation satellites and in situ sensors such as ground stations, airborne and sea-borne sensors. Its processes these data and provides users with reliable and up-to-date information through a set of services related to environmental and security issues: marine, atmosphere, land, climate change, and emergency and security services.

In November 2014, Mercator Ocean, the French center for ocean monitoring and forecasting was entrusted by the European Commission to operate and implement the Copernicus Marine Service. The service provides free and fully open access to regular and systematic reference information on the physical state and marine ecosystems for the global ocean and the European regional seas (temperature, currents, salinity, sea surface height, sea ice, marine optics, nutrients, etc.). The Copernicus Marine Service offers all comers, freely and openly, simply and instantaneously, information on the physical and biogeochemical state of the global ocean and six regional seas in Europe: temperature, currents, salinity, sea surface height, sea ice, marine optics, nutrients, etc. These digital data are scientifically qualified and regularly updated. This capacity encompasses satellite and in-situ observation derived products, the description of the current situation (analysis), the prediction of the situation a few days ahead (forecast), and the provision of consistent retrospective data records for recent years (re-analysis). Subscribers to the Copernicus Marine Service have access to a catalogue of ocean-related products and information that they can download from a single point of entry, a web-portal.

THE EU COPERNICUS PAN-EUROPEAN AND LOCAL LAND COMPONENT OF THE LAND SERVICE - LAND MONITORING USE CASES

Sousa A[1], Langanke T[1], Dufourmont H[1]
[1]*European Environment Agency*

Copernicus, Europe's Earth Observation programme, provides data and services for six core thematic areas. The Copernicus Land Monitoring Service (CMLS) delivers products on a local, continental and global scales. The European Environment Agency (EEA) plays a key role in the implementation of Copernicus, in particular in the technical coordination of CLMS.

Copernicus, even though in its operational phase, has not yet reached its full use potential. In fact, some of the land products and services, with an update frequency ranging from 3 to 6 years, are still at the beginning of producing time series, which will indeed add a great value in terms of change monitoring. The current ongoing production, addressing the 2015 reference year, enters the era of big data and multi-temporal image processing, which incorporates large data volumes from different sensors now including Sentinel 1 and Sentinel 2 twin satellites.

The presentation will address some current use cases in the land domain. Some of them take the format of indicators such as: land take and intensity of land take; imperviousness and imperviousness change indicator; urban sprawl 'indicator' based in different metrics such as weighted urban proliferation, dispersion of built-up areas, land uptake per person, utilisation density and urban permeation, among others. Other use cases on the basis of Copernicus data and services have assessed landscape fragmentation; land cover flows (from mapping to accounting); land recycling estimates; proximity of urban green areas; accessibility to public transport; forest distribution and forest connectivity in Europe; likelihood of High Nature Value forest area derived from HRL combined with other layers; and use of Sentinel products for mapping abandoned arable land.

KEY WORDS: Copernicus, land monitoring, use cases

THE EU COPERNICUS PAN-EUROPEAN AND LOCAL LAND COMPONENT OF THE LAND SERVICE - OBSERVING EUROPE: THE SERVICE'S PRODUCTS

Dufourmont H[1], Langanke T[1], Sousa A[1]
[1]*European Environment Agency*

The European Environment Agency (EEA) plays a key role in the implementation of Copernicus, Europe's Earth Observation programme, in particular in the technical coordination of the Copernicus Land Monitoring Service (CLMS).

The pan-European component of CLMS produces land cover/land use information i.e. the Corine Land Cover (CLC) dataset and a set of High Resolution Layers (HRL). The CLC is a vector-based dataset and provides a time-series (from 1990 until 2012) and a change layer. The HRL are raster-based datasets providing information on land cover characteristics and as such complementary to CLC. The 2015 update is ongoing and includes: an update of the HRL imperviousness and HRL "forest", including a re-analysis of their respective time-series;, a modified HRL "grassland"; a modified HRL on "water and wetness", and a new HRL on "small woody features" (SWF), the latter based on Very High Resolution (VHR) imagery.

The local component aims to provide specific and more detailed information on different hotspots, i.e. areas that are prone to specific environmental challenges. It is based on very VHR imagery in combination with other available datasets (high and medium resolution). The current products are: Urban Atlas providing comparable land use and land cover data covering major Functional Urban Areas (FUA); Riparian Zones, which addresses land cover and land use in areas along rivers, to support biodiversity monitoring and the improvement of the "green" and "blue" infrastructures in Europe; and the mapping of Natura 2000 (N2K) areas, for which the mapping at 2 timestamps (2006-2012) enables assessing the effectiveness of the N2K instrument as a conservation policy measure.

KEY WORDS: Copernicus, land monitoring, land cover, land use, Corine, Urban Atlas, Riparian zones, Natura2000 sites

THE EUROPEAN MARS CROP YIELD FORECASTING AND MONITORING SYSTEMS - CONTRIBUTIONS TO GEOGLAM

Baruth B[1], Rembold F, Van der Velde M, Niemeyer S
[1]*European Commission - JRC*

Transparency is nowadays recognised as a key factor for stabilising markets and avoiding excessive volatility. The timely publication of accurate crop yield estimates could help to anticipate fluctuations in expected crop production levels. In order to provide this information, the European Commission's Joint Research Centre (JRC) has been making operational forecasts of national-level crop yields across all EU Member States and neighbouring countries since 1993. The crop-yield forecasts are publically available and published in the monthly MARS Bulletin along with a description of dominant agro-meteorological conditions. These forecasts are based on information from the MARS Crop Yield Forecasting System (M-CYFS), which provides country-level analyses by dedicated analysts and integrates meteorological and remote sensing information as well as crop growth model results.

The M-CYFS is progressively refined and updated, thus keeping it state-of-the-art. In this context, research and development in crop growth modelling to improve the simulation and estimation of crop yield and agricultural production, but also in developing and testing new forecasting methods from remote sensing and long-range meteorological forecasts and the inclusion of regional statistics is performed. The results of the monthly analysis are shared with the GEOGLAM community and form an integral part of the Crop Monitor analysis to serve the needs of the AMIS Market Monitor.

For areas outside Europe and more in a food security context, the JRC has recently developed the ASAP system (Anomaly hot Spots of Agricultural Production), which proposes a rapid two-step analysis to provide timely warnings of production deficits in water-limited agricultural systems every month. Affected countries are flagged as agricultural hotspots and this information, accompanied by short narratives, is then made available on a website and is also shared with the partners contributing to the GEOGLAM Crop Monitor for Early Warning (CM4EW).

KEYWORDS: agricultural monitoring, crop yield forecasting, early warning

THE EXTENT OF MANGROVE CHANGE AND POTENTIAL FOR RECOVERY FOLLOWING CYCLONE YASI AT HINCHINBROOK ISLAND, QUEENSLAND, AUSTRALIA

Asbridge E[1], **Lucas R[1]**, Rogers K[2], Accad A[3]
[1]University of New South Wales, [2]University of Wollongong, [3]Queensland Herbarium Brisbane Botanic Gardens Mt Coot-tha

Large storm events, such as cyclones, are significant drivers of change within mangrove ecosystems with the extent of initial damage determined by storm severity, location and distribution (exposure), and mangrove species composition and structure (height). The long-term recovery of mangrove is often dependent upon hydrological regimes, as well as the frequency of storm events. On 3rd February, 2011, Tropical Cyclone Yasi (Category 5) struck the coast of north Queensland Australia with its path crossing the extensive mangroves within and surrounding Hinchinbrook Island National Park. Based on a combination of Landsat-derived Foliage Projective Cover (FPC), Queensland Globe aerial imagery and RapidEye imagery (2009-2015), 16 % of the 13,795 ha of mangroves (combination of Hinchinbrook Island, the island within the Hinchinbrook Channel and the Queensland coastline) experienced severe windthrow during the storm. The damage from the cyclone was primarily inflicted on mangroves dominated by Rhizophora stylosa, whose large prop roots were unable to support them as wind speeds exceeded 280 km hr-1. The classification of 2016 RapidEye data indicated that many areas of damage had experienced no or very limited recovery, with this confirmed by a rapid decline in Landsat-derived FPC (from > 90 % from 1986 to just prior to the cyclone to < 20 %) and no subsequent increase. Advanced Land Observing Satellite (ALOS-1) Phased Arrayed L-band Synthetic Aperture Radar (SAR) L-band HH backscatter also increased with this attributed to both a reduction in foliage cover and an increase in woody debris within affected areas. The low ability of mangroves to recover from Cyclone Yasi could be attributed to an inability of mangrove species, particularly R. stylosa, to resprout from remaining plant material and recover from hydrological regimes that previously allowed the establishment of mangroves species in previous decades but then became less suitable for recovery and colonisation. This study indicates that increases in storm intensity and frequency predicted with changes in global climate may lead to a reduction in the area, diversity and abundance of mangroves surrounding Hinchinbrook Island.

THE FUTURE AIRBUS DEFENCE AND SPACE OPTICAL AND SAR SATELLITE CONSTELLATION, A VISION FOR 2020 AND BEYOND

Eloff C[1]
[1]*Airbus Defence And Space*

Airbus Defence and Space is one of the global leaders in providing Earth Observation satellite products and services since its establishment in 1982. The engineering and operational excellence acheived during the inception of SPOT 1 (1986) until today with the SPOT 6 and SPOT 7 satellites, paved the way for optical along-track scanning systems with off-nadir viewing and linear arrays of charge-coupled detectors. The success of the highly agile Pléiades very high resolution twin satellite constellation further complimented the SPOT series since 2011. Furthermore, the two high precision X-Band satellites TerraSAR-X and TanDEM-X were used in a unique helix formation orbit to collect the Earth-surface to generate the highest resolution gloabl elevation model known as WorldDEM.

The focus for this article is not to dwell in the past, but to introduce the exciting futuristic optical and SAR constellation missions of Airbus Defence and Space as from 2020. The new optical constellation will consist of four satellites, each one positioned in a perfect orbit to complement each other to produce the ability to serve high temporal applications as well as to image any target per day over the globe. This quadruple constellation will provide world-class resolution with several multi-spectral bands to ensure its relevance to the market as well as to serve a broad range of geoinformatic applications. It is foreseen that the first satellite will be launched during 2020 with a vision to complete this constellation by 2021.

In 2015, the World Radiocommunication Conference, has doubled the allocated frequency bandwidths for Earth Observation Synthetic Aperture Radar (SAR) systems. This decision opens the door to an unprecedented image resolution quality in colour of future SAR satellites. Airbus Defence and Space, already working on the next generation of SAR satellites for several years - as a follow-on mission to the TerraSAR-X and TanDEM-X, can now move forward in offering new applications based on this unprecedently high-quality SAR imagery with a resolution better than 25cm.

All these new comers will grow the ranks of the actual Airbus Defence and Space's constellation to more than 10 satellites, optical and radar sensors combined.

THE GEO BLUE PLANET INITIATIVE: OCEAN AND COASTAL OBSERVATIONS FOR SOCIETAL BENEFIT

Cripe D[4], Seeyave S[1], DiGiacomo P[2], Smail E[2], Steven A[3]
[1]Partnership for Observation of the Global Oceans (POGO), [2]National Oceanic and Atmospheric Administration (NOAA), [3]Commonwealth Scientific and Industrial Research Organisation (CSIRO), [4]Group on Earth Observation

We live on a Blue Planet, and Earth's waters benefit many sectors of society. In 2015, through development of a United Nations Sustainable Development Goal explicitly targeted at the oceans (SDG 14), the global community has prioritised the need for concerted action to ensure sustainable growth and management of blue economies across the planet. Sustainable Development Goals 13 and 15 (climate action and life on land) further recognise that the future of our Blue Planet is increasingly reliant on the services delivered by marine, coastal and inland waters and on the advancement of effective, evidence-based decisions on sustainable development.

The overall goal of Blue Planet Initiative is to ensure the sustained development and use of ocean and coastal observations for the benefit of society. The objectives to achieve this goal are: (1) Increase integration of and access to in situ and remote sensing ocean observation data; (2) Conceptualize, promote and facilitate the development of end-to-end ocean information services; (3) Improve connections between the producers and providers of ocean observation data, products and information and the end users; and (4) Increase societal support and build capacity for ocean observations.

As a Group on Earth Observations (GEO) Initiative, Blue Planet will allow for direct linkages between the ocean observing community and users within an agreed, yet flexible framework. Blue Planet will support GEO's work to promote open and accessible data and will work to facilitate, promote and conceptualize prototype/pilot services that address the GEO societal benefit areas. Blue Planet will also work to facilitate the development of prototype/pilot services that address specific policy mandates into GEO Flagships. Blue Planet activities are organized into four components: 1) data integration and informatics, 2) information services, 3) user engagement and 4) capacity building and advocacy.

KEYWORDS: GEO, ocean observations, user engagement, SDG-14, capacity building

THE GEO CARBON AND GHG INITIATIVE: TOWARD POLICY-RELEVANT GLOBAL CARBON CYCLE OBSERVATION AND ANALYSIS

Heiskanen J[13], Bombelli A[1], Butler J[2], Canadell J[3], Ciais P[4], DeCola P[5], Dolman A[6], Duren R[7], Fujimoto T[8], Houweling S[9], Jackson R[10], Key R[11], Kim D[12], Kutsch W[13], Lavric J[13], Loescher H[14], Muraoka H[15], Obregon A[16], Pfeil B[17], Plummer S[18], Saigusa N[19], Shiomi K[20], Scholes R[21], Shvidenko A[22], Sivakumar V[23], Suzuki R[24], Tanhua T[25], Telszewski M[26], Vermeulen A[27], Yi L[28]

[1]CMCC, [2]NOAA, [3]CSIRO, [4]LSCE, [5]Sigma Space Corp., [6]VU Amsterdam, [7]JPL-NASA, [8]JMA, [9]SRON, [10]Stanford University, [11]Princeton University, [12]Wondo Genet College, [13]ICOS Head Office, [14]NEON, [15]Gifu University, [16]GEO-Sec, [17]UIB, [18]ESA Climate Office / CEOS , [19]NIES, [20]JAXA/EORC, [21]Witwatersrand University, [22]IIASA, [23]UKZN, [24]JAMSTEC, [25]GEOMAR, [26]IOCCP, [27]ICOS Carbon Portal, [28]IAP/CAS

To better understand the carbon cycle and the climate system, and to address society's efforts to mitigate and adapt to climate change, we require long-term, high-quality observation systems covering the atmospheric, oceanic, terrestrial, and anthropogenic domains. Budget uncertainties of carbon dioxide and other greenhouse gases (GHGs) make climate change mitigation and adaptation strategies difficult.

Currently, several initiatives and services are developing at global and regional scale, but a globally coordinated service for delivering timely and reliable domain-overarching information on the carbon cycle and GHG fluxes to the decision-makers, the scientific community, and the general public is still missing. To fill this gap, we are developing in the framework of GEO a global Carbon and GHG Initiative that builds on existing initiatives and networks, with the objective to ensure their continuity and coherence, and obtain a comprehensive, globally coordinated service for timely delivery of relevant and traceable data and data products.

The GEO Carbon and GHG initiative shall address policy agendas and will operate as a common and open platform to plan and implement strategies and joint activities at the global level from science to policy. Our objective is to develop an operational and independent system for monitoring and evaluating variations in the carbon cycle and GHG emissions as they relate to human activities and climate change, and to provide decision makers with timely and reliable policy-relevant information, recommendations, and services.

THE GEOSS SOCIO-ECONOMIC AND ENVIRONMENTAL INFORMATION NEEDS KNOWLEDGE BASE (SEE-IN KB): LINKING THE SUSTAINABLE DEVELOPMENT GOALS TO EARTH OBSERVATIONS, MODELS, AND CAPACITY BUILDING

Plag H[1], Jules-Plag S[2]
[1]Old Dominion University, [2]Tiwah UG

The "Socio-Economic and Environmental Information Needs Knowledge Base" (SEE-IN KB) is developed as a collaborative platform for the co-creation of knowledge in support of societal goals. By bringing providers and users together, it is possible to "learn" how Earth observations and models are used to create practice-relevant knowledge and to identify gaps. The design of such a collaborative platform attractive to both users and providers has to be innovative to achieving the necessary revolution in how Earth observations are informing decisions.

A core function of the SEE-IN KB is the linking of societal goals and targets to Essential Variables (EVs). Focusing on the use case of the Sustainable Development Goals (SDGs), the SEE-IN KB uses a goal-based approach to identify those EVs that are essential for the quantification of SDG indicators and the assessment of policy impacts on SDG targets. The approach aims to be consistent with the UN's System of Environmental-Economic Accounting (SEEA). The SEE-IN KB includes the rules for defining the observational needs and how to link them to user knowledge needs. Existing knowledge repositories and databases are leveraged as far as possible for populating the SEE-IN KB. User feedback concerning the fitness for purpose of both data and processes is collected.

The SEE-IN KB is a component of the GEOSS Knowledge Base. It documents the relationships between the data and the processes (models, workflows, algorithms) needed to develop the information and knowledge needed by societal users. The concept of networks inherent in the conceptual model of the SEE-IN KB allows for the construction of business processes to answer "What if?" questions. This functions supports the planning of policies and activities facilitating progress towards the SDGs. Increasingly, user types, applications and requirements are linked to actual users, models and dataset, respectively, and this allows execution of business processes.

THE GERMAN EARTH OBSERVATION PROGRAMME

Schaadt P[1]
[1]*German Space Administration - DLR*

Global change, a sustainable development of our habitats, an efficient use of resources, securing our mobility and our competitive position in the world of advanced technology, the need to deal with crises and to minimize the risks imposed on us by natural and technological hazards, all this puts mankind before huge challenges. Earth observation by satellite can help to stay on top of these tasks. Earth observation is a strategic benefit for policymakers, industry, and citizens.

Germany's Earth observation program covers the entire spectrum of these capabilities. Distinguished task of the German Space Agency is defining the German space planning on behalf of the federal government.
The German earth observation activities are carried out in two main areas.
First area is the International Program with contributions programs such as the Earth Observation Envelope Program at ESA, COPERNICUS at the EU or the operational EUMETSAT Programs.
Second area is the National Program, which is complementary and is supporting missions, technologies, data exploitation and routine utilization. Hereby all disciplines are covered: Radar, Optics, Spectrometry and Lidar. In the field of X-band radar technology, the TerraSAR-X and TanDEM-X missions are among the world's best. With the Hyperspectral mission ENMAP for the first time in the world data with high spectral and geometric resolution, ranging from the Visible to Shortwave Infrared spectrum, will serve the global user community. METimage will be the European continuation of the NOAA-AVHRR mission while the French-German mission MERLIN will measure atmospheric Methane with a Lidar.

The activities of the national Earth observation program, the main achievements and the plans for the future will be introduced.

THE GERMAN HYPERSPECTRAL ENVIRONMENTAL MAPPING AND ANALYSIS PROGRAM (ENMAP). CURRENT STATUS AND APPLICATION EXAMPLES FOR THE GEOSCIENTIFIC COMMUNITY

Mielke C[1], Bösche N[1], Rogaß C[1], Segl K[1], Guanter L[1]
[1]*Helmholtz Center Potsdam German Research Center For Geoscience Gfz*

The Environmental Mapping and Analysis Program (EnMAP) is a spaceborne imaging spectroscopy mission under development by a consortium of German Earth Observation research institutes. The core payload of the EnMAP satellite is a combination of two prism-based imaging spectrometers covering the 420-2450 nm spectral window with 242 contiguous spectral bands, with a ground sampling distance of 30 m and a swath of 30 km. The mission is now in phase D (construction). The expected mission lifetime is 5 years. Top-of-atmosphere radiance and geometrically-corrected surface reflectance products will be produced and distributed by the ground segment.

As part of the on-going mission preparatory activities, the EnMAP Box, an open source software tool provided by the EnMAP science team, is being implemented. It includes a collection of algorithms dedicated to the specific science themes within EnMAP, which are for example: forest, water, soil and geology. Preprocessing and visualization are also part of the tools offered by the EnMAP Box.
Here we present an overview and a status update on the EnMAP mission, followed by results from the EnMAP preparatory phase in the field of geology. For this purpose, we used simulated EnMAP data analyzed by the EnMAP Geological Mapper (EnGeoMAP).

This module includes two algorithms: EnGeoMAP REE and EnGeoMAP BASE, for surface material and mineral characterization in geology. The main aim of EnGeoMAP is to provide a software tool to the geoscience community for a better understanding of the mining landscape, which includes mineral exploration, mine site, and post-mining-landscape mapping. The dynamic nature of the mineral exploration and extraction process itself will in future require synergetic applications between EnMAP and a large global mapping mission, such as Sentinel-2. Here Sentinel-2 will be able to deliver the necessary multitemporal detail and rapid coverage to the focused and detailed material analysis offered by EnMAP.

THE GLOBAL FOREST OBSERVATIONS INITIATIVE IN SUPPORT OF DEVELOPING COUNTRIES SETTING-UP THEIR NATIONAL FOREST MONITORING SYSTEM

Seifert F[1], Herold M[2], Rosenqvist A[3], Harvey T[4]
[1]European Space Agency, [2]Wageningen University, [3]soloEO, [4]GFOI Office, [5]NASA, [6]ENVIRONMENTAL ACCOUNTING SERVICES

The United Nations Framework Convention on Climate Change has recognised the important role that deforestation and forest degradation have in contribution to global anthropogenic emissions of carbon dioxide. In order to address this issue, the UNFCCC has adopted in the Paris Agreement a mechanism for reducing emissions from deforestation and forest degradation and the role of conservation, sustainable management of forests and enhancement of forest carbon stocks in developing countries (REDD+), which would ultimately provide financial incentives for emission reductions. One important requirement needed in order to implement REDD+ is that countries need to establish national forest monitoring systems (NFMS) that provide annual, national, estimates of emissions of greenhouse gases and changes in forest carbon stocks that are reported biannually.

The Global Forest Observations Initiative (GFOI) was established to assist countries establish such systems, based on time-series of wall-to-wall Earth Observations. GFOI provides methodological guidance and building capacity in the implementation of improved forest mapping methods. Related research like integrating multiple, complementary satellite observations, monitoring of forest degradation and estimating carbon stock and uncertainties in the mapping should assist more operational monitoring systems. This paper summarises the GFOI approach and identifies issues that need urgent attention in order to allow operational monitoring systems to be implemented.

KEYWORDS: UNFCCC Paris Agreement, REDD+, GFOI, MRV, NFMS

THE GLOBAL GEODETIC OBSERVING SYSTEM

Gross R[1]
[1]*Jet Propulsion Laboratory, California Institute of Technology*

Earth observations are needed not only for scientific research but also for societal applications such as disaster prevention and mitigation, managing resources like energy, water, and food, mitigating the effects of climate change, and protecting the biosphere, the environment, and human health. Geodetic observations provide the metrological foundation for Earth observations and provide the means to determine mass transport in the Earth system. Geodetic observations are therefore a cornerstone of the Earth observing systems needed for scientific research and societal applications. Geodetic observations also provide the basis for realizing the reference systems that are required in order to assign coordinates to points and objects in space and time and to describe the motion of the Earth in space. The International Terrestrial Reference Frame (ITRF) determined by geodetic observations is the indispensable foundation for all the sustainable Earth observations that are used by science and society for so many purposes, including navigation, mapping, surveying, construction, land development, natural resource management and conservation—in fact, all decision-making activities that have a geo-related component. It allows different spatial information, such as imagery from different space and airborne platforms, to be geo-referenced and aligned with each other. And it plays a key role in modeling and estimating the motion of the Earth in space, in measuring change and deformation of all components of the Earth system, and in providing the ability to connect measurements made at the same place at different times, a critical requirement for understanding global, regional and local change.

KEYWORDS: geodetic observations, terrestrial reference frame

THE HIGH RESOLUTION WIDE-SWATH MISSION AND WORLDSAR. THE NEXT GENERATION OF X- BAND SAR SERVICES

Janoth J[1], Kaptein A[1], Jochum M[1], Joumel P[1]
[1]*Airbus Defence And Space - Intelligence*

Launched in 2007, the TerraSAR-X Mission provides X-Band data and services on an operational basis. Recent improvements and evolutions of the program comprise the introduction of new SAR imaging modes and the upcoming constellation with the Spanish PAZ satellite. The "HRWS" (High Resolution Wide-Swath) Mission is intended to constitute the next National German civilian X-band SAR program for institutional, scientific and commercial use, and is designed to guarantee the TerraSAR-X data and service continuity for institutional, scientific and commercial end-users well beyond the year 2030. Designed for an operational system lifetime of 10 years, HRWS will bring improved system capabilities compared to the current Mission and will be an in-orbit demonstrator of new Digital Beam Forming SAR technologies. The Phase 0/A Study has been kicked off in May 2016 with the overall goal to identify the driving user requirements and to assess the overall feasibility of the HRWS Mission. As part of the Phase 0 a Mission Concept, a preliminary Customer Technical Requirements Specification and a preliminary System Concept will be established. After successful Mission Definition Review the HRWS Study will proceed with the Phase A focusing on the assessment of the technical feasibility and on system definition. The HRWS Mission and potential extensions will be subject to a partnership model, "WorldSAR", in which partners can participate through co-investment, subscription, and ownership of additional satellites operated in constellation.

KEYWORDS: TerraSAR-X Follow-on Mission, X-Band Continuity, HRWS (High Resolution Wide Swath), Digital beam Forming.

THE HYPERSPECTRAL SENSOR DESIS ON MUSES: PROCESSING AND APPLICATIONS

Cerra D[1], Carmona E[1], Müller R[1], Alonso Gonzalez K[1], Bachmann M[1], Gerasch B[1], Krawczyk H[1]
[1]DLR

Teledyne Brown Engineering (TBE) located in Huntsville, Alabama, USA and the German Aerospace Center (DLR), Germany, will develop and operate a hyperspectral instrument integrated in the Multi-User-System for Earth Sensing (MUSES) platform installed on the International Space Station (ISS). The DESIS hyperspectral instrument is realized as a pushbroom imaging spectrometer spectrally sensitive over the VNIR range from 400 to 1000 nm with a spectral sampling distance of about 3 nm. The instrument allows, besides standard Earth data products, acquisition of stereo or Bidirectional Reflectance Distribution Function (BRDF) products and continuous observations of the same targets on ground (using forward motion compensation mode). DESIS level 1A (raw data) products will be long-term archived while level 1B products (systematically and radiometrically corrected data), level 1C products (geometrically corrected data), level 2A products (atmospherically corrected data) will be processed on demand. The level 1B processor corrects the hyperspectral image cube for systematic effects of the focal plane detector array, e.g. radiometric non-uniformities, and converts the system corrected data to physical at-sensor radiance values based on the currently valid calibration tables. The level 1C processor creates orthoimages based on Direct Georeferencing techniques implementing a line-of-sight model, which uses on-board measurements for orbit and attitude determinations as well as the sensor look direction vectors based on the currently valid calibration values. Furthermore it is foreseen to automatically extract ground control points from existing reference data sets by image matching techniques to improve the geometric accuracy better than one pixel size. The level 2A processor performs atmospheric and haze correction of the images by estimating the aerosol optical thickness and the columnar water vapour.

THE JAXA KYOTO AND CARBON (K&C) INITIATIVE GLOBAL MANGROVE WATCH (GMW)

Lucas R[1], Rosenqvist A[2], Bunting P[3]
[1]University Of New South Wales, [2]Solo EO, [3]Aberystwyth University

The Global Mangrove Watch (GMW) has been developed as part of the Japanese Space Exploration Agency's (JAXA) Kyoto and Carbon (K&C) Initiative and is establishing 25 m baselines of global mangrove extent from the mid 1990s onwards. More specifically, the GMW is updating existing baseline maps of mangrove extent, initially using freely available historical Japanese Environmental Resource Satellite (JERS-1) SAR data (1996) and Advanced Land Observing Satellite (ALOS) Phased Arrayed L-band SAR data (2007-2010). For key areas, reference is being made to Landsat sensor data acquired on an annual basis from 1985. Changes from the latter (2010) baseline are being tracked using time-series of ALOS-2 PALSAR-2 data, with consideration given to Landsat sensor and Sentinel-1/2 SAR data. From 2018, annual updates to the baseline of extent will be put in place to provide timely information on changes in mangroves and the causes, drivers and consequences of these changes. All changes will be associated with quantitative information on the nature of change (e.g., loss of biomass, structure). The changes observed, including detected warnings, will be placed in the context of previous change, with focus on causes, drivers and consequences. Reference to retrieved biophysical parameters (e.g., biomass, height, if available) will also be used to better convey the nature of change. The historical and near real time observations of mangroves will be validated using a range of ground truth data. The intention is to prevent or limit further loss of mangroves and facilitate their restoration. Hotspots for conservation, intervention management or restoration will also be identified. Changes from the baselines are being used to routinely monitor mangroves, thereby providing timely forewarning of adverse events (e.g., climate-related dieback, deforestation) and processes (regeneration and colonization). The datasets are intended for use in management planning and policy development and supporting environmental impact assessments.

THE LAND POTENTIAL KNOWLEDGE SYSTEM: APPLICATION OF EARTH OBSERVATION DATA FOR SUSTAINABLE LAND MANAGEMENT

Ndungu L[1]
[1]*RCMRD/SERVIR E&SA*

Kenya's 2030 agenda for sustainable development not only includes meeting the 16 Sustainable Development goals, but also in implementing the Addis Ababa Action Agenda and the Paris Outcome of COP21 on climate change. To meet these challenges, there is a need for data collection, aggregation and dissemination at both local decision making levels, but also at policy levels. The role of technology in data collection and dissemination in Africa has gained momentum with the advent of affordable smart phones that can bridge the gap between information producers and scientists and information users at cross cutting levels. However, more important is the development of tools that can allow for dis-aggregation of information for consumption by different users.

The Land potential knowledge system aims to provide tools that will not only promote sustainable agriculture, while ensuring sustainable consumption and production patterns, but will also provide critical information to assist in adapting to climate change and mitigating land degradation. It is hoped that uptake of the tools will assist in combating desertification and reversing land degradation, through identification of areas with the ability to recover and where investments will yield some returns. LandPKS has utilized elevation and elevation derivatives(slope and curvature), soil texture, climate data and land use as inputs to predictive models that provide indicators of the land's productivity and erosion risk. Geo-information allows predictions to be made from any point globally providing a universal method that allows for the land potential to be evaluated and will in future form a platform where users facing similar land management problems can share local working knowledge on indigenous knowledge that they have successfully applied to sustainably manage their land.

THE NASA-ISRO SYNTHETIC APERTURE RADAR (NISAR) MISSION

Rosen P[1], Kumar R[2]
[1]Jet Propulsion Laboratory, California Institute of Technology, [2]Space Applications Center, Indian Space Research Organization

NASA has joined forces with the Indian Space Research Organisation (ISRO) to measure Earth change through the eyes of a dual-frequency synthetic aperture radar, the NASA-ISRO SAR (NISAR) mission now in development for a launch in 2021. The mission's primary science objectives are codified in a set of science requirements to study Earth land and ice deformation, and ecosystems, globally with 12-day sampling over all land and ice-covered surfaces throughout the mission life. The US and Indian science teams share global science objectives; in addition, India has developed a set of local objectives in agricultural biomass estimation, Himalayan glacier characterization, and coastal ocean measurements in and around India. Both the US and India have identified agricultural and infrastructure monitoring, and disaster response as high priority applications for the mission.

With this range of science and applications objectives, NISAR has demanding coverage, sampling, and accuracy requirements. The system requires a swath of over 240 km at 3-10 m SAR imaging resolution, using full polarimetry where needed. NISAR carries two radars, one operating at L-band (24 cm wavelength) and the other at S-band (10 cm wavelength); the radars can operate simultaneously. The L-band system is being designed to operate up to 50 minutes per orbit, and the S-band system up to 10 minutes per orbit. The orbit will be controlled to within 300 m for repeat-pass interferometry measurements. This unprecedented coverage in space, time, polarimetry, and frequency, will add a new and rich data set to the international constellation of sensors studying Earth surface change.

This work was partially performed at the Jet Propulsion Laboratory, California Institute of Technology, under contract with the National Aeronautics and Space Administration.

THE NEED FOR A GLOBAL REFERENCE DATABASE FOR LAND COVER AND LAND USE MAPPING

Fritz S[1], See L[1], Geller G[2], Obregón A[2], Tsendbazar N[3], Herold M[3]
[1]International Institute For Applied Systems Analysis, [2]Group on Earth Observations (GEO), [3]Wageningen University

Greater amounts of calibration and validation data are needed to improve land cover and land use maps derived from remote sensing. Although there have been efforts in compiling existing reference datasets and making them openly available, e.g. through the GOFC-GOLD portal, they represent a fraction of the reference data collected and stored locally around the world. Crowdsourced data, e.g. through Geo-Wiki and initiatives such as OpenStreetMap, represent another growing source of in-situ data that have the potential to benefit remote sensing. Although there are quality concerns surrounding crowdsourced information, new developments in quality assurance can provide the necessary confidence for use as reference data. What is needed is a coordinated effort by community members, which could be supported by the Group on Earth Observations (GEO), to create a global reference database that could be used for different purposes, i.e. calibration, validation and for the generation of hybrid maps. For calibration and hybrid map generation, the proposed database could bring together existing datasets. For validation, which requires higher accuracy data and probability sampling, there are already different sampling schemes proposed, e.g. from Tsinghua University, Boston University, and the Global Grid proposed by Theobald (2016, Remote Sensing). A specific portal, which could be linked to the Global Common Infrastructure of GEOSS, could be developed by GEO members for access to the reference database. Visualization, data publishing and legend harmonization tools would need to be developed along with data standards and the schema for labelling features, e.g. the Land Cover Modelling Language (LCML) or another ontology. Validation activities in the CGLOPS project are currently producing a flexible validation dataset that could be used for different purposes. The Horizon2020 LandSense Citizen Observatory will develop a distributed model for reference data, and may provide good practice for scaling up this concept globally.

THE NORTHERN VOICE: LISTENING TO INDIGENOUS AND NORTHERN PERSPECTIVES ON MANAGEMENT OF DATA IN CANADA

Ledrew E[1], Friddell J[1], Church D[1], Alix G[1]
[1]*University Of Waterloo*

The Canadian Cryospheric Information Network and Polar Data Catalogue (CCIN/PDC) were originally developed to provide a trusted archive to store data from Canadian cryospheric research and to provide a public access portal to this information. The CCIN/PDC has since expanded its collection to include a wealth of other research including the biological and social sciences. Since its inception CCIN/PDC has engaged Indigenous and northern Canadians to ensure that their information needs are being met. Feedback from these communities led to the development of PDC Lite: a version of the full PDC Search application that accommodates slower internet speeds and allows one to search by particular northern communities. The PDC Lite continues to be improved by input from the communities that it serves.

A major knowledge gap in the polar data management community is the degree to which Indigenous people wish to store, manage, and protect Traditional and Local Knowledge (TLK). To facilitate discussion and strengthen collaborative relationships, CCIN/PDC co-hosted two major meetings in 2015. Emerging from both these events was a need to prioritize "human interoperability" and the need to have indigenous and northern community involvement at all levels of data management.

Future plans for CCIN/PDC include more effective partnerships in which we work with and listen to northern and Indigenous Canadians to better understand their requirements for data management services and expertise. The ultimate goal is to provide data and information that meets their needs and enables and supports their individual data management goals.

THE POSSIBILITY OF USING 5-METER RAPIDEYE DATA FOR MONITORING MAIZE PLANTS INFESTED BY STEM BORER

Abdel-Rahman E[1,2], Kyalo R[1], Landmann T[1], Le Ru B[1,3]

[1]*International Center For Insect Physiology and Ecology (icipe)*, [2]*Department of Agronomy, Faculty of Agriculture, University of Khartoum*, [3]*IRD/CRNS UMR IRD 247 EGCE, Laboratoire Evolution Génomes Comportement et Ecologie, CNRS*

Maize (Zea mays L.) is the most important staple crop in East Africa. However, the production of the crop is significantly reduced by many biotic and abiotic constraints. Among the biotic constraints, stem borer is regarded as a major insect pest that attacks the crop causing up to 21% yield loss. In this study, we tested the hypothesis that stem borer-infested maize plants can be estimated using 5-m RapidEye observation. RapidEye image was acquired on the 3rd of January 2015 for a test site in Kenya. Three RapidEye bands (red, red edge and near infrared) and six spectral vegetation indices were used as predictor variables to estimate stem borer-infested maize plants. Data on stem borer infestation were collected by inspecting 100 maize plants within an area of 25 by 25 meter in each sample field (n = 64); and number of infested plants was recorded. Two generalized linear models (i.e. negative binomial: NB and zero-inflated negative binomial: ZINB) were employed to develop stem borer-infested maize plants predictive models. The models were evaluated using root mean square error (RMSE) and ratio prediction to deviation (RPD) metrics based on a leave one-out cross-validation method. Results showed accurate models for monitoring stem borer-infested maize plants and ZINB model outperformed NB one. We conclude that 5-m RapidEye data, NB and ZINB models can be used to predict site-specific stem borer-infested maize plants. Our results can be used to understand linkages between maize stem bore-infested sites and underlying environmental factors as well as cropping systems for better knowledge on stem borer propagation and spread in Kenya.

KEYWORDS: RapidEye, Maize, Stem borers, GLM, Agricultural productivity

THE POSSIBILITY OF USING MODIS-BASED PHENOMETRICS AND GENERALIZED LINEAR MODELING (GLM) FOR MAPPING LAND DEGRADATION PROCESSES IN SOMALILAND

Landmann T[1], Dubovyk O[2], Ghazaryan G[2], Kimani J[1], Abdel-Rahman E[1,3]
[1]International Centre Of Insect Physiology And Ecology (icipe), [2]Center for Remote Sensing of Land Surfaces (ZFL), University of Bonn, [3]Department of Agronomy, Faculty of Agriculture, University of Khartoum

Invasive species and deforestation are significant land degradation factors in eastern Africa that in unity severely impede rangeland and cropland productivity with dire consequences for livelihoods of agro-pastoralist communities. Using 250-meter Moderate Resolution Imaging Spectroradiometer (MODIS) satellite data and trends computed from Enhanced Vegetation Index (EVI) vegetation productivity and phenology metrics, we produced the first comprehensive data set on invasive species occurrence (Prosopis juliflora and Parthenium hysterophorus) and deforestation for a study area in western Somaliland (eastern Africa). A 30-meter Landsat reference map showing 'Deforestation' and areas of propagation of the two invasive species between 2001 and 2015 was produced with the help of field observations collected multi-seasonally during the years 2014 and 2015. Generalized Linear Modeling (GLM) with a binomial logistics link function was used to link the reference data (Landsat-based map) to the MODIS-based vegetation trends. A probability of occurrence map for each of the three land degradation factors was derived. The GLM parameters showed that both 'amplitude' and 'small integral' trends were uniquely and highly significant for Parthenium hysterophorus ($p<0.001$ and regression coefficients $>|5|$). 'EVI trend' was highly significant and relevant for both Prosopis juliflora and 'Deforestation'. The probability maps showed that cropland areas in the south-western part of the study area as well as urban zones are highly infested with Parthenium hysterophorus and Prosopis juliflora which is confirmed by recent reports. Interventions that address land rehabilitation and invasive species management could well profit from spatial information feeds on propagation zones for invasive species and where deforestation is particularly severe.

KEYWORDS: MODIS EVI, Landsat, land degradation, bush encroachment, Logistic Regression.

THE POTENTIAL OF SENTINEL-DATA FOR THE OBSERVATION OF WETLAND DYNAMICS IN POYANG AND DONGTING LAKE, CHINA

Huth J[1], Wang Y[2], Cheng Y[3], Yesou H[4], Kuenzer C[1], Clauss K[5]
[1]German Aerospace Center (DLR); Earth Observation Center (EOC), [2]Jiangxi Normal University, [3]Max Planck Institute for Ornithology, [4]University of Strasbourg - ICube, [5]University of Wuerzburg

The Poyang Lake and Dongting Lake are the two largest freshwater lakes in China. Located in the middle part of the Yangtze River catchment these large floodplain lakes are home to four Ramsar sites initiated by the UNESCO, to foster wetland conservation. The wetlands in and around the lakes provide numerous important ecosystem services for human well-being, e.g. freshwater resources, retention area for Yangtze River floods, buffer for drought events, natural habitats for millions of migratory birds, etc. Particularly since the last two decades, however, anthropogenic influences and related degradation of these wetlands have dramatically increased. Major threats can be attributed to upstream hydropower development, lake regulation, urbanization, industrialization, and agricultural intensification, all having serious impacts on the natural functions of the wetlands at Poyang and Dongting Lake.

In the presented study multi-sensor and multi-scale remote sensing data are used for a comprehensive monitoring of the wetland dynamics at Poyang and Dongting Lake. The study will demonstrate the potential of newly available Sentinel-data in combination with historic satellite data from Landsat, TerraSAR-X, Envisat and other missions for the generation of longer time series. Up-to-date information as well as information for the past decade on annual water and land surface changes will be presented. Furthermore, based on a comparison of the two lakes different examples for similarities and differences in wetland development in China will be discussed.

Finally, qualitative and quantitative analyses will provide insights into the rapid changes on the Earth's surface that have taken place due to human influence: In the last 10 to 15 years both lakes have shown a shrinking trend in water surface extent. Wetlands are increasingly used for resource exploitation, e.g. sand extraction and the cultivation of economically important plants.

KEYWORDS: Wetlands, Floodplain Lakes, Poyang Lake, Dongting Lake, China, Optical Data, Radar Data, Sentinel-1, Sentinel-2.

THE PRODUCTION OF PAST DEFORESTATION MAPS OF THE BRAZILIAN CERRADO TO SUPPORT A PARTIAL FREL SUBMISSION WITHIN THE BRAZILIAN REDD+ STRATEGY

Valeriano D[1]
[1]*National Institut for Space Research*

This work presents the methodology adopted for the cartographic basis for the Forest Reference Emission Level for the Brazilian Savannah, the Cerrado, which is part of the Brazilian REDD+ policy strategy. The Cerrado domain was divided into groups as ecoregions and classification criteria were established for each one of them. A base map of natural and anthropic areas was then made for the year of 2000 which was followed by biennial deforestation maps for the period 2000-2010. The natural areas map was discriminated into vegetation formations of grassland, savannah and forest areas as a basis for the determination of relevant REDD+ land cover change.

THE RESPONSES OF DRY-WET CONDITIONS TO EL NINO SOUTHERN OSCILLATION: A CASE STUDY IN CHINA

Sun Z[1], Wang X[2], Tani H[2], Zhong G[1], Yin S[1]
[1]Graduate School of Agriculture,Hokkaido University, [2]Research Faculty of Agriculture, Hokkaido University

Dry and wet conditions are the basic reasons why there will be drought and flood which are amongst the most serious forms of meteorological hazards. Whilst a growing number of studies have investigated how dry-wet conditions' regime influenced by natural climatic variations, the spatial patterns forms of dry-wet conditions influenced by climatic variations are poorly understood. Since dry-wet conditions are the response of internal interactions in the atmosphere and feedback from oceans and land surface, and the El Nino Southern Oscillation (ENSO) is one of the most important atmosphere-ocean coupled event of climate variability, it is very necessary to understand the relationship between ENSO with dry-wet conditions, and the differences of spatial responses of dry-wet conditions to ENSO. In order to achieve the above object, in the present study, we proposed a dry-wet index calculated by using the reanalysis precipitation and temperature data, modelled soil moisture, sensible heat flux and latent heat flux, and remote sensing evaporation data to assess the dry-wet levels. We also used the Granger causality test to examine the impacts of ENSO, depended on geostatistics methods and spatial analysis to find out the distributions of spatial responses patterns. Through the analysis of spatiotemporal data series, the results show that the dry-wet conditions can be well indicated by using the dry-wet index proposed in this study; the dry-wet conditions in China are significantly affected by ENSO, but with the regional differences.

THE ROLE OF BAND SELECTION IN PAN SHARPENING FOR YOUNG WOODY VEGETATION MAPPING

Fundisi E[1], Tesfamichael S[1]
[1]*University Of Johannesburg*

Expansion of woody vegetation results in the transformation of an ecosystem, and thus it is critical to have efficient monitoring and management strategies. In this regard spatial resolution of monitoring is critical particularly in semi-arid environments. This study investigated the utility of remote sensing to characterize the dynamics young woody vegetation in a semi-arid area. Specifically, the study explored the effect of band selection during pansharpening on the ability to discriminate woody vegetation from coexisting land cover types. Red-green-blue spectral bands (30 m resolution) of Landsat 8 imagery was pansharpened using the panchromatic band (15 m resolution) of the same image to improve spatial resolution. Near-infrared (0.85 - 0.88) μm, shortwave infrared 1 (1.57 - 1.65) μm and shortwave infrared 2 (2.11 - 2.29) μm bands were each used as the fourth spectral band during pansharpening, resulting in three pansharpened images. Unsupervised classification was performed on each pansharpened image as well as on non-pansharpened multispectral image (made up of six spectral bands in the visible to near-infrared regions of the electromagnetic spectrum). Comparison of the classified images showed the influence of spectral bands used for pansharpening in assessing woody vegetation. This study shows the importance factoring in bands when pansharpening a multispectral image to enhance spatial resolution in an effort to characterize woody vegetation.

KEYWORDS: woody vegetation, Landsat, pan sharpening, unsupervised classification

THE ROLE OF SATELLITE OBSERVATIONS FOR MEASURING AND MONITORING THE URBAN SDG TARGETS AND INDICATORS

Paganini M[1]
[1]*European Space Agency*

With the adoption of the 2030 Agenda for Sustainable Development, all countries agree to undertake some collective and transformative efforts to integrate the three pillars of sustainable development (economic growth, social development and environmental protection) into their national policies and programmes. When adopting the 17 Sustainable Development Goals (SDGs) and the 169 SDG Targets, the United Nations recognised the importance of satellite observations to ensure access to high quality, timely and disaggregated geospatial data, in particular for least developed countries where data is often incomplete, inaccessible, or simply inaccurate.

More than half of the global human population lives in urban environments and the dynamic trend of urbanization is growing at an unprecedented speed. In emerging economies, the urban population is expected to double in 30 years, adding 2 billion more people, particularly in South Asia and Africa. To meet the ambitious objective of SDG 11 to make cities more inclusive, safe, resilient and sustainable while leaving nobody behind, all countries and cities will soon face the hard task of implementing the targets of SDG 11 and reporting on the indicators that have been defined to monitor progress. How urban indicators will be collected, with which methods, and by whom need to be urgently addressed.

With the emergence of government-funded satellites with open and free data policies and long-term observation continuity, such as the Sentinel missions of the European Copernicus Program, Earth Observation is becoming an essential and cost-effective instrument to understand and monitor the complexity of urban environments.

The presentation will show how Earth Observation satellite can support the provision of timely and accurate information necessary to measure and monitor several targets and indicators of SDG 11 such as 11.3 on sustainable urbanization, 11.6 on urban air quality and 11.7 on urban green areas or 11.1 on slums.

THE ROLE OF SPACE AGENCIES IN REMOTELY SENSED ESSENTIAL BIODIVERSITY VARIABLES

Paganini M[1], Leidner A[2], Turner W[2], Wegmann M[3], Geller G[4]
[1]European Space Agency, [2]NASA, [3]University Würzburg, [4]Group on Earth Observations

The Group on Earth Observations Biodiversity Observation Network (GEO BON) is developing the Essential Biodiversity Variables (EBVs) as the key variables that are fundamental to be collected globally and regularly to understand and monitor changes in the Earth's biodiversity. The monitoring of a limited number of essential observations on the structural, functional and compositional aspects of biodiversity is seen as the most cost effective and efficient framework to develop a global biodiversity knowledge system. The EBVs are based on the integration of remotely sensed observations that can be measured systematically and globally by satellites, with field observations from local sampling schemes integrated into large-scale generalisations. The subset of EBVs that can be derived from space-based remote sensing are often referred to as remotely sensed EBVs (RS-EBVs). Satellite remote sensing allows wide scale, repeatable and cost effective measurements, yet their application in global biodiversity monitoring is still insufficiently developed. The obstacles that hindered greater use of satellite data in biodiversity monitoring were mainly restrictive data access policies, insufficient time series to capture the temporal dynamics of ecosystems at appropriate scales, or uncertainties on the long term availability of satellite observations The emergence of government-funded satellite continuity missions with open and free data policies, such as the Sentinel missions of the European Copernicus Program or the US Landsat series, offer an unprecedented ensemble of satellite observations that collectively enable the development of satellite-based monitoring systems of the Earth environment. The long-term availability of continuous observations make satellite remote sensing an essential instrument for the development of the EBVs. The presentation will explain the role space agencies should play in the EBVs development. As an example the approach followed by the European Space Agency for the engineering of High Resolution RS-EBVs on the structure and function of terrestrial ecosystems will be showed.

THE SOUTH AFRICAN NATIONAL SPACE AGENCY ENABLING ACCESS TO COMMERCIAL EO DATA

Saloojee I[1], Kekana A[1], Lavhengwa T[1]
[1]*SANSA*

This paper explores the implementation and impact of a single licence multi user approach in South Africa to provide access to high resolution commercial satellite data to support public sector science and decision making processes. The paper explores the extent to which the availability of data, which would have otherwise been prohibitively expensive, has stimulated science and the development of decision making applications within the public sector. The paper further argues that extending such an approach to the private sector would be a stimulus for innovation. Lastly, the paper showcases SANSA's EODC as a portal for accessing its data holdings in the Southern African Region.

THE SUSTAINABLE DEVELOPMENT GOALS A BASIS TO DEVELOP DIGITAL EARTH

Hernandez M[1]
[1]*Future Earth Engagement Commitee*

Agreeing to the Sustainable Development Goals, the world will need to evaluate progress and look ahead to emerging challenges and eventual associated solutions.

Data is and will be one of the most important issues. Jointly with the importance of data is the necessary digital platform that will be required to host all required data and what is more important, a Digital Earth platform that should enable all the different audiences of users (e.g. scientists, decision makers, and the civil society) to store and easily retrieve and visualize the data. Such a Digital Earth platform will be a key element in order for the national and international community to be able to access the vast amount of data that will be used.

"Future Earth" is a global environmental change research platform with the aim of providing knowledge and support to accelerate transformations to a sustainable world. It will integrate as different disciplines from the natural and social sciences (including economic, legal and behavioural research), engineering and humanities. Future Earth will require large amount of data coming from many different scientific research disciplines. How to store and retrieve such data is something that could be addressed by the Digital Earth platform that International Society for Digital Earth (ISDE) aims to develop.

A common web-based distributed ICT platform to enable a access, visualization and modeling of all this data is missing. This could be the niche for ISDE in order to develop Digital Earth.

This paper will address the needs to support the SDGs and will aim to open an overall discussion for ISDE to develop the desired Digital Earth platform based on the SDGs needs.

KEYWORDS: Sustainable Development Goals, Digital Earth, Future Earth

THE USE OF DLR FIREBIRD MISSION DATA FOR HOT SPOT DETECTION AND FIRE CHARACTERIZATION

Klein D[1], Strobl C[1], Fuermann M[1,3], Fischer C[1], Frauenberger O[1], Lorenz E[2], Halle W[2]
[1]German Aerospace Center (DLR), Earth Observation Center (EOC), [2]German Aerospace Center (DLR), Institute of Optical Sensor Systems, [3]University Stockholm, Dept. of Physical Geography

The successful launch of the BIROS satellite in June 2016 followed the Technology Experiments Carrier - TET-1 satellite system, launched in 2012, and thus completed the FireBIRD mission. Based on DLR's Bi-Spectral and Infrared Remote Detection (BIRD) sensor system (2001-2004) and its thermal infrared sensor concept, the scientific small-satellite mission aims at detecting high temperature anomalies with a special focus on wild fire detection and fire characterization.

The main advantage of FireBIRD data is the higher spatial resolution and extended dynamic range of the mid-wave (MWIR) and long-wave thermal (LWIR) channels with 320 m pixel size compared to other thermal satellite data like MODIS or Sentinel 3 SLTSR with 1 km spatial resolution. The repetition rate for both satellites is on average 3 days, due to off-nadir viewing capabilities of the satellites also several image acquisitions are possible on consecutive days for the same location.

At DLR processing chains have been implemented for near-real-time processing L1B and L2 level data. While the Level 1B data sets represent Top-of-Atmosphere radiances, the L2 data show hot spots as well as derived fire radiative power and fire temperature of the detected fire clusters. The implemented fire detection algorithm adapts the method proposed by Zhukov e. al. [2006] for BIRD, which makes use of the bi-spectral method using MWIR and TIR channel.

The paper will present the FireBIRD mission, data access and results, focusing on the FireBIRD data capabilities for hot spot detection and fire characterization on different test sites, ranging from volcanoes (Mt. Aetna) to active fire events as the ones from last summer in Portugal. Results are compared to other satellite data, e.g. MODIS, AVHRR, SLSTR, etc..

KEYWORDS: FireBIRD, MODIS, Hot Spot Detection, fire radiative power

THE USE OF GIS DATA IN THE DESERTIFICATION RISK'S CARTOGRAPHY - CASE OF THE AURÈS REGION (ALGERIA)

Hassen B[1]
[1] *University of Batna2*

The risk's cartography is a primordial step for the valuation and management of desertification phenomenon but it is a complicated spot, which necessitate a big amount of spatial and statistic's data. The use of GIS permits to manage and use these data efficiency.

The objective of our study is the realization of the sensitivity to the desertification map of south of the Aurès region by means of Geographical information system in accordance with the MEDALUS method (Mediterranean Desertification and Land Use), which uses the qualitative indices to define the sensitive environment zones to the desertification.

The creation of the database consist of four information layer (soil quality, vegetation quality,, climate quality and the socioeconomic state) when the articulation in the space and in the time is submit to the validation on ground.

Once the database has corrected it help to the elaboration of the sensitivity to the desertification map with calculation of the indices of the sensitivity to the desertification (ISD).

The result is a risk's map at a middle scale which presents a big efficacy in word of synthesis of a desertification phenomenon. The maps make a tool of help to decision as far as the protection of natural resources is concerned in regions stroke by the aridity.

KEYWORDS: Desertification, Indices, risk's cartography, Geographical Information System, MEDALUS,Aurès.

THE USE OF REMOTE SENSING TECHNOLOGY TO UNDERSTAND SNOWFALL RISK IN SOUTH AFRICA

Pillay D[1]
[1]National Disaster Management Centre

During the 2013/14 financial year, the National Disaster Management Centre (NDMC) partnered with the CSIR Meraka Institute to map snowfall incidents suing satellite imagery and quantitative data. Snowfall incidents present a considerable hazard to selected areas of South Africa affecting vulnerable communities and major transportation routes. The alternative effect is that snowfall incidents incur a local cost in terms of disaster management funding and local municipalities must plan ahead for the cleaning and treatment of snow affected areas.

Using multi temporal medium resolution data and quantitative techniques the project was able to provide an indication of the selected areas in south Africa that have a high risk of snowfall hazards. Statistical methods were used to refine the satellite imagery results to ensure that selected geographical areas in South Africa were highlighted.

In addition, the hazard profile was subjected to a GIS modelling processes in which the hazard was combined with coping capacity and areas of vulnerability to highlight arrears in South Africa that have a high versus low risk rating in terms of the a complete snowfall risk product. The modelling also provided a means to increase capacity and understand the nature of vulnerability in the selected areas.

The NDMC concluded this product and made the results available for national planning in relation to snowfall incidents. The data also informed planning in terms of budgetary allocations to manage different local municipalities in terms of future snowfall occurrences. satellite imagery provided a suitable base of information in which an indicative risk product could be generated outside of any other sources of information.

THE WOODY AND HERBACEOUS CONSTITUENTS OF MODIS LAI AND THEIR APPLICATION IN UNDERSTANDING FIRE REGIMES AND HERBIVORY IN AFRICA

Kahiu M[1], Hanan N[1]
[1]*South Dakota State University*

Savannas are expansive tree-grass or shrub-grass systems covering about 20% of terrestrial ecosystems. The vegetation structure of savanna systems plays a critical role in regulating terrestrial carbon cycle, ecosystem productivity, and the hydrological cycle. Savannas are also important for human livelihoods and biodiversity conservation as they influence provision of important products such as fuel-wood, wild foods and forage for herbivores. Ecosystem services provided by the herbaceous and woody layer are distinct in magnitude and seasonality e.g. provision of fodder, changes in carbon sequestration. Hence, it is important to understand the distinct seasonality and the extent of herbaceous and woody layers to better understand and model the impacts on hydrological and biogeochemical cycles, fire, herbivory and net primary production. Although the herbaceous and woody cover components play distinct roles in earth system processes and human well-being, they are not well represented in earth observation and ecosystem modeling. In remote sensing data, savanna ecosystems remain a challenge due to the presence of mixed woody-herbaceous components at scales much finer than most medium and coarse resolution products.

In this paper we will present the partitioning of woody and herbaceous components of savanna ecosystems using MODIS leaf area index from 2003 to 2015. Additionally, we will use the partitioned leaf area index estimates to explore the distinct roles played by woody and herbaceous cover in fire and herbivory in sub-Saharan Africa.

THINKING REMOTELY, ACTING LOCALLY - VERYIFYING THE UTILITY OF SENTINEL-2 AND LANDSAT 8 FOR SUSTAINABLE AGRICULTURAL DECISION MANAGEMENT IN KWA-ZULU NATAL

Atkinson J[1], Barichievy K[1], Adjorlolo C[2]
[1]*Kzn Dept. Agriculture & Rural Dev.*, [2]*South African National Space Agency*

Agricultural sustainability in South Africa is in the process of being redefined. One of the biggest threats facing the sector is the loss of agricultural land to non-productive land uses. Legislative prescripts such as the Spatial Planning and Land Use Management Act as well as the Draft Policy on the Preservation and Development of Agricultural Land seek to provide the necessary strategic framework to encourage holistic and integrated socio-economic and agricultural development in the country. We argue that there's presently a disjunct between the policies that are expected to underpin the growth of the sector and the scientific mechanisms used to evaluate and verify the potential and capability of the natural resources base within each Province. Presently, KZN relies on the Bio-Resource Classification to derive long-term average yield data for a variety of crops and ultimately govern agricultural development and sustainability. However many of these crop yield estimates are antiquated and require re-evulation and remote sensing therefore offers great potential for extending the investigative reach of resource managers and decision makers in the Province.

Consequently, top on the research agenda is the ability to intergrade field-based measurements of crop performance with readily available synoptic sensor platforms to enable near real-time assessment of the commodity base from a farm to framework (Provincial and National) level. The objective of this study is to determine the utility of accessing the suite of Sentinel-2 as well as LANDSAT 8 derived biophysical top of canopy (TOC) products at farm level by calibrating the vegetation characteristics of the sensors with precision-level (2.5 m x 6 m) in-field yield data. By evaluating the relationship between sensor and ground data, we hope to show that the application of Sentinel-2 and LANDSAT 8 offer real potential for improving decision making at farm level.

KEYWORDS: Sentinel-2, LANDSAT 8, KZN, Precision Agriculture

TIME SERIES ANALYSIS OF GLOBAL DAILY SNOW COVER EXTENTS DERIVED FROM MEDIUM RESOLUTION REMOTE SENSING DATA - GLOBAL SNOWPACK

Dietz A[1]
[1]*German Aerospace Center (DLR)*

Global snow cover is an important environmental parameter, as it influences hydrology, vegetation, radiation balance, and the living space of humans and animals. Snow is an essential source for freshwater in many regions of the world and at the same time, snow cover depends on precipitation and temperature during the snow season. As climate and weather varies, also the amount as well as the onset, duration, and offset of snow cover changes throughout the years. It is important to analyze this variability in order to identify possible trends, but also to predict the impact of the snow cover situation on local freshwater availability, floods, or the influences on vegetation.

The presentation will focus on the production of daily, global snow cover extent datasets derived from medium resolution remote sensing data and their application in terms of time series analyses. Data sources include AVHRR, MODIS, and Sentinel 3. The presented products include the overall snow cover duration, early season snow cover duration, and late season snow cover duration. Together these products form the Global SnowPack: a set of snow cover parameters mage available on a global scale by DLR.

The presentation will give a quick overview of how the processing steps are implemented before some examples are given on how the Global SnowPack datasets look like and what they can be used for. A long time series of daily snow cover datasets will be discussed in terms of possible trends, extreme events, and mean longterm conditions.

TOWARDS A NATIONAL OCEANS AND COASTAL INFORMATION MANAGEMENT SYSTEM FOR SOUTH AFRICA: SYSTEM SPECIFICATION AND DESIGN METHODOLOGY

McAlister B[1], McFerren G[1]
[1]*CSIR*

This paper highlights the need for a national oceans and coastal information management system for South Africa and how such a system could support the oceans economy of South Africa. This system will entail earth observation based monitoring of ocean and coastal environmental variables and socio-economic activities in service to decision-support-tools for various stakeholders. The methodology used in the specification and design of a National Oceans and Coastal Information Management System for South Africa is described. Based on the expressed needs, a business function model, system context and subsequent system decomposition is defined. This is followed by a detailed architectural description of all major system elements guided by the RM-ODP framework. A technology roadmap is also provided that shows current solution identification and effectiveness evaluation. Finally, current implementation progress, lessons-learnt and way-forward are discussed.

TOWARDS BEHAVIOR DEPENDENT SPECIES DISTRIBUTION MODELING

Remelgado R[1], Wegmann M[1]
[1]*University Of Würzburg*

Movement is a crucial element in animal ecology. It offers important information on individual and collective behavior and providing the basis for efficient conservation planning. In order to fully understand movement one needs to consider the underlying environmental conditions that guide it. Its perception by the animal influences its decision process and, consequently, its movement. In this context, remote sensing offers a unique opportunity. It supports the monitoring of surface conditions providing a unique overview on natural and human driven processes on multiple spatial and temporal scales as well as on their impact on animal movement. However, connecting movement and remote sensing data is not an easy task. Movement occurs on a small temporal scales (e.g. minutes, hours) and varying spatial scales (e.g. meters to kilometers) making a direct assimilation of movement and remote sensing near impossible when considering the spatial and temporal constraints of both datasets. Additionally, while movement data provides information on the presence of an individual describing true absences (i.e. low suitability) for species distribution modeling is a challenge. Existing models such as MaxEnt attempt to tackle this issue by performing random background sampling. However, this assumes very specific environmental requirements from the animal and disregards the element of choice as well as the underlying complexity of the landscape. Within this presentation we discuss the impact of behavior dependent sample selection strategies from movement data on the accuracy of species distribution modeling. In this analysis we used White Stork (ciconia ciconia) tracking data from 13 individuals collected during the summer of 2013 within Germany.

KEYWORDS: Biodiversity, Multi-disciplinary Applications

TOWARDS DEVELOPMENT OF A NATIONAL HUMAN SETTLEMENT LAYER USING HIGH RESOLUTION IMAGERY

Mudau N[1]
[1]*Sansa*

Information on the spatial distribution of building structures is important for urbanisation studies, spatial planning, service delivery and environmental management. High resolution imagery provides information on the built-up elements including buildings, roads and open spaces. This study investigates the extraction of building structures using SPOT 6 satellite imagery. The proposed methodology uses object based classification technique to extract building structures from other land use features. This method uses textural information derived from 1.5m panchromatic band to classify building structures from segmented image objects. Soil index derived from the multispectral bands was used to eliminate the confusion between building structures and open areas. The methodology was tested in different areas including urban, rural and informal settlements. The results show the spatial distribution of building structures within different human settlement types. This paper presents the results and assesses detection accuracy within these different settlement types. The results of this study will contribute towards development of a high resolution national human settlement layer which can be used to support sustainable human settlement development.

KEYWORDS: satellite imagery, building extraction, SPOT 6, Object based classification

TOWARDS P-BAND PASSIVE MICROWAVE SENSING OF SOIL MOISTURE

Walker J[1], Ye N[1], Yeo I[2], Jackson T[3], Kerr Y[4], Kim E[5], McGrath A[6]
[1]Monash University, [2]University of Newcastle, [3]United States Department of Agriculture, [4]Centre National d'Etudes Spatiales, [5]NASA Goddard Space Flight Center, [6]Airborne Research Australia

Economic, social and environmental planning for a water-limited future requires a capacity to provide information on soil moisture content in a way that is useful for such applications. Of particular importance is meeting the world's growing demand for food production, which is limited primarily by the soil moisture available for germination and growth of crops and pasture. Timely soil moisture status and forecasts are therefore critical for (i) grain growers to make informed decisions on what and when to plant based on likely germination rates and crop yield, (ii) graziers to be proactive in their management of stocking rates based on likely pasture growth, and (iii) dairy and other high water use agriculture to undertake more efficient irrigation scheduling practices. A fundamental limitation is that current remote sensing technology can only provide moisture information on the top 5 cm layer of soil at most, being one-tenth to one-quarter of the wavelength (21 cm at L-band; 1.4 GHz) using the current SMAP and SMOS soil moisture dedicated missions of NASA and ESA. Consequently, we have developed an airborne passive microwave sensing capability at P-band to develop a new state-of-the-art satellite concept that will provide soil moisture data for the top 15 cm layer of soil using radiometer observations at P-band (40 cm; 750 MHz). Not only would P-band provide soil moisture information on a soil layer thickness that more closely relates to that affecting crop and pasture growth, but it is expected to produce greater spatial coverage with improved accuracy to that from L-band. This is because P-band should be less affected by surface roughness conditions and have a reduced attenuation by the overlaying vegetation. This paper will present some early results from initial trial flights.

TOWARDS PROTOTYPING A GLOBAL LANDSAT-8 SENTINEL-2 BURNED AREA PRODUCT

Roy D[1], Huang H[1], Kumar S[1], Zhang H[1], Yan L[1], Li J[1], Boschetti L[1]
[1]*South Dakota State University*

Fire products derived from coarse spatial resolution satellite data have become an important source of information for the multiple user communities involved in fire science and applications. The advent of the MODIS on NASA's Terra and Aqua satellites enabled systematic production of 500m global burned area maps, and more recently other European coarse spatial resolution data have been used to derive burned area products. There is, however, an unequivocal demand for systematically generated higher spatial resolution burned area products. Moderate spatial resolution contemporaneous satellite data from Landsat-8 and the Sentinel-2 sensors provide the opportunity for detailed spatial mapping of burned areas. Combined, these polar-orbiting systems may provide 10m to 30m multi-spectral global coverage up to every 3 days. Combination of the sensor data is complex especially due to their different spatial and spectral resolution. This NASA funded research presents results to prototype a combined Landsat-8 Sentinel-2 burned area product. The processing to combine the data and preliminary results for Southern Africa are presented and implications for future research discussed.

TOWARDS UNDERSTANDING HUMAN SETTLEMENT GROWTH IN SOUTH AFRICA

Mudau N[1], Kemper T[2], Tsoeleng T[1], Mashalane M[1]
[1]Sansa, [2]European Commission, Joint Research Centre

South Africa is urbanizing rapidly. It is estimated that by 2050 nearly 80% of South Africa's population will be living in urban areas. About two-third of South Africa 50 million population currently live in urban areas. This figure increased from 52% in 1990. The increase of people living in urban areas in South Africa is attributed to population growth, migration of people from rural to urban areas and abolishment of apartheid spatial planning laws. Population increase results in increased demand of land for building houses, food and services. Unmanaged urbanization causes various problems including overload on existing infrastructure and services and degraded environment. Understanding urban spatial growth is vital for sustainable development. Monitoring human settlement developments can reduce the negative impact of urbanization. Advancement in remote sensing technology provides high resolution imagery suitable for mapping and monitoring of human settlement over larger areas. In this study we investigated the use of SPOT 5 imagery to detect human settlement growth between 2007 and 2014 in South Africa. A South Africa - Global Human Settlement Layer (SA- GHSL), system, was used to automatically detect human settlement data from SPOT imagery. A post classification technique was applied to detect human settlement growth across the country. The results highlight areas where human settlement growth took place between 2007 and 2014. The results can be used by authorities to support spatial planning and environmental management.

KEYWORDS: Urbanization, urban spatial growth, SPOT 5, remote sensing, human settlements

TRACKING FARM MANAGEMENT PRACTICES WITH REMOTE SENSING

Stals J[1], Ferreira S[1]
[1]*Geoterraimage*

Earth observation data is effective in monitoring agricultural cropping activity over large areas. An example of such an application is the GeoTerraImage crop type classification for the South African Crop Estimates Committee (CEC). The satellite based classification of crop types in South Africa provides a large scale, spatial and historical record of agricultural practices in the main crop growing areas. The results from these classifications provides data for the analysis of trends over time, in order to extract valuable information that can aid decision making in the agricultural sector. Crop cultivation practices change over time as farmers adapt to demand, exchange rate and new technology. Through the use of remote sensing, grain crop types have been identified at field level since 2008, providing a historical data set of cropping activity for the three most important grain producing provinces of Mpumalanga, Freestate and North West province in South Africa. This historical information allows the analysis of farm management practices to identify changes and trends in crop rotation and irrigation practices. Analysis of crop type classification over time highlighted practices such as: frequency of cultivation of the same crop on a field, intensified cultivation on center pivot irrigated fields with double cropping of a winter grain followed by a summer grain in the same year and increasing cultivation of certain types of crops over time such as soybeans. All these practices can be analyzed in a quantitative spatial and temporal manner through the use of the remote sensing based crop type classifications.

KEYWORDS: farm management, crop types, crop statistics, classifications, spatial analysis

TRIPLESAT CONSTELLATION AND PROTECTING ENVIRONMENT USING ITS SATELLITE IMAGERY

Lloyd D[1]
[1]*Scsgi*

Title:
TripleSat Constellation and Protecting Environment Using Its Satellite Imagery

Abstract

To measure the time sensitive key parameters for the effective management of rapid urbanization, requires regular monitoring in high resolution. To date the challenge have been address by aerial surveys done once per year, once per quarter in the highest revisit time. The opportunity offered by TripleSat Constellation allows new sets of urban processes to be monitored on a regular basis. The Triplesat Constellation comprises three sub-meter satellites phased 33 minutes/ 120 degree after each other that enable the unique daily targeting capability among very high resolution satellites. The operator of TripleSat Constellation is Twenty First Century Aerospace Technology Co. Ltd (21AT) that is the first commercial EO satellite operator in China. Its first satellite, Beijing-1 small satellite, was launched in 2005 and has been used successfully to reduce the environment impact of the fast urbanization of the Capital of China.

With the launch of TripleSat Constellation (named Beijing-2 in China) on 10 July 2015, 21AT, in cooperation with its partner SCSgi in Africa, now is able and is very keen to share its experience in one hand to manage the fast development and in the other hand to reduce the impact on the environment for sustainable development.

The paper first gives brief introduction of the newly launched TripleSat Constellation and then describes in detail its experience in eliminating the problems associated with the messy urbanization based on changes detected by regular monitoring from space.

UAV-BASED SURVEILLANCE IN SUPPORT OF WEED MANAGEMENT IN PROTECTED AREAS: THE CASE OF LUDWIGIA PERUVIANA INFESTATION IN URBAN WETLANDS OF SYDNEY

Metternicht G[1], Zhou Y[1], Bergmarks J[1], Giles S[1], Lo K[1], Puckeridge A[1], Roods B[1]
[1]*University of New South Wales*

Ludwigia peruviana is a small shrub classified as top priority weed for eradication in wetlands around the city of Sydney. Infestations clog wetlands, reducing biodiversity of the area and destroying habitats. Dense stands of Ludwigia can intercept almost all incident light, leading to the loss of native plants and animals. Since the 1980s, the government of NSW has developed management plans with the Regional Weeds Advisory Committees and local councils for its containment and/or local eradication. Strategies include early detection, early intervention to control individuals reaching maturity and prevention of spread via stormwater runoff. Containment and eradication approaches require access to up-to-date, reliable information on the spatial extent and location of infestations. Remote sensing through unmanned aerial vehicles (UAVs) has the potential to provide such information: for baseline mapping of infestation, and to assess success of interventions.

UAVs have become accessible, relatively inexpensive and flexible means for near real time data acquisition. Using the Warriewood wetlands as a test site, this paper explores effectiveness of using a RGB camera on a UAV for rapid mapping of Ludwigia occurrence. It analyses the effect of seasonality, spatial and spectral resolution for a "quick-and-dirty" mapping of weed occurrence. Two components are used for evaluating cost-effectiveness: per-class accuracy assessment and cost associated with mapping, including image turnaround, time spent on data collection and image processing.

Initial results indicate that where rapid response is required for mapping current or new incursions or to provide evidence of success in containment through treatments undertaken by contractors, UAVs are cost-effective to locate the spread of the incursion, and assist in rapid assessment of success in subsequent treatment of the area. The benefits are economic and environmental, as chemical treatments (ie. herbicide application) could be more targeted.

UAV-BORNE AND AIRBORNE REMOTE SENSING FOR FOREST HEALTH

Smigaj M[1], Gaulton R[1], Barr S[1], Suarez J[2]
[1]Newcastle University, [2]Forest Research, Northern Research Station

Climate change has a major influence on forest health by indirectly affecting the distribution and abundance of pathogens, as well as the severity of tree diseases. Changing weather conditions may also result in the introduction of non-native invasive pest species. The detection and robust monitoring of affected forest stands is therefore crucial for allowing management interventions to reduce the spread of infections.

Stress induced by an invasion of insects or onset of disease manifests itself in tree foliage, and may result in a variety of changes to plant's physiological processes. When a plant is under stress, stomatal closure occurs to help reduce water losses and prevent the entry of microbes and host tissue colonisation. This mechanism can cause an increase in leaf and canopy temperature.

Nevertheless, there has been limited research into the use of thermal remote sensing for tree health monitoring as required high spatial resolution data is usually obtained with low temporal frequency. Newly emerging technologies, such as unmanned aerial vehicles (UAVs), could supplement aerial data acquisition. This project investigates the use of airborne and UAV-borne sensors for detection of disease symptoms, in particular low-cost UAV-borne microbolometer thermal system for monitoring disease-induced canopy temperature rise.
The research is based in Queen Elizabeth Forest Park, Scotland, where research plots were established in pine stands, exhibiting various stages of stress. Extensive structural measurements of sample trees were collected, including visual estimation of red band needle blight infection level. These measurements were accompanied by airborne hyperspectral, thermal and LiDAR data, as well as a thermal UAV-borne imagery.

The presentation will show results of UAV-borne thermal imaging for detection of disease-induced canopy temperature increase, as well as analysis of the acquired airborne data.

UNDERSTANDING TEMPORAL AND SPATIAL VARIATION OF SOIL AVAILABLE NUTRIENTS UNDER UNIFORM FERTILIZATION CONDITION BY INTEGRATING WOFOST MODEL AND TIME SERIES SATELLITE DATA

Meng J[1], Cheng Z[2]
[1]Institute Of Remote Sensing And Digital Earth, Chinese Academy Of Sciences, [2]College of Resources and Environment, University of Chinese Academy of Sciences

Soil available nutrients, including available nitrogen (N), available phosphorus (P) and available potassium (K), are the key determinants in crop growth, field stable output and ecological balance. Mapping temporal trends and spatial variation characteristics of soil available nutrients is necessary to understand the nutrients flux mechanism. However, the soil available nutrients are usually variable between and within growth seasons, making it difficult for common methods to simulate the temporal and spatial variation with the required precision. Thus in this study, we first introduce a new method to map soil available nutrients by integrating the crop model and time series remote sensing data, then we maped the soil available nutrients of the farm from 2012 to 2016 and analyzed its variation within and between years. Because the farm is implementing uniform fertilization management through last 5 years, the temporal variation can also be simulated using transformation algorithms. The soil available nutrients contents were simulated and analyzed in two ways. Firstly, we applied the simulation method to the study area in the five years to monitor the interannual variation and analyzed the influencing factor. The results indicated that crop growth and precipitation intensity can both bring variations between years and plots. Then the within-year variation was also analyzed by estimating the soil available nutrients contents of a whole growth season with the temporal step of five days which showed that the nutrients contents presented downtrend basically while the variation can be quite different among plots. In general, the temporal and spatial variation of soil available nutrients under uniform fertilization condition can be simulated with high stability and accuracy. Appropriate fertilization measures can also be developed to protect field ecological environment based on the analysis of simulation results.

KEYWORDS: soil available nutrients, spatial and temporal variation, WOFOST, fertilization, HJ-1 CCD

UNIVERSAL AND UNIFIED REQUIREMENTS FOR CORE CERTIFICATION OF TRUSTWORTHY DATA REPOSITORIES

Mokrane M[1], Edmunds R[1], L'Hours H[3], Dillo I[2]
[1]ICSU World Data System, [2]Data Archiving and Networked Service, [3]UK Data Archive

Data created and used by scientists—including Earth observations—should be managed, curated, and archived in such a way to preserve the initial investment in collecting them. Funding authorities increasingly require continued access to data produced by the projects they fund, and have made this an important element in Data Management Plans. Likewise, the Group on Earth Observations (GEO) adopted Data Sharing and Data Management Principles to ascertain that data made available through the Global Earth Observation System of System (GEOSS) remain useful and meaningful into the future.

Sustainability of repositories raises several challenging issues in different areas: organizational, technical, financial, legal, etc. Certification can be an important contribution to ensuring the reliability and durability of data repositories and hence the potential for sharing data over a long period. By becoming certified, repositories can demonstrate to both their users and their funders that an independent authority has evaluated them and endorsed their trustworthiness.

Core certification involves a minimally intensive process whereby data repositories supply evidence that they are sustainable and trustworthy. A repository first conducts an internal self-assessment, which is then reviewed by community peers. Such assessments help data communities—producers, repositories, and consumers—to improve the quality and transparency of their processes, and to increase awareness of and compliance with established standards. This community approach guarantees an inclusive atmosphere in which the candidate repository and the reviewers closely interact.

The Core Trustworthy Data Repository Requirements were developed by a Partnership Working Group between ICSU World Data System (ICSU-WDS) and Data Seal of Approval (DSA) within the Research Data Alliance. This set of universal and unified requirements draws from criteria previously used by ICSU-WDS and DSA. The two organizations are now developing common procedures to be implemented jointly and offered to the community including to GEO data providers.

UNLOCKING THE VALUE OF HIGH RESOLUTION, ACCURACY, SPECTRUM AND REVISIT

Fortescue A[1]
[1] *DigitalGlobe*

Content: Presentation of companies plans with new Earth Observation programs. Heads of these companies are asked to contribute to the sessions with a talk of about 15 minutes highlighting a few outstanding achievements of their EO activities and to give an outlook of their future EO program. This will give presenters the opportunity to advertise their activities to a broad forum of scientists, remote sensing experts from research and industry, environmental and resource managers and engineers. In addition they will meet your colleagues from the other companies.

DigitalGlobe owns and operates the most agile and sophisticated constellation of high-resolution commercial earth imaging satellites. WorldView-1, GeoEye-1, WorldView-2, and WorldView-3 together are capable of collecting over one billion square kilometers of quality imagery per year and offering intraday revisits around the globe. On November the 11th 2016 DigitalGlobe launched WorldView-4, expanding our constellation further, doubling our 30.cm resolution capability and adding an additional 680,000km2 per day in collection capacity. This presentation will showcase the unrivalled capability of leveraging this constellation and discuss how DigitalGlobe will enter the Small Imaging Satellite domain with its joint venture with TAQNIA & KACST, developing six or more sub-meter resolution imaging satellites to be launched in late 2018 or early 2019.

UNMIXING THE PHENOLOGY OF TREES AND GRASSES IN SEMI-ARID SAVANNAS

Scholes R[1], van den Hoof C[1]
[1]*University Of The Witwatersrand*

High-frequency multi-temporal remote sensing is typically at a spatial resolution such that each pixel observes many individual plants, each with their own phenological pattern. This is especially obvious in savannas, which their characteristic mix of an open tree canopy and an underlying grass cover. We use our understanding of the different temporal pattern of tree and grass growth, supplemented with additional data on tree cover, to unmix a long-term savanna FAPAR record into the temporally-resolved grass and tree components.

USE OF EARTH OBSERVATION DATA FOR PRACTICAL TARGET SETTING AND MONITORING OF PROGRESS TOWARDS LAND DEGRADATION NEUTRALITY

Minelli S[1], Alexander S[1], Castillo V[1], Metternicht G
[1]*United Nations Convention To Combat Desertification (unccd)*

One year after the adoption of the Sustainable Development Goals (SDGs), more than 100 Parties to the United Nations Convention to Combat Desertification (UNCCD) have already mobilized to set and monitor voluntary targets to achieve land degradation neutrality (LDN), one of the targets under SDG 15, life on land. Depending on domestic circumstances, LDN target setting will be a top-down, politically driven or a bottom-up, technically driven process - or rather a combination of both. The bottom line is that both political support and sufficient knowledge and information will be necessary to set baselines and ambitious, yet realistic targets.

Based on the experience gained through the LDN pilot project, the UNCCD concludes that: 1) Establishing baselines for the extent of land degradation is a priority for each country wishing to set an LDN target; and 2) A balance needs to be struck between comparability across countries and the contextual nature of land degradation. Land cover, land productivity and carbon stocks – the UNCCD land-based indicators – can represent a minimum common denominator for all countries. However, additional national indicators, information and expert knowledge are needed for context specific interpretations to define trade-offs and identify measures to halt and reverse land degradation in a participatory approach. Earth observation data offers a viable and cost-effective means of measuring progress. While in the long-term, all countries should be enabled to independently perform relevant data collection and analysis on land degradation, global data sources can bridge data gaps and ultimately decrease the reporting burden on countries.

Monitoring challenges are no excuse for inaction. The pilot project and recent efforts by the UNCCD Science-Policy Interface to develop a conceptual framework for LDN demonstrate that Earth observation, monitoring capacity and LDN research can evolve in parallel with the political process.

KEYWORDS: land degradation neutrality, SDG, indicators

USE OF EARTH OBSERVATION DATA TO MAP INVASIVE SPECIES IN KENYA'S WILDLIFE CONSERVANCIES

Kiema J[1]
[1]*RCMRD*

RCMRD through SERVIR E&SA, in partnership with Northern Rangeland Trust, (NRT) and Laikipia Wildlife Forum, (LWF), is mapping current extents of invasive plant species (specifically the Acacia reficiens and Opuntia spp.) in Northern Kenya wildlife conservancies. Heavy infestation by invasive plant species shrinks forage space available for both livestock and wildlife. While MODIS NDVI data provide an indicative index of green vegetation in an area, the actual forage that is food for wildlife is less if an area is heavily invaded by sometimes drought resistant unpalatable plants.

The small size of the conservancies present a challenge in using coarser resolution EO data such as MODIS and Landsat in mapping the actual extents of the invasive plants as they do not provide enough details to discriminate vegetation to a species level. Both Sentinel-2 and Digital Globe's WorldView2 are being explored as an alternative to improve the mapping accuracy while data collected from a mobile app developed by SERVIR E&SA team is being used for validation.

SERVIR E&SA will also model the future occurrences of the invasive species using predictive technology based on the known suitable environmental conditions to characterize areas of possible invasion. For sustainable landscapes and ecosystems this information is key to policy decision makers.

KEYWORDS: invasive species, earth observation, geospatial technologies, mapping, modelling

USE OF RADAR AND OPTICAL DATA TO SUPPORT CONSERVATION AND SUSTAINABLE UTILISATION OF MANGROVES

<u>Lucas R[1]</u>, Lucas R[1], Rosenqvist A[2], McOwen C[3], Hillarides L[4]
[1]University Of New South Wales, [2]Solo Earth Observation (soloEO), [3]United Nations Environment Programme-World Conservation Monitoring Centre, [4]Wetlands International

Though small in terms of relative area (0.4% of global forest cover), mangroves are disproportionally important to humans because of the multitude of ecosystem services they provide, their economic value and rich biodiversity. Nevertheless, their current deforestation rates are 3-5 times higher than the global average for forests. Reducing current mangrove loss, degradation and fragmentation as well as ensuring their sustainable utilisation, restoration and conservation is therefore important and relates directly to the aims of Aichi Biodiversity Targets 5, 7, 11, 14 and 15 of the Convention of Biological Diversity's (CBD) Strategic Plan. However, at present there are no suitable indicators to determine the status and trends of mangroves. In this paper, we describe how the Global Mangrove Watch (GMW) of the Japan Aerospace Exploration Agency (JAXA) can contribute to this gap in knowledge. The main product anticipated from the GMW is an updated mangrove baseline map of extent; in turn, annual updates to this baseline are being produced with particular focus on hotspots of change. By combining these products with additional information such as the structure (height) of the mangrove forest and the diversity of mangroves species links can be made to carbon dynamics and biodiversity change. Ultimately, this information supports a range of policies including the Ramsar Convention's Global Wetlands Observing System (GWOS), the United Nations Framework Convention on Climate Change (UNFCCC), Reducing Emissions from Deforestation and Degradation (REDD+) and the Intergovernmental Science-Policy Platform on Biodiversity (IPBES).

USER NEEDS AND GAP ANALYSIS: FROM A USER REQUIREMENTS REGISTRY TO A GEOSS KNOWLEDGE BASE LINKING USERS TO KNOWLEDGE

Plag H[1]
[1]*Old Dominion University*

The Global Earth Observation System of Systems (GEOSS) is focused on the creation of knowledge required to address societal issues. The experience with the development of the User Requirements Registry (URR) during recent years provides a basis for the development of a collaborative platform, the GEOSS Knowledge Base, for co-creation and co-usage of knowledge derived from Earth observations, socio-economic data and models. Designing a platform attractive to both users, providers and those engaged in the creation of system and transition knowledge requires innovative and foreward looking thinking anticipating the future technological capabilities. Artificial intelligence helps in extracting user needs from existing resources and from user activities on the collaborative platform, and it facilitates the linkage between these needs and those that can meet them. Providing an engaging decision-preparation platform coupled to a diverse social ecosystem of users, knowledge experts, and Earth observation providers creates a "crowd-sourcing" environment for the identification of knowledge needs and the determination of gaps in the layered system of knowledge creation down to the level of individual data services and sensors.

An initial step is the Socio-Economic and Environmental Information Needs Knowledge Base (SEE-IN KB), which documents relationships between Earth observations, socio-economic data and the processes (models, work flows, algorithms) for creating knowledge serving a wide range of user needs. The SEE-IN KB includes the rules for defining observational needs and linking them to knowledge needs. Existing knowledge repositories are leveraged as far as possible. User feedback concerning the fitness for purpose of data and processes is collected. The networks model allows for the construction of business processes to answer "What if?" questions in support of policy development facilitating progress towards societal goals. User types, applications and requirements are linked to actual users, models and dataset and this allows execution of business processes.

USING LIDAR DERIVATIVES TO ESTIMATE SEDIMENT GRAIN SIZE ON BEACHES IN FALSE BAY

Burns J[1], Lück-Vogel M[1,2]
[1]University Of Stellenbosch, [2]Council for Scientific and Industrial Research (CSIR)

Coastal environments form where the land, sea and atmosphere meet and interact in unique and dynamic ways and a state of dynamic equilibrium is ideally maintained by the natural ocean processes of wind, waves and tides. Human interference, leading to global and climate change in particular, continues to have a big impact on coastal environments. The vulnerability of the sandy coast is dependent on the physical characteristics such as orientation/exposure, beach slope, and sand grain size. These characteristics are also dependent on each other; for calm beaches, flatter slopes are prevalent with a finer grain size; and conversely with steeper beaches which are usually exposed to higher wave energy, grain size tends to be larger. Knowledge about one of these parameters (beach slope, grain size, wave energy) can therefore theoretically be used as a proxy to predict the other factors. This information would be of great interest for coastal protection and disaster risk management.

Field assessments and surveys are, however, expensive and often impossible in a country such as South Africa with a long and largely inaccessible coast. Remote sensing technologies and LiDAR (light detection and ranging) in particular, hold much potential for the assessment of relevant physical beach parameters. The intensity characteristic of LiDAR scanning essentially represents the amount of energy reflected from the target of the scan and is affected by the composition and roughness of the surface.

The work that will be presented coupled field measurements of slope and grain size to LiDAR derived slope and intensity to determine whether these can be used as a proxy to predict beach grain size. This would provide a valuable tool for the spatial assessment of beach vulnerability on a national scale. The initial findings of this study will be presented at this conference.

KEYWORDS: coast, grain size, LiDAR, intensity

USING LONG-TERM WATER CYCLE AND VEGETATION SATELLITE PRODUCTS FOR IMPROVED MODELLING OF TERRESTRIAL ECOSYSTEM DYNAMICS

Dorigo W[1], Forkel M[1], Demuzere M[2], Miralles D[2], Papagiannopoulou C[3], Waegeman W[3], Teubner I[1]
[1]Vienna University Of Technology, Remote Sensing Research Group, [2]Ghent University, Laboratory of Hydrology and Water Management, [3]Ghent University, Research Unit Knowledge-based Systems

Global warming is expected to impact the global water cycle, leading to a increase in the frequency and severity of extreme events like floods and droughts. The anticipated changes in moisture availability are expected to alter ecosystem dynamics and composition through multiple interacting pathways. This would in turn affect vegetation production and ecosystem composition and dynamics. However, the impacts of changes in the global water cycle on ecosystems as predicted by models are uncertain, as the link between water variability and vegetation is only poorly represented at the scale of the models.

We used long-term, satellite products of different water cycle components (e.g. ESA CCI Soil Moisture, GLEAM root zone soil moisture) and vegetation dynamics (e.g. Vegetation Optical Depth, GIMMS NDVI 3g, GFED burned area) in combination with various climatic datasets (radiance, temperature, ocean oscillation indices) to assess short-term and long-term controls of ecosystem dynamics. To disentangle the various potential controls of ecosystems dynamics we used state-of-the-art data-driven approaches like random forest and ridge regression.

Our work shows that, even though at the global scale the drivers are mostly linear, regionally, extreme events like droughts and wildfires may have a considerable impact on vegetation dynamics in a non-linear way. Besides, these impacts may be indirect and counter-intuitive, e.g. a wet period in semi-arid regions leading to higher fire fuel accumulation and, hence, to larger wildfire probability. The main control variables identified by the data-driven approaches will be used to benchmark and optimise the performance of the state-of-the-art ecosystem model LPJml.

This study is funded by the Science Award of the Vienna University of Technology (http://eowave.geo.tuwien.ac.at/), the Living Planet Fellowship program of the European Space Agency, and the Belgian Science Police Office SAT-EX project (http://www.sat-ex.ugent.be/).

KEYWORDS: carbon cycle, water, climate, ecosystem, machine learning

USING MACHINE LEARNING AND GEOBIA FOR THE AUTOMATION OF LAND COVER CLASSIFICATION WITH A TIME SERIES OF LANDSAT AND SENTINEL 2 DATA

Lueck W[1]
[1]*PCI Geomatics*

The past decade has seen an advancement of Geographic Object Based Image Analysis (GEOBIA), quantitative image analysis on high resolution optical satellite imagery, machine learning image classification techniques and time series analysis of satellite imagery. The availability of frequent observations from sensors and satellite missions, designed for the monitoring of natural resources and land cover such as Sentinel 2 and Landsat 8, has driven the operationalization of these remote sensing techniques.

The author's hereby present a methodology and highly automated workflow implemented in the PCI Geomatica 2017 software package to achieve the generation of a highly accurate land cover classification, using imagery acquired over the eastern part of South Africa. A time series of Sentinel 2 and Landsat 8 imagery acquired over a period of one year is used to demonstrate the workflow. Imagery is automatically pre-processed to provide radiometrically normalized and quality flagged data that can then be reliably used in a data stack for the time series analysis. This pre-processing entails the atmospheric correction of imagery using radiative transfer modelling, spectral pre-classification for the masking of cloud, cloud shadow, water and surface classes with distinct BRDF characteristics. This classification layer is then used for a topographic normalization and BRDF correction over an area with high topographic relief. A set of indices and spectral features are calculated for each dataset to be used in a time series analysis. The data is subsequently segmented using the Sentinel 2, 10m GSD bands for this purpose. Features are calculated from the entire time series data-stack consisting of both Landsat 8 OLI and Sentinel 2 MSI sensor for pure pixels only, disregarding sensor measurements along the boundary of segments. Thematic features are calculated from temporal matrices that are then used in a Support Vector Machine classification leading to classification accuracies around 90%.

USING MULTI-ANGULAR REMOTE SENSING TO ANALYSE CHANGES IN DRY WOODLAND STRUCTURE

Knox N[1], Strohbach B[1], de Cauwer V[1]
[1]*Namibia University Of Science And Technology*

Dry woodlands in Namibia are restricted to the north of the country. This is an ecosystem experiencing a great deal of utilization pressure. The population in the region live in a matrix of both rural and urban centres. The population makes extensive use of these woodland resources for firewood, building and construction, crafts, etc. As a result of this limited distribution, and utilization pressure, it is necessary to be able to monitor the changes taking place in this ecosystem. Monitoring changes that occur in this ecosystem are challenging, because utilization of the resources is usually not as a result of extensive ground clearing, but rather localised removal of individual trees or even harvesting sections of individual trees. In this research we have looked at using two scales of optical imagery for analysing how multi-angular data can be used to analyse changes in dry woodland structure. High resolution imagery acquired from UAV's are used to analyse vegetation structure to determine whether this system can be used to model and monitor small scale structural changes (at the 1km2 range). At the larger regional landscape scale we investigated the use of multi-angular imagery acquired from the MISR-HR (multi-angular imaging spectro-radiometer – high resolution) sensor to analyse regional structural changes to the dry woodlands. The multi-angular properties were investigated over the period of 2000-2012 to ground based studies which have recorded the changes to the vegetation composition, but not specifically to changes in the structure of the woodlands. Recently conducted fieldwork (and the UAV imagery) will provide the basis for generating the algorithms to link the actual structure of these woodlands too the lower resolution multi-angular imagery, and these will then be used to estimate the extent of structural change from the preceding time frame.

USING SATELLITE DATA FOR IMPROVED URBAN DEVELOPMENT IN DEVELOPING COUNTRIES

Haeusler T[1], Gomez S[1], Enssle F[1]
[1]*GAF AG*

Since 2008 the European Space Agency (ESA) has worked closely together with International Financing Institutions (IFIs) and their client countries to harness the benefits of Earth Observation (EO) in their operations and resources management. The EO for Sustainable Development (EO4SD) programme is a new initiative of the ESA which aims to achieve an increase in the uptake of satellite based information into the regional and global programmes of the IFIs. In May 2016 the EO4SD Urban project was initiated (with a duration of 3 years) which has the overall aim to integrate the application of satellite data for urban development programmes being implemented by the IFIs. The project is applying a variety of geo-spatial products from baseline land use/land cover data, urban green areas to more specific products such as building footprints and informal settlement mapping for the implementation of projects in about 40 different cities in Asia, Africa and Latin America. The cities for the current project have been identified in consultation with the IFIs, and encompass both mega-cities as well as small to medium sized cities. Additionally as one of the key Urban policy frameworks that Governments have committed to implement is the United Nations Sustainable Development Goal 11 "Make cities and human settlements inclusive, safe, resilient and sustainable", the project also addresses which geo spatial products can support the assessment of Indicators and their methodologies. The application of the geo-spatial products for urban planning in the 40 cities, illustrate that EO data has the unique ability to allow the evaluation of past and present spatial features and structures on the ground with frequent, detailed global coverage, allowing comprehensive analyses of the development and trends of spatial urban patterns.

USING SPATIAL CONTEXT TO INTEGRATE LANDSAT AND SENTINEL-2 DATA FOR FOREST CHANGE MONITORING

Hamunyela E[1], Reiche J[1], Pratihast A[1], Verbesselt J[1], Herold M[1]
[1]*Wageningen University And Research*

Free and open access satellite data streams from Landsat and Sentinels sensors offer new opportunity to monitor forest ecosystems at unprecedented spatial and temporal scales. However, these sensors have different radiometric calibrations, spatial and spectral resolutions. These differences lead to temporally inconsistent multi-sensor time series. To use these multi-sensor time series for forest disturbance monitoring, the inter-sensor differences should be addressed to avoid amplified false detections. In this work, we investigated whether inter-sensor differences in Landsat-Sentinel-2 time series can be addressed by using spatial context. Using normalised difference vegetation index (NDVI) time series derived from Landsat-7/ETM+ and Landsat-8 OLI and Sentinel-2A data, we show that normalising the time series spatially results into temporally consistent multi-sensor time series. We also show that both spatial and temporal detection of forest disturbances improve significantly when data from both Landsat and Sentinel-2 sensors are combined. The spatial normalisation approach we proposed here would allow for multi-sensor data harmonization to monitor forest changes in near real-time.

KEYWORDS: Deforestation, Landsat, Sentinel-2, time series, spatial contex

USING SPATIO-TEMPORALLY EXPLICIT MODIS EVI TIME SERIES TO INFORM HERBIVORE MANAGEMENT DECISION MAKING IN SOUTH AFRICAN NATIONAL PARKS

Smit I[1], Simms C[1], Wessels K[2]
[1]South African National Parks, [2]CSIR

Small or fragmented protected areas (PAs) often lack intact ecological processes like predation and migration. These missing processes, compounded by tourism expectations of game viewing opportunities and grazing condition, often result in PAs having to actively manage herbivore populations. This is also the case in many of the smaller National Parks in South Africa where "excess" herbivores need to be removed from time to time, often through live capture. Stable carrying capacity approaches to determine off-takes, although traditionally popular and simple to apply, has fallen out of favour in conservation circles in recent years due to the recognition of the importance of spatio-temporal variability in herbivore numbers and associated impact on system functionality and resilience. As such, South African National Parks (SANParks) has substituted stocking rate models with a decision-making process that integrates various data sources to inform herbivore off-takes in many of the smaller National Parks. In this presentation I will introduce the data needs and the steps involved in this decision-making process, with a specific emphasis on the remote sensing datasets (MODIS Enhanced Vegetation Index) used in the decision-making process. The long-term (17 years) and spatially explicit nature (500m2) of the EVI time series has proved valuable in informing herbivore management decisions. Some real-world examples of how this data was used for herbivore decision-making in SANParks will be illustrated as case studies.

USING THE GLOBCURRENT DATA PORTAL FOR MONITORING AND STUDYING MESOSCALE PROCESSES AND VARIABILITY

Johannessen J[1], Chapron B[2], Collard F[3], Rio M[4], Donlan C[5], Gaultier L[3], Korosov A[1], Hansen M[1], Quartly G[6], Escola R[7], Roca M[7], Nardelli F[6], Piollé J[2], Danielson R[1]
[1]Nansen Environmental And Remote Sensing Center, [2]Ifremer, [3]OceanDataLab, [4]CLS, [5]ESA, [6]Plymouth Marine Laboratory-PML, [7]isardSAT

The GlobCurrent project (http://www.globcurrent.org) funded under the ESA Data User Element (DUE) from 2014-2017 aims to: (i) advance the quantitative estimation of ocean surface currents from satellite sensor synergy; and (ii) demonstrate impact in user-led scientific, operational and commercial applications that, in turn, will improve and strengthen the uptake of satellite measurements.

Within the GlobCurrent project global and regional observation-based surface currents from 1992-2016 are now available at a spatial resolution of 25 km and temporal resolution of 1 day (geostrophic) and 3 hours (Ekman). These fields are combined with high resolution snapshot satellite products from Sentinel-1 surface features, Sentinel-2 sunglint and Sentinel-3 ocean color and SST retrievals. In addition in-situ observations are available including Argo profiling floats and drift, surface drifting buoys. These satellite- and in-situ based fields and products are moreover complemented and extended by a range of validation studies pursued by independent user led teams. In this presentation we demonstrate how the GlobCurrent multisensor synergy based analyses approach can stimulate and strengthen research and development to better understand the 2-dimensional surface expressions in the context of the upper ocean 3-dimensional mesoscale variability and dynamics.

VALIDATION OF SAR-BASED LAND COVER AND LAND COVER CHANGE MAPS AND DETECTABILITY OF SLASH-AND-BURN ACTIVITIES IN THE KWAMOUTH REGION, MAI-NDOMBE DISTRICT, DRC

Haarpaintner J[1], Mazinga A[2], Mane L[2]
[1]*Norut - Northern Research Institute,* [2]*OSFAC - Observatoire Satellital des Forêts d'Afrique Centrale*

As part of the ESA DUE Innovator III project "SAR for REDD", a 6-day field mission was carried out in the Kwamouth region in the south-west of the Mai-Ndombe district in the Democratic Republic of Congo (DRC). The aim of this fieldwork was to validate forest and forest change products based on C- and L-band synthetic aperture radar satellite imagery. A transect was driven with a 4x4 along the road from Masia-Mbio to Kwamouth. Along the transect, GPS position and ground photography were taken at forest/non-forest transitions and at characteristic land cover types and land cover changes. In addition, 20 flights with a remotely piloted aerial system, a DJI Phantom-3 quadrocopter, were performed to collect aerial photos, that were then stitched together into aerial mosaics covering up to 50 ha each. Land covers were dominantly primary and secondary forest, savannahs, dry and wet grasslands and small scale agriculture of mainly cassava. Observed land cover changes were deforestation, forest degradation and an intensive slash and burn activity of degraded forests or agriculture.

In this presentation, (a) we verify forest loss areas detected from multi-year SAR imagery prior to the field work and (b) we investigate the detectability of slash-and-burn activities in near-real time imagery taken shortly before and after those activities have taken place. Slash-and-burn activities in agricultural areas seem to be better detectable in dense time series of C-band SAR imagery whereas deforestation is more clearly visible in yearly L-and SAR forest change products. Aerial and satellite imagery are compared and changes in SAR backscatter signatures are specified.

VALIDATION OF THE FINAL TANDEM-X DEM IN THE LOWVELD SAVANNA, SOUTH AFRICA, USING HIGHLY ACCURATE DIFFERENTIAL GNSS AND TLS GROUND MEASUREMENTS

Baade J[1], Schmullius C[2]
[1]Department of Geography, Physical Geography, Friedrich-Schiller-University Jena, [2]Department of Geography, Earth Observation, Friedrich-Schiller-University Jena

The German X-SAR TanDEM-X mission acquired data for a new and truly global Digital Elevation Model (DEM) from January 2010 to December 2015. Since October 2016, the final DEM is available and first results suggest an accuracy of about 1 m; an order of magnitude higher than the initially targeted benchmark for the linear error (LE90 < 10 m). Being sensible not only to buildings and other infrastructure but as well to canopy cover, the TanDEM-X DEM actually represents a Digital Surface Model (DSM) as compared to a Digital Terrain Model (DTM).

In open terrain highly accurate ($\sigma < 0.05$ m) differential Global Navigation Satellite System (dGNSS) based ground measurements provide a reliable measure to assess the accuracy of the TanDEM-X DEM. However, in vegetated terrain, ground point measures just provide evidence for the sensibility of the DEM to canopy cover but tell little about the accuracy of the TanDEM-X height readings. Here, up to date and accurate DSMs are needed to assess its accuracy and to understand where in the canopy the X-SAR scattering centre is located. In quite open vegetated terrain, like the Lowveld Savanna in South Africa, it is further of interest to identify the lower canopy cover threshold biasing the TanDEM-X DEM height readings.

Here we present the results of a local scale TanDEM-X DEM height accuracy validation utilizing over 10,000 dGNSS-based ground survey points from fourteen sites across the Kruger National Park (KNP), South Africa, characterized by mainly pristine Savanna vegetation. Additional investigations into the representation of savanna vegetation in the TanDEM-X DEM are based on five high resolution DSMs derived by Terrestrial Laser Scanning (TLS). The DSMs cover about 1 km^2 each and provide a selection of typical Lowveld savanna vegetation types.

KEYWORDS: accuracy assessment, ground based measurements, canopy cover, Kruger National Park

VERY HIGH RESOLUTION MAPPING OF SPEKBOOM CANOPY COVER

Harris D[1], Vlok J[2], van Niekerk A[3]
[1]Department of Geography and Environmental Studies, Stellenbosch University, [2]Nelson Mandela Metropolitan University, [3]Centre for Geographical Analysis, Stellenbosch University

Very high resolution canopy cover maps of Spekboom (Portulacaria afra) are required to assist with restoration of degraded habitat in the Little Karoo, a large semi-arid region in South Africa. There have been relatively few studies addressing high resolution remote sensing of vegetation over large spatial extents, especially in arid areas. Variations in habitat and level of degradation, in addition to radiometric variations in the imagery, make this a challenging problem. Here we present a per-pixel classification approach for canopy cover mapping of Spekboom. Multi-spectral 0.5m resolution aerial imagery, comprising 2228 scenes covering a 5893 km2 area, was acquired from Chief Directorate: National Geospatial Information. The imagery was corrected with a novel technique for the extraction of surface reflectance by calibration with satellite data. Ground truth was produced by selecting and hand labelling polygons in the imagery. A feature clustering and ranking procedure that is robust to feature redundancy was applied to select an informative feature subset from a typical set of spectral, textural and vegetation index features. Support Vector Machine (SVM), random forest, decision tree, K-Nearest Neighbour (kNN) and Bayes normal classifiers were evaluated against both the hand labelled image ground truth and canopy cover ground truth acquired at 20 field sites. All classifiers except the Bayes normal classifier performed similarly well. The decision tree produced the best results on the field ground truth and was chosen as the final classifier. It produced a mean absolute canopy cover error of 5.85% with a standard deviation of 4.65% on the field ground truth.

KEYWORDS: Very high resolution, aerial, vegetation mapping, classification, radiometric correction

VESSEL DETECTION CAPABILITIES IN VIIRS SATELLITE IMAGERY

Lebona B[1,2], Kleynhans W[1,3], Celik T[2]
[1]*Csir Meraka Institute,* [2]*University of the Witwatersrand,* [3]*University of Pretoria*

Actively monitoring the designated Economic Exclusive Zone, designated and regulated by a particular coastal country, can be a resource intensive task; with South Africa's spanning 1,535,538 square kilometers. Remote sensing offers an efficient and cost effective solution for maritime domain awareness with an array of Earth Observation tools currently being used including transponder tracking, radar, infrared and optical satellite imagery. It has been observed that developing a robust target detection tool involves incorporating a combination of Earth Observation tools and object detection methods within those tools to give a final classification with a defined degree of accuracy. In our research we investigate the potential of vessel detection using satellite imagery from the Visible Infrared Imaging Radiometer Suite. An adaptive threshold will be applied to the low level light sensing Day-Night band imagery to detect nocturnal light emission sources, which are expected to be very visible against the contrast of the dark ocean background and being far from major anthropogenic light sources. Detections will be validated against transponder data from the Automatic Identification System collected by the Maritime Safety and Security Information System. False alarms will be investigate; as lunar illumination, lightning and gas flares are a few of the factors observed to cause false autonomous detections in these images. Information from the transponder system will allow us to analyse the type of ships capable of being detected by nocturnal light emission sensors and the advantages it could contribute to a maritime domain awareness system.

VIIRS ACTIVE FIRE DATA TO SUPPORT REGIONAL ENVIRONMENTAL APPLICATIONS

Csiszar I[1], Schroeder W[2], Giglio L[2], Kondragunta S[1], Tsidulko M[3], Ellicott E[2], Justice C[2]
[1]NOAA/NESDIS Center for Satellite Applications and Research, [2]University of Maryland, Department of Geographical Sciences, [3]I.M. Systems Group, Inc.

The Visible Infrared Imaging Spectroradiometer (VIIRS) on the Suomi National Polar-orbiting Partnership satellite has been providing fire information since early 2012. Availability of VIIRS data is planned for four additional satellites of the Joint Polar Satellite System, extending continuous observations into the 2030s. VIIRS represents continuity with the Moderate Resolution Imaging Spectroradiometer (MODIS) on the NASA Earth Observing System Terra and Aqua satellites. Improvements over MODIS include full daily spatial coverage at low latitudes, which is a critical feature for most of the African continent; higher spatial resolution, allowing for the detection of smaller fires typical of many regions in Africa; and more compatible measurements along the satellite swath, enabling better monitoring of daily variability of small fires at a regional level. Following fire product development to take advantage of the enhanced capabilities provided by VIIRS, a suite of mature fire products is now available for the user community through global near-real-time production and distribution systems. Algorithms to generate VIIRS fire products are also included in direct broadcast processing packages. VIIRS fire information is now used to support various societal benefit areas. For example, VIIRS fire radiative power information, generated at NOAA, is used as input for air quality applications. NASA's VIIRS fire products support various real-time fire management and science applications. Coordination with regional users is facilitated by the Global Observation of Forest and Land Cover Dynamics Fire theme, including regional networks in Southern Africa, Central Africa and West Africa. This also includes validation activities, continuing previous efforts for the Advanced Very High Resolution Radiometer and MODIS. Reprocessing of the VIIRS data record with the most complete satellite input and with the highest quality algorithms is ongoing. This effort will allow for correcting of deficiencies in particular at the beginning of the data record and for rigorous retrospective analyses.

VISIBLE AND INFRARED SPECTRAL CHARACTERISATION OF CHINESE CABBAGE (BRASSICA RAPA L. SUBSPECIES CHINENSIS), GROWN UNDER DIFFERENT NITROGEN, POTASSIUM AND PHOSPHORUS CONCENTRATIONS

Mokoatsi B[1], Tesfamichael S[1], Araya H[2], Mofokeng M[2]
[1]University Of Johannesburg, [2]Agricultural Research Council, Roodeplaat Vegetable and Ornamental Plant Institute

There is a need to intensify research efforts on improving productivity of indigenous vegetables in South Africa. One research avenue is operationalising remote sensing techniques to monitor crop health status. This study aimed at characterising the spectral properties of an indigenous vegetable, Chinese cabbage (Brassica Rapa L. subspecies Chinensis) grown under varying fertiliser treatments: nitrogen (0 kg/ha, 75 kg/ha, 150 kg/ha, 225 kg/ha and 300 kg/ha), phosphorus (0 kg/ha, 50 kg/ha, 100 kg/ha and 150 kg/ha) and potassium (0 kg/ha, 30 kg/ha and 60 kg/ha). Visible and infrared spectral measurements were taken from a total of 120 samples inside the laboratory. Contiguous spectral regions were extracted and assessed to compare spectral response of the different fertiliser treatments using analysis of variance and t-tests. The results showed significant difference among treatment levels at different spectral regions for each nutrient. Therefore, the regions extracted from the visible and near-infrared portions of the spectrum can be used to discriminate various nutrient concentrations in crops. This indicates the potential for the use of spectroscopy in monitoring food quality parameters, thereby reducing the cost of traditional methods. Further research using advanced statistical analysis techniques are needed to accurately quantify fertiliser concentrations found in the crops.

KEYWORDS: Chinese cabbage, spectral signatures, fertilisers, classification.

VULNERABILITY ASSESSMENT OF A WETLAND TO A DROUGHT: CASE STUDY OF DIE VLEI WETLAND, EASTERN CAPE PROVINCE– SOUTH AFRICA

Marembo K[1]
[1]*University Of Fort Hare*

South Africa, has been experiencing a drought over the 2014 to 2016 period and impacts have been seen in the particularly dry areas such as the Eastern Cape Province. Despite the presence of Die Vlei Wetland in the Hogsback area, the decrease in rainfall totals during the drought has posed tremendous pressure and threat to the environment let alone the sustainable viability of the wetland. This is a cause of concern as Die Vlei wetland, is a major water source for both commercial and domestic activities in the town. Whilst the impact is not clear, officials have hypothesized that the recent droughts have caused a decline in the agricultural productivity of the farmers and forestry company that use the wetland for purposes such as irrigation. This study applies geospatial modelling tools in the form of Spot Imagery to monitor using time-series analysis the wetlands' physical state change before and during the drought and its impact on farmers productivity. Findings aids in measuring the degradation extent caused by the drought and suggest preventive measures the authorities should consider in order to minimize impacts in the future. The evidence based results aid decision makers with suggestive solutions towards wetland protection and management of natural resources.

WATER ACCOUNTING +: REMOTE SENSING BASED ASSESSMENTS OF WATER USE AND AVAILABILITY

Rebelo L[1]
[1]*International Water Management Institute*

Currently, water resource monitoring is well below the levels needed across much of the world. In many countries, water monitoring networks have declined over the past few decades, there are only scattered examples of water quality monitoring, and few countries have adopted sound water accounting mechanisms or reporting systems. Water accounting quantifies how much water is in a system, where, when and in what quality it is available, how much is demanded and consumed in time and place, and how well it is currently managed with respect to meeting those demands. Water accounting has emerged as an important tool to understand water availability and use and is practiced in some form in every managed watershed, although frameworks and available information vary hugely from one country to another. Incomplete and partially accessible water flow data in ungauged or poorly gauged basins is a fundamental problem in understanding hydrological processes and managed water flows in many parts of the world, and is one of the main reasons for the absence of operational national level water accounting systems. In addition, the integration of data and information across sectors that depend on access to water remains a challenge. In order to overcome the difficulty in measuring all water flows and fluxes in a river basin with multiple water users, a new framework, referred to as Water Accounting Plus (WA+) has been developed over the past few years. This approach uses new, public domain remote sensing datasets to analyze the water flows, fluxes, stocks, consumption, and services from complex river basins or at the national scale. This paper presents the WA+ framework, the input datasets used (precipitation, soil moisture, evapotranspiration, land use) and annual water accounts for the Nile Basin derived from these.

KEYWORDS: water accounting, water resource monitoring, reporting

WATER RELATED ACTIVITIES IN GEO (GROUP ON EARTH OBSERVATIONS)

Aellen V[1]
[1]*GEO*

Water, one of the most essential components of our planet, is not evenly distributed over the globe. When it is missing (droughts) or is too abundant (floods), or when its quality is degraded, it is a source of problems for the populations depending on it. Further its global distribution is being threatened by climate change, with cold regions being the most sensitive to it. In GEO, the Group on Earth Observations, the water issue is being tackled by a number of Community Activities, such as AquaWatch, dealing with water quality monitoring, and Global Initiatives such as GDIS (Global Drought Information System), GEOCRI (GEO Cold Region Initiative), or GEOGLOWS (GEO Global Water Sustainability).

AquaWatch intends to deliver, on a routine and sustained basis, timely, consistent, accurate and fit-for-purpose water quality data products and information to support water resource management and decision making in coastal and inland waters.

The purpose of the Global Drought Information System (GDIS) is to assist in ensuring the sustainability of the global water supply and to carry out global monitoring of the variability of water as it relates to drought and water scarcity.

Lastly GEOGLOWS intends to provide a coordination framework for all water related initiatives under the GEO programme. It presently facilitates the use of Earth observation assets to contribute to mitigating water shortages, excesses and degraded water quality arising from population growth, climate change and industrial development on a world-wide basis. It develops knowledge based on the analysis of Essential Water Variables (EWVs) and uses this knowledge to guide applications related to minimizing basin and/or regional risk; to inform policies related to enhancing global water sustainability, and to support capacity building through regional programs and alliances such as AmeriGEOSS, with the need to take hold in Asia or Africa or even Europe.

WAVE TRANSFORMATION ON OCEAN CURRENTS

Chapron B[1], Collard F[2], Kudryavtsev V[3]
[1]Ifremer, [2]OceanDataLab, [3]SOLAB/RSHU

Under favourable imaging conditions, the Sentinel-2 Multi-Spectral Instrument (MSI) can provide spectacular and novel quantitative ocean surface wave directional measurements in satellite Sun Glitter Imagery (SSGI). Owing to a relatively large-swath with high spatial resolution (10 m) ocean surface roughness mapping capabilities, changes in ocean wave energy and propagation direction can be precisely quantified at very high resolution, across spatial distances of 10 km and more. This provides unique opportunities to study ocean wave refraction induced by spatial varying bottom depth and surface currents. As expected and demonstrated over the Grand Agulhas current area, the mesoscale variability of near-surface currents, documented and reconstructed from satellite altimetry, can significantly deflect in-coming south-western swell systems. Based on ray-tracing calculations, and unambiguously revealed from the analysis of Sentinel-2 MSI SSGI measurements, the variability of the near-surface current explains significant wave-current refraction, leading to wave-trapping phenomenon and strong local enhancement of the total wave energy. In addition to its importance for wave modelling and hazard prediction, these results open new possibilities to combine different satellite measurements and greatly improve the determination of the upper ocean mesoscale vorticity motions.

WAVELET DECOMPOSITION OF SEA ICE TRENDS IN THE BEAUFORT SEA AND EASTERN CANADIAN ARCTIC

LeDrew E[1], Liao R[1]
[1]*University Of Waterloo*

Arctic Sea ice has exhibited considerable interannual variability superimposed upon a general downward temporal trend. This is evident in passive microwave imagery collected over the past four decades, and confirmed through operational airborne survey. Only recently evident in this record is the ongoing loss of multi-year ice which may have tremendous consequences for global climate through the impact on the thermohaline circulation.

Of tremendous interest in understanding the atmosphere cryosphere feedback processes that may drive this are the variances from this general trend that may be related to extreme synoptic events. In this study we examine the NSIDC NASA team record of sea ice in the Eastern Canadian Arctic and the Beaufort Sea using wavelet decomposition of the major spatial patterns. Such decomposition signals may provide clear demarcation of the anomalous events. When this analysis is coupled with study of atmospheric typing patterns, we may gain insight into processes accounting for interannual variability in the long term trends, and how they vary by region.

WETLAND HYDROLOGICAL INTEGRITY ASSESSMENT WITH UNMANNED AERIAL SYSTEMS (UAS)

Boon M[1], Tesfamichael S[2]
[1]Department of Zoology University Of Johannesburg, [2]Department of Geography, Environmental Management and Energy Studies University Of Johannesburg

A key component of the health status of a wetland ecosystem is the present hydrological integrity. The analysis of on-site activities that impact on the hydrology of a particular wetland was undertaken using an Unmanned Aerial System (UAS) in combination with field studies. The WET-Health methodology was followed for the hydrological assessment, where wetland health is a measure of the deviation of a wetland's structure and function from its natural reference condition. The extent and nature of activities within the wetland such as impoundments, excavations and indicators of visible damage such as erosion gullies was determined through high resolution UAS mapping using a commercial off-the-shelf digital camera. Structure from Motion (SfM) computer vision techniques were used to derive ultra-high resolution point clouds, orthophotos, digital elevation models (DEMs) and digital terrain models (DTMs). Visual and statistical analyses of the point clouds and surface models were used to derive detailed information useful for enhanced hydrological health assessment of the wetland. The WET-Health hydrology module completed with the aid of the UAS products still indicated that the hydrology of the wetland is completely modified as indicated by the "F" Present Ecological State (PES) category and that the hydrological state of the wetland will deteriorate (change score). However a higher impact score were determined through the enhanced visualisation and completion of scale-appropriate measurements of hydrological features. In conclusion, the use of UAS can significantly enhance the assessment of surface hydrology of wetlands and thereby allow for more effective management, decision making and conservation of wetland ecosystems.

KEYWORDS: Wetland hydrology, Unmanned Aerial Vehicle, Structure-from-motion, 3D point clouds, DEM

WETLAND MAPPING, INVENTORY AND ASSESSMENT IN THE CONTEXT OF THE SDGS

Rebelo L[1]
[1]*International Water Management Institute*

Wetlands are the source of multiple societal benefits, and global policy agreements concur that wetlands are vital ecosystems for human development and well being. But despite this, 40% of wetlands have been lost in the last 40 years, along with a 76% loss in wetland species populations. With the focus of the global agenda on a new set of Sustainable Development Goals, the importance of wetlands is increasingly recognised. Implicitly or explicitly mentioned in 7 goals and 27 targets, wetlands are critical to achieving sustainable food production systems, ensuring a supplies of freshwater, providing protection to societies against inundation and flooding. Despite their importance, information on wetland ecosystems and their services is often scattered, difficult to find, and hard to integrate into decision making. Regional scale maps of wetland extents and baseline conditions do not exist, and national level inventories are scarce.

Earth observations and geospatial information is critical to support the implementation of the SDGs at national, regional and local levels, as well as for monitoring and reporting against the global indicator framework. The guidance for Goal 6.4 in particular recommends the use of EO data to establish a baseline for water related ecosystems including wetlands. This paper investigates recent advances in data, tools and applications available for countries looking to implement an Earth Observation based approach to determine wetland extent at the national level, as well as to track progress towards the SDG Goal 6 and associated targets.

KEYWORDS: Wetland mapping, extent, sustainable development

WETLAND VEGETATION MAPPING USING REMOTE SENSING: THE CASE OF KLIPRIVIERSBERG NATURE RESERVE, SOUTH AFRICA

Mosime M[1], Tesfamichael S[1]

[1]*Department of Geography, Environmental Management and Energy Studies, University of Johannesburg, PO Box 524, Auckland Park 2006*

The aim of this study was to compare the performance of Landsat and SPOT imagery to map wetland vegetation types in the Klipsriviersberg Nature Reserve, South Africa. The Gauteng Conservation Plan 3.3 (C-Plan 3) was used to delineate the boundaries of the wetlands in the study area. C-Plan 3 is a component of Gauteng Department of Agriculture and Rural Development (GDARD) that identifies sites that are critical for maintaining biodiversity. According to the plan, the proposed study area falls within the Critical Biodiversity Areas (CBA) and Ecological Support Areas (ESA). Limited field data were collected within the boundaries of the wetlands during summer 2015 when the vegetation cover was relatively high. These data identified features including sparse vegetation, dense vegetation, grassland and bare land in the study area. Additional samples were added from Google Earth image to increase sample size. Both the field data and Google Earth data were used as reference against which the performances of SPOT and Landsat product were compared. Unsupervised classification was used to classify SPOT and Landsat images acquired in summer 2015. The results showed that overall accuracy of SPOT images is higher than Landsat images. This is attributed to its high spatial resolution of 1.5 m compared to 30 m spatial resolution of Landsat imagery. These indicate that SPOT imagery is recommended to map wetland vegetation in localised area such as the study area. The current high temporal resolution of the image has also an added advantage that conversationalist should exploit.

KEYWORDS: wetland vegetation, SPOT, Landsat, Klipriviersberg Nature Reserve

WHERE, WHEN AND WHY ARE THERE ELEPHANT POACHING HOTSPOTS IN KENYA?

Ouko E[1]
[1] *Rcmrd*

Poaching for elephant tusks is a major short-run threat to the African elephant with land fragmentation a threat in the longer run. Due to difficulties in distinguishing poached ivory and ivory purchased from legal sources, the Kenyan government decided not to trade in ivory confiscated from poachers. This decision was announced to the world on 18th July 1989. Kenya burned 2,000 confiscated elephant tusks to show its effort and commitment to saving the elephant from eminent extinction. This study identifies the spatial and temporal clusters of elephant poaching incidences in Kenya and the associated biophysical and human factors using geographical information systems, spatial scan statistic-SaTScan, and boosted regression trees. The spatial scan statistic detected most likely significant clusters (hotspots) for time window of 1, 6, and 12 months. Similarly, significant secondary clusters were also simulated from the analysis. More elephant poaching crimes were confirmed to be repeated next to the protected areas boundaries, at lowlands and at mean altitude of 1300 meters above sea level. Areas closer to roads and rivers contributed more to poaching cases. High income regions recorded more elephant related crimes. Regions dominated by kaolin clay soils, bush-lands, forests, plantations and grasslands are main targets of the poachers. This study provides evidence of the existence of statistically significant poaching hotspots/clusters in Kenya and also identifies the associated factors explaining such patterns. The applied methods demonstrated their relevance and applicability in analysing elephant crime data to identify hotspots.

KEYWORDS: SaTScan, spatial and temporal clusters, boosted regression trees, most likely clusters, secondary clusters, variables.

WIND CHANGES ABOVE WARM AGULHAS CURRENT EDDIES

Rouault M[1]
[1]*Nansen Tutu Center For Marine Environmental Research*

Sea surface temperature (SST) estimated from the Advanced Microwave Scanning Radiometer onboard the Aqua satellite, altimetry derived sea level height as well as GlobCurrent surface ocean velocity are used south of the Agulhas Current to identify warm-core mesoscale eddies presenting a distinct SST perturbation superior to 1 C to the surrounding ocean. The analysis of twice daily instantaneous charts of equivalent stability neutral wind speed estimates from the SeaWinds scatterometer onboard the QuikScat satellite collocated with SST during the lifespan of for those six identified eddies show stronger wind speed above those warm eddies than surrounding water at all wind directions as was found in previous studies. However, only half of the case show higher wind above the eddies at the instantaneous scale. 20 % of the cases had incomplete data due to partial global coverage of the scatterometer for one path. For cases where the wind is stronger above warm eddies, there is no relationship between the increase in surface wind speed and the SST perturbation but we do find a linear relationship between the decrease in wind speed from the center to the eddy border downstream and the SST perturbation. SST perturbations range from 1 C to 6 C for a mean eddy SST of 15.9 C and mean SST perturbation of 2.65 C. Diameter of eddies range from 100 to 250 km. Mean background wind speed is about 12 m.s-1 mostly southwesterly to northwesterly and ranging mainly from 4 m/s to 16 m/s. Mean wind increase is about 15 % at 1.8 m.s-1. Wind speed increase of 4 to 7 m.s-1 above warm eddies is not uncommon.

WOODY RESOURCE PATTERNS AND LAND MANAGEMENT IN THE SOUTH AFRICAN LOWVELD WITH L-BAND SAR AND LIDAR IMAGERY

Mathieu R[1], Main R[1], Naidoo L[1], Wessels K[2], Mograbi P[3], Smit I[4], Asner G[5], Linden H
[1]CSIR-NRE, [2]CSIR-Meraka, [3]Department of Environmental Science, Rhodes University, [4]Scientific Services SANParks, [5]Department of Global Ecology, Stanford University

Savannahs and woodlands account for 35% of the land in southern Africa. Harvesting of woody plants and land use changes can threaten the sustainability of the provision of raw materials to poor rural communities. In addition, bush densification is increasingly seen as a major regional threat for food security and biodiversity. In South Africa tree cover is believed to increase at a rate of 5-6% per decade and to encroach into grasslands; bush encroachment affects 10-20 million ha. Despite these dynamics there are yet limited information on spatial patterns of woody vegetation in the country. We mapped woody cover over the northern section of the South African Lowveld (6.5M ha) with Synthetic Aperture Radar and LiDAR airborne data. Field plots were used to calibrate and validate extensive LiDAR-based maps of structural metrics, which were then used to upscale woody cover at satellite level (individually processed scenes - ALOS PALSAR 1 2010) using random forest. The maps were produced at 1ha pixel size, produce good accuracy, and are a significant improvement on global products which are the only available datasets in the region (e.g. LandSat VCF). Woodland sand woody cover patterns will be reported and analysed in the region considering land management (conservation, communal, commercial use) and environmental conditions. The detailed SAR & LIDAR-based map provides a unique opportunity to revisit our understanding of woody cover patterns in the region.

KEYWORDS: savannahs, woody cover, SAR, LiDAR

WORLD ATLAS OF DESERTIFICATION AND THE CONCEPT OF LAND PRODUCTIVITY DYNAMICS: A CONTRIBUTION TO THE UNCCD'S GLOBAL LAND OUTLOOK

Cherlet M[1], Sommer S[1], Weynants M[1], Kutnjak H[2]
[1]European Commission, [2]University of Zagreb

Increase in human population and changes in consumption patterns have created enormous pressures on the Earth's natural resources. A multitude of global drivers or anthropogenic land change processes are interacting simultaneously around the globe and include processes such as demographic changes, globalized economies, intensified agriculture. Interactions are however different and are defined by regional and local socio-economic and biophysical conditions, as do final impacts on the local land resources. The World atlas of Desertification considers that these processes can lead to land degradation on their own or combined. And therefor builds a framework of providing a convergence of evidence of human-environment interactions to identify areas of increased stress where pathways towards land degradation are or need to be identified. Earth observation now offers ideal data and downstream products that can be used as proxies for these processes. Land productivity is an essential concept for monitoring land degradation. Based on 15 years of Copernicus vegetation related satellite observations, long-term seasonal dynamics and current 'arrival' state can be mapped. As land productivity alone at that scale can tell only a small part of the land change story, this needs to be complemented with other contextual information, much of which can be derived from Earth Observation, in order to provide insight on the underlying causes and trends of other processes to further evaluate and assess the land status. Meaningful global overviews of combined processes are nevertheless essential and need to be 'viewed' from different thematic angles to yield focused information. Routine processing and integration of those datasets can provide a good base-monitoring scheme to the Global Land Outlook initiative from the UNCCD.

AUTHOR INDEX

AUTHOR INDEX

AUTHOR INDEX

AUTHOR INDEX

AUTHOR INDEX

AUTHOR INDEX

AUTHOR INDEX

AUTHOR INDEX

AUTHOR INDEX

AUTHOR INDEX

AUTHOR INDEX

AUTHOR INDEX

AUTHOR INDEX